Fluid Catalytic Cracking III

ACS SYMPOSIUM SERIES **571**

Fluid Catalytic Cracking III

Materials and Processes

Mario L. Occelli, EDITOR
Georgia Institute of Technology

Paul O'Connor, EDITOR
Akzo Nobel Chemicals

Developed from a symposium sponsored
by the Division of Petroleum Chemistry, Inc.,
at the 206th National Meeting
of the American Chemical Society,
Chicago, Illinois,
August 22–27, 1993

American Chemical Society, Washington, DC 1994

Library of Congress Cataloging-in-Publication Data

Fluid catalytic cracking III: materials and processes / Mario L. Occelli,
Paul O'Connor, editors

   p.   cm.—(ACS symposium series, ISSN 0097–6156; 571)

   Includes bibliographical references and indexes.

   ISBN 0–8412–2996–1

   1. Catalytic cracking—Congresses.

   I. Occelli, Mario L., 1942-   . II. O'Connor, Paul. III. American
Chemical Society. Division of Petroleum Chemistry. IV. Title: Fluid
catalytic cracking 3.   V.   Title: Fluid catalytic cracking three.
VI. Series.

TP690.4.F563   1994
665.5'33—dc20   94–33553
                                                      CIP

Foreword

THE ACS SYMPOSIUM SERIES was first published in 1974 to provide a mechanism for publishing symposia quickly in book form. The purpose of this series is to publish comprehensive books developed from symposia, which are usually "snapshots in time" of the current research being done on a topic, plus some review material on the topic. For this reason, it is necessary that the papers be published as quickly as possible.

Before a symposium-based book is put under contract, the proposed table of contents is reviewed for appropriateness to the topic and for comprehensiveness of the collection. Some papers are excluded at this point, and others are added to round out the scope of the volume. In addition, a draft of each paper is peer-reviewed prior to final acceptance or rejection. This anonymous review process is supervised by the organizer(s) of the symposium, who become the editor(s) of the book. The authors then revise their papers according to the recommendations of both the reviewers and the editors, prepare camera-ready copy, and submit the final papers to the editors, who check that all necessary revisions have been made.

As a rule, only original research papers and original review papers are included in the volumes. Verbatim reproductions of previously published papers are not accepted.

M. Joan Comstock
Series Editor

Contents

INDEXES

Preface

CATALYSTS, AND THEREFORE CATALYSIS, ARE CRITICAL to the chemical and petrochemical industries, two of the largest and most important industries in the United States, with employment exceeding 1.1 and 0.75 million, respectively. As reflected in two other ACS Symposium Series companion books, volumes 375 and 452, increasing public concern with the effects of chemicals and industrial emissions on the environment and the concern with the supply of oil are still powerful forces that provide a strong incentive for increasing research efforts aimed at the discovery of novel fluid cracking catalysts (FCC) and catalytic processes.

The chapters in this volume reconfirm the idea that the industrial development of FCC is and will remain dominated largely by experimental studies. Formulation, characterization, microactivity testing, pilot-plant testing, and product analysis cannot be avoided. Microreactor and pilot-plant testing remain the preferred methods for the establishment of working correlations between theoretical calculations, physicochemical properties, microactivity testing, and commercial results.

Modern spectroscopic techniques are needed to analyze and understand the science of catalyst design and catalysis. The emphasis remains on analytical techniques such as infrared spectroscopy, solid-state nuclear magnetic resonance spectroscopy, and X-ray photoelectron spectroscopy. In addition, the atomic force microscope can describe the surface architecture of FCC microspheres with an unprecedented resolution. Scanning probe microscopes are emerging as new tools in catalyst characterization.

Acknowledgments

The views and conclusions expressed in this book are those of the authors. We thank them, too, for the effort they have made, both in presenting their work at the symposium and in preparing the camera-ready manuscripts used in this volume. We also express our gratitude to all our colleagues for acting as technical referees.

MARIO L. OCCELLI   PAUL O'CONNOR
Georgia Institute of Technology   Akzo Nobel Chemicals
Atlanta, GA 30332   3800 AZ Amersfoort, Netherlands

August 9, 1994

Chapter 1

Role of Carbocations in Hydrocarbon Reactions Catalyzed by Strong Acids

G. K. Surya Prakash

Loker Hydrocarbon Research Institute and Department of Chemistry, University of Southern California, Los Angeles, CA 90089–1661

This review describes the role of strong acids, particularly superacids (both Bronsted and Lewis type) in hydrocarbon transformation processes. The key to the chemistry is the electrophilc activation of C-H and C-C σ-bonds. These reactions involve both trivalent carbocations (carbenium ions) as well higher coordinate carbocations (carbonium ions) as distinct intermediates. Mechanism of many industially important processes such as isomerization, cracking, alkylation and related electrophilic reactions are also discussed.

Before the turn of last century, saturated hydrocarbons [paraffins] played only a minor role in industrial chemistry. They were mainly used as a source of paraffin wax as well as for heating and lighting purposes. Aromatic compounds such as benzene, toluene and related compounds obtained from the destructive distillation of coal were the main source of industrial organic materials used in the preparation of dyestuffs, pharmaceutical products, etc. Acetylene, obtained by the hydrolysis of calcium carbide, was the key starting material for the emerging organic chemical based industry. It was the ever increasing demand for motor fuel [gasoline] after the first world war that led to the study of isomerization and cracking reactions of petroleum fractions. After the second world war, rapid economic expansion required more and more abundant and cheap sources for chemicals which resulted in industry switching over to petroleum based ethylene as the main chemical feed-stock.

One of the major difficulties that had to be overcome is the low reactivity of some of the major components of petroleum. The lower boiling fractions [up to 250°C] are mainly straight-chain saturated hydrocarbons or paraffins [parum affinis: slight reactivity] which have little reactivity. Consequently, lower fractions were cracked to give alkenes [mainly ethylene, propylene and butylenes]. The straight-chain liquid hydrocarbons have also very low octane numbers which make them less desirable as gasoline components. To transform these paraffins into useful components for gasoline and other chemical applications they have to undergo diverse reactions such as isomerization, cracking and alkylation. These reactions, which are large scale industrial processes, require acidic catalysts [at temperatures around 100°C] or noble metal catalysts [at higher temperatures, 200 - 500°C] capable of activating strong covalent C-H and C- C σ-bonds (1).

0097–6156/94/0571–0001$08.00/0
© 1994 American Chemical Society

Since the early 1960s, superacids (2) [Bronsted acids stronger than 100 % sulfuric acid and Lewis acid stronger than AlCl$_3$ in its reactivity] are known to react not only with π and n-donor bases but also with saturated hydrocarbons, even at room temperature and much below 0 °C involving carbocationic intermediates. Superacids not only encompass liquid systems but also solids such as perfluroalkanesulfonic acids, metal oxides, sulfides, halides, sulfates etc. as well as mixed oxides including natural clays, minerals and zeolites (2). The liquid or solid superacids can also be immobilized with inert supports (2). This discovery initiated extensive studies devoted to electrophilic reactions and conversions of saturated hydrocarbons including methane. Many of these reactions are prototypical of industrial hydrocarbon conversions which occur at much higher temperatures over acid catalysts. This paper will discuss the role of carbocations in strong acid catalyzed reactions of hydrocarbons, particularly, saturated ones.

Direct Observation of Carbocations.

The most important step in the acid catalyzed hydrocarbon transformation processes is the formation of the intermediate carbocation, R$^+$ [equation 1]. Whereas all investigations involving isomerization, cracking, and alkylation reactions under acidic conditions seem to suggest that a trivalent carbocation [carbenium ion] as the key intermediate, no direct evidence for its existence was available till the end of 1950.

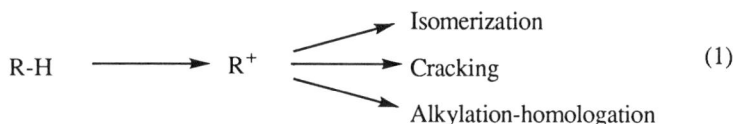

$$R\text{-}H \longrightarrow R^+ \begin{array}{l} \nearrow \text{Isomerization} \\ \longrightarrow \text{Cracking} \\ \searrow \text{Alkylation-homologation} \end{array} \qquad (1)$$

Until the early 1960s simple alkyl cations were considered only as transient species. The only isolable carbocationic species until that time were of the triarylcarbenium type (2). The intermediacy of simple alkyl cations has been inferred from the indirect kinetic and stereochemical studies of reactions, particularly, solvolytic reactions. No reliable measurements, other than electron impact measurements in the gas phase (mass spectrometry), were known. The formation of gaseous organic cations under electron bombardment of alkenes, alkyl halides, and other precursors had been widely investigated in mass spectrometric studies (3). No direct observation of alkyl carbocations in the condensed phase was achieved prior to the early 1960s.

The observation of alkyl cations such as *tert*-butyl cation **1** and isopropyl cation **2** in solution was a long standing challenge. The existence of alkyl cations in systems containing alkyl halides and Lewis acid halides has been inferred from a variety of observations, such as vapor pressure depressions of CH$_3$Cl and C$_2$H$_5$Cl in the presence of GaCl$_3$ (4), conductivity of AlCl$_3$ in alkyl chlorides (5), and of alkyl fluorides in BF$_3$ (6), as well as the effect of ethyl bromide on the dipole moment of AlBr$_3$ (7). However, in no case had well-defined, stable alkyl cation complexes been established even at very low temperatures. Prior to 1960, Symons and coworkers (8) and Deno *et al* (9) had investigated the fate of a variety of aliphatic alcohols and alkenes in neat H$_2$SO$_4$ and oleum. However, all these studies failed to provide convincing evidence for the existence of free alkyl cations.

In 1962, Olah *et al* (10) first directly observed alkyl cations in solution. They obtained *tert*-butyl cation **1** when *tert*-butyl fluoride was dissolved in excess antimony pentafluoride, which served both as the Lewis acid and the solvent.

Subsequently, the counter ion was found to be, under these conditions, primarily the dimeric $Sb_2F_{11}^-$ anion (11) whereas in $SbF_5:SO_2$ and $SbF_5:SO_2ClF$ solutions, both SbF_6^- and $Sb_2F_{11}^-$ anions are formed.

$$(CH_3)_3C\text{-}F \ + \ 2 \ SbF_5 \longrightarrow (CH_3)_3C^+ \ Sb_2F_{11}^-$$

1

The possibility of obtaining stable alkyl fluoroantimonate salts from alkyl fluorides and subseqently alkyl chlorides in antimony pentafluoride solution (neat or diluted with sulfur dioxide, sulfuryl chloride fluoride, sulfuryl fluoride, Freon 11 and Freon 113) or in other superacids (2) such as $HSO_3F:SbF_5$ (Magic Acid), $HF:SbF_5$ (fluoroantimonic acid) $HF:TaF_5$ (fluorotantalic acid), $CF_3SO_3H:B(OSO_2CF_3)_3$ and the like has been evaluated in detail, extending studies to all isomeric C_3-C_8 alkyl halides as well as to a number of higher homologs (2). Using Bronsted superacids stable carbocations can now be generated from precursors such as alcohols, alkenes, ethers, sulfides, amines and even alkanes (*vide infra*). Substrates which possess non-bonded electron pair donors initially form onium ions.

Isomeric propyl, butyl, and pentyl fluorides with antimony pentafluoride gave the isopropyl, *tert*-butyl, and *tert*-amyl cations (as their fluoroantimonate salts) **2**, **1**, and **3**.

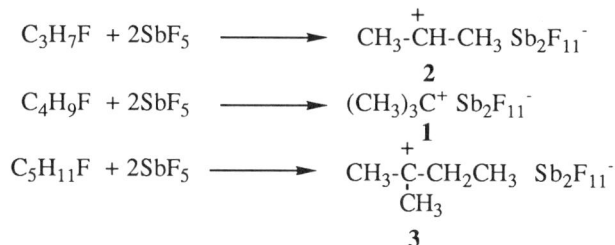

$$C_3H_7F \ + \ 2SbF_5 \longrightarrow CH_3\overset{+}{-}CH\text{-}CH_3 \ Sb_2F_{11}^-$$

2

$$C_4H_9F \ + \ 2SbF_5 \longrightarrow (CH_3)_3C^+ \ Sb_2F_{11}^-$$

1

$$C_5H_{11}F \ + \ 2SbF_5 \longrightarrow \underset{CH_3}{CH_3\overset{+}{\underset{|}{C}}\text{-}CH_2CH_3} \ Sb_2F_{11}^-$$

3

The secondary butyl and amyl cations can be observed only at very low temperatures (less than -95 °C), and they readily rearrange to more stable tertiary cations. Generally, primary alkyl cations are not observed under the superacid conditions.

To prove that stable alkyl cations, and not exchanging donor-acceptor complexes, were obtained, Olah *et al* in addition to ^{1}H NMR studies also investigated as early as 1962 the ^{13}C NMR of potentially electropositive carbenium carbon atoms in alkyl cations (12). The ^{13}C NMR chemical shift of the carbocationic center in *tert*-butyl cation in $SbF_5:SO_2ClF$ at -20 °C is δ^{13}C 335.2 (referenced from the tetramethylsilane signal) with a long range ^{13}C-^{1}H NMR coupling constant of 3.6 Hz. The ^{13}C NMR chemical shift of the isopropyl cation is δ^{13}C 320.6 with a long range [13]C-^{1}H coupling of 3.3 Hz. The direct [13]C-^{1}H coupling at the carbocationic center is 169 Hz (indicating sp[2] hybridization of the cationic center). Thus the substitution of the methyl group in the *tert*-butyl cation by hydrogen causes an upfield shift of 14.6 ppm. Such deshielding upon methyl substitution has been attributed to methyl substituent effect. The *tert*-butyl cation is much more stable than the secondary isopropyl cation due to additional C-H hyperconjugation. This effect has been recently demonstrated by the crystal structure analysis of *tert*-butyl cation salt with $Sb_2F_{11}^-$ anion (13a).

    Since 1960s a wide variety of structurally diverse carbocations (13b) have been prepared under so called long-lived stable ion conditions and characterized primarily by low temeprature ^{1}H and ^{13}C NMR spectroscopy (2). In addition to solution NMR studies, solid state NMR, UV-Visible, infrared, Raman, X-ray photoelectron spectroscopy and X-ray crystallography have been employed to delineate carbocationic structures, their mode of charge delocalization and stabilization (2). Some of the representative trivalent tricoordinate carbocations that have been characterized under stable ion conditions are shown in **Scheme-1** (only the counter anions are not shown).

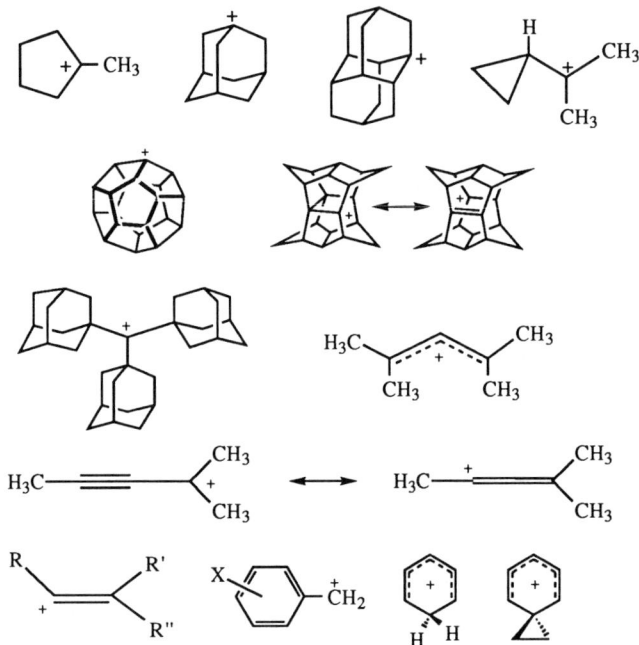

Scheme-1

    Not only cyclic, polycyclic and bridge-head carbocations but also cyclopropyl carbinyl, allylic, propargylic, vinyl, and dienylic carbocations have been investigated (2). These also include the cationic intermediates of the Friedel-Crafts reactions of aromatic compounds. In addition to trivalent carbomonocations a series of carbodications and polycations have been studied (**Scheme-2**). These encompass acyclic, cyclic and polycyclic systems as well as aromatically stabilized dications (2).
    Some carbocations tend to undergo fast degenerate rearrangements through intramolecular hydrogen or alkyl shifts to the degenerate structures. The question arises whether these processes involve equilibrations between limiting classical ion intermediates (trivalent, tricoordinate carbocations, also called carbenium ions) whose structures could be adequately described by two-electron, two center bonds separated by low energy transition states, or whether nonclassical, hydrogen or alkyl bridged carbocations (high coordinate carbocations, also called carbonium ions) are involved, which also require the presence of two-electron bonds between three or

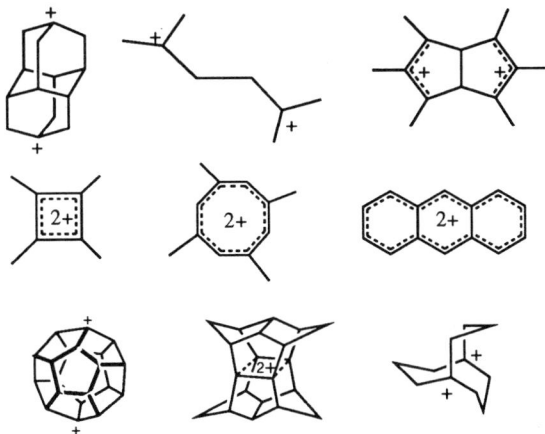

Scheme-2

more centers for their description (2). It is generally difficult to answer this question by NMR spectroscopy alone because of its relatively slow time scale; however, NMR has been used to deduce structures where degenerate rearrangements lead to average coupling constants and chemical shifts. Solid state ^{13}C NMR (using cross-polarization magic angle spinning techniques at low temperatures) (14a), isotopic substitution (14b) as well as faster methods such as infrared, Raman, UV-Visible, X-ray photoelectron spectroscopy are (ESCA) are particularly useful (2). Some typical

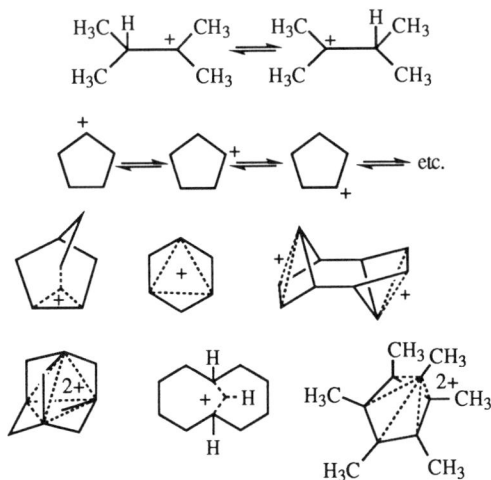

Scheme-3

examples are shown in **Scheme-3**. These high coordinate carbonium ions are prototypical of intermediates of strong acid catalyzed hydrocarbon reactions.

C-H and C-C σ-Bond Protolysis, Hydrogen- Deuterium Exchange

In 1946 Bloch, Pines, and Schmerling (15) observed that n-butane would isomerize to isobutane under the influence of pure aluminum chloride only in the presence of HCl. They proposed that ionization occurs through initial protolysis of the alkane to *sec.*-butyl cation [which subsquently isomerizes to *tert*-butyl cation] as indicated by the formation of minor amounts of hydrogen gas in the initial stage of the reaction [equation 2].

$$\text{n-C}_4\text{H}_{10} + \text{HCl} \xrightarrow{\text{AlCl}_3} \text{sec-C}_4\text{H}_9^+ \text{ AlCl}_4^- + \text{H}_2 \qquad (2)$$

The first direct evidence of protonation of alkanes under superacid conditions was reported independently by Olah and Lukas (16) as well as Hogeveen and coworkers (17). When n-butane or isobutane was reacted with HSO_3F+SbF_5 (1:1, Magic Acid), *tert*-butyl cation was formed [eqation 3] exclusively. Both *sec*-butyl cation and *tert*-butyl cation [the former easily rearranges to the latter] have been characterized by a host of spectroscopic techniques including low temperature NMR spectroscopy (*vide supra*). In fact, many different carbocations have been formed by direct protolysis (2).

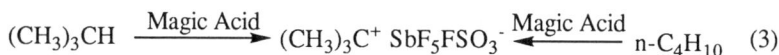

$$(\text{CH}_3)_3\text{CH} \xrightarrow{\text{Magic Acid}} (\text{CH}_3)_3\text{C}^+ \text{ SbF}_5\text{FSO}_3^- \xleftarrow{\text{Magic Acid}} \text{n-C}_4\text{H}_{10} \qquad (3)$$

It was shown (18) that even *tert*-butyl cation undergoes degenerate carbon scrambling at higher temperatures in superacids (**Scheme-4**). A lower limit of E_a of 30 kcal mol^{-1} was estimated for the scrambling process which could correspond to the energy difference between *tert*-butyl and primary isobutyl cation [the latter described as a partially delocalized, 'protonated cyclopropane' intermediate].

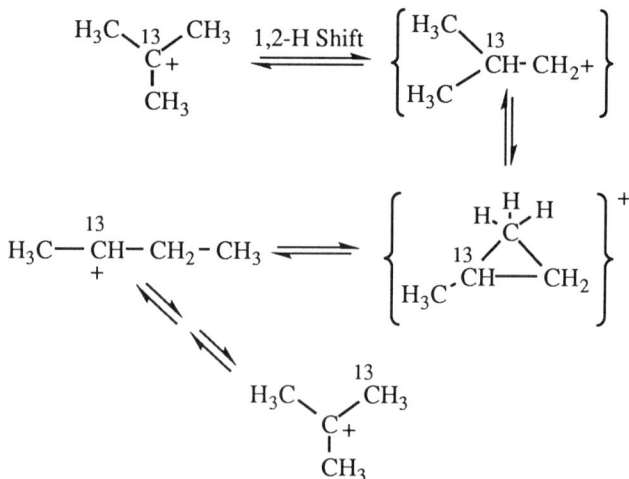

Scheme-4

In compounds containing only primary hydrogen atoms such as neopentane and 2,2,3,3-tetramethylbutane, a carbon-carbon bond [C-C protolysis] is broken rather than a carbon-hydrogen bond [equation 4].

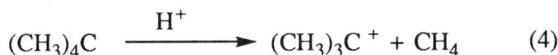

$$(CH_3)_4C \xrightarrow{\quad H^+ \quad} (CH_3)_3C^+ + CH_4 \qquad (4)$$

Hogeveen and coworkers have suggested a linear transition state for protolytic ionization of hydrocarbons. This, however, may be the case only for sterically congested systems. Results of protolytic reactions of hydrocarbons in superacidic media were interpreted by Olah as an indication for the general reactivity of covalent C-H and C-C σ-bonds of saturated hydrocarbons [equation 5]. The reactivity is primarily due to the σ-donor ability of a shared electron pair (of σ-bond) via two electron, three center bond formation. The transition state of the reactions are consequently of a three center bound pentacoordinate carbonium ion nature (2). Many such bridged carbonium ions have been characterized in superacid solutions (*vide supra*). Strong indication for the mode of protolytic attack was obtained from hydrogen-deuterium exchange studies. Monodeuteriomethane was reported (19) to undergo clean deuterium-hydrogen exchange without side reactions in HF-SbF$_5$ system [equation 6].

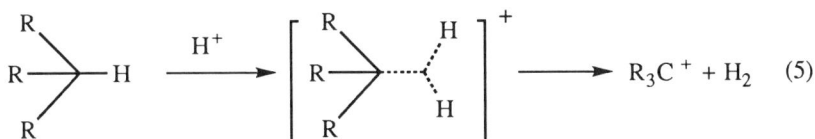

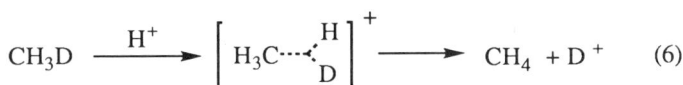

$$R_3C^+ + H_2 \longrightarrow \left[R_3C \cdots \begin{array}{c} H \\ H \end{array} \right]^+ \longrightarrow R_3CH + H^+ \quad (7)$$

The reverse reaction of protolytic ionization of hydrocarbons to trivalent carbenium ions can be considered as alkylation of molecular hydrogen by the electrophilic carbenium ion through the pentacoordinate carbonium ion (20a,b) [equation 7]. Indeed, Hogeveen has experimentally obtained support for this point by reacting stable alkyl cations in superacids with molecular hydrogen (20a). The use of molecular hydrogen to reduce cracking over a variety of heterogeneous catalysts is an industrially important reaction.

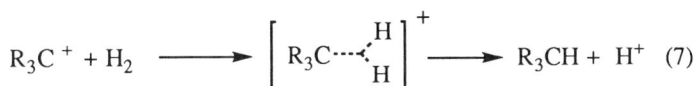

Further evidence for the pentcoordinate carbonium ion mechanism for the protolysis of alkane was obtained in the H-D exchange reactions observed with isobutane. When isobutane is treated with deuterated superacids [DSO$_3$F:SbF$_5$ or DF:SbF$_5$] at low temperature and atmospheric pressure hydrogen-deuterium exchange is observed both at tertiary C-H and primary C-H bonds, although the rate of exchange at the tertiary C-H bond is significantly faster due to its higher basicity (21). These results are best explained as proceeding through a two-electron, three center bound pentacoordinate carbonium ion. The H-D exchange in isobutane in superacid media is fundamentally different from the H-D exchange observed by

Otovos and coworkers (22) in D_2SO_4, who found the eventual exchange of all the nine methyl hydrogens but not the methine hydrogen [equation 8].

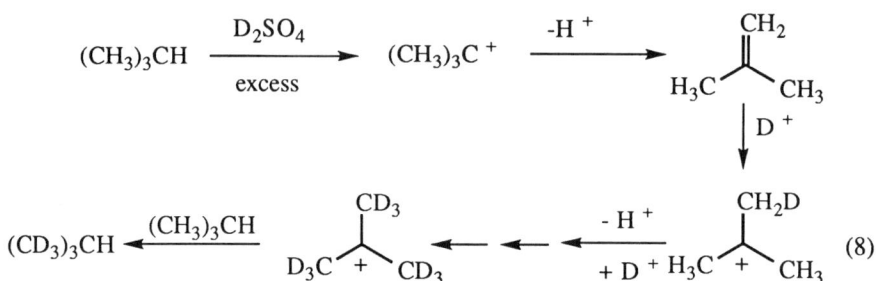

(8)

Otvos et al. suggested (22) that under the reaction conditions a small amount of *tert*-butyl cation is formed from isobutane in an oxidative step which then deprotonates to isobutylene. The reversible deprotonation and deuteriation of isobutylene is responsible for the H-D exchange of methyl hydrogens, whereas tertiary hydrogen is involved in intermolecular hydride transfer from unlabeled isobutane [at the tertiary C-H position]. Under superacidic conditions, where no olefin formation occurs, the reversible isobutylene protonation cannot be involved in the exchange reaction. More recently, even hydrogen- deuterium exchange has been observed (23a) in the *tert*-butyl cation in $DF-SbF_5$ solution indicating the intermediacy of protio-*tert*-butyl dication [equation 9]. Such protolytic activation leading to *"superelectrophiles"* in strong Bronsted superacids has been recently discussed (23b). Such activations involving strong Lewis acids are also possible (23b).

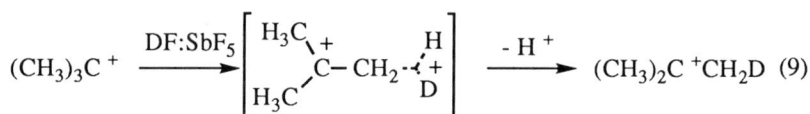

(9)

One of the main initial difficulties in comprehending the carbocationic nature of acid-catalyzed transformations of alkanes via hydride abstraction mechanism was that no clear evidence for stoichiometric amount of hydrogen gas evolution was observed in the reaction mixture. For this reason, a complimentary mechanism was proposed (24) involving direct hydride abstraction by the Lewis acid such as SbF_5 [equation 10].

$$R\text{-}H + 2\ SbF_5 \longrightarrow R^+\ SbF_6^- + SbF_3 + HF \quad (10)$$

It has been pointed out by Olah (21a) that if SbF_5 would directly abstract an hydride, it would need to form SbF_5H^- ion involving a weak Sb-H bond compared to the strong C-H bond being broken. Thermodynamic calculations also show that direct oxidation of alkanes by SbF_5 is not feasible. However, it is generally assumed that the hydrogen produced is partially consumed in the reduction of one of the acid components. The direct reduction of SbF_5 in absence of hydrocarbons with molecular hydrogen requires more forcing conditions [50 atm, high temperature] which suggests that the protolytic ionization of alkanes proceeds probably via solvation of protonated alkane by SbF_5 and concurrent ionization-reduction [equation 11].

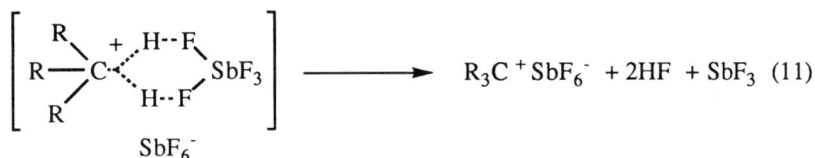

$$\left[\begin{array}{c} R \\ R-C \cdots \\ R \end{array} \overset{+}{\cdots} \begin{array}{c} H\cdots F \\ \diagdown \\ SbF_3 \\ \diagup \\ H\cdots F \end{array} \right]_{SbF_6^-} \longrightarrow R_3C^+ SbF_6^- + 2HF + SbF_3 \quad (11)$$

Culman and Sommer have reinvestigated (25, 26) the mechanism of SbF_5 oxidation of saturated hydrocarbons and have observed stoichiometric evolution of hydrogen. They found efficient oxidation of isobutane to *tert*-butyl cation by neat SbF_5, free of protic acid, at - 80 °C in the presence of proton trap like acetone [equation 12]. Thus, they concluded that proton is not essential for the C-H bond oxidation by SbF_5, contrary to what was previously suggested.

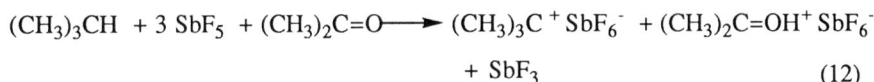

$$(CH_3)_3CH + 3\ SbF_5 + (CH_3)_2C=O \longrightarrow (CH_3)_3C^+ SbF_6^- + (CH_3)_2C=OH^+ SbF_6^-$$
$$+ SbF_3 \quad (12)$$

However, it must be realized that in the demonstration of the oxidizing ability of pure SbF_5 towards C-H bonds still produces protic acid in the system. In the absence of proton trap like acetone, after the initial stage of the reaction, the *in situ* formed HF-SbF_5 conjugate superacid will itself act as a protolytic ionizing agent. The situation is similar to the question of the Lewis acid initiators in the polymerization of isobutylene. Even if the Lewis acids initiate the reaction, the conjugate Bronsted-Lewis acid produced in the initial stage will promote protic initiation. Regardless, the study of protic acid free SbF_5 in initiating reactions of alkanes such as isobutane, much furthered the knowledge in the field (26).

In studies involving solid acid-catalyzed hydrocarbon cracking reactions using HZSM-5 zeolite, Haag and Dessau (27) were able to account nearly quantitatively for H_2 formed in the protolytic ionization step. This is a consequence of the solid zeolite not being easily reduced by molecular hydrogen.

Acid catalyzed alkane transformations under oxidative conditions can also involve radical cations leading to the initial carbenium ions. In the context of the present discussions only protolytic path is considered.

Isomerization

Isomerization of hydrocarbons was first reported by Nentzescu and Dragan (28). They found that when n-hexane was refluxed in $AlCl_3$, it was converted to branched isomers. This reaction is of major economic importance as the straight chain C_5- C_8 alkanes are the main constituents of fraction obtained by the refining of crude oil. Since branched hydrocarbons have considerably higher octane numbers than their linear conterparts, the combustion properties of gasoline can be substantially improved by the isomerization. A number of methods involving liquid and solid acid catalysts have been developed (29). The isomerization again involves carbenium and carbonium ion intermediates. Classic examples of isomerization are n-butane iomerization to isobutane and the rearrangement of tetrahydrocyclopentadiene to adamantane [equation 13] (30, 31).

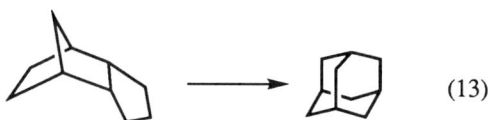

(13)

Cracking and β-Scission

The reduction in molecular weight of various crude oil fractions is an important reaction in petroleum chemistry. The process is called cracking. Catalytic cracking is usually achieved by passing the hydrocarbons over metallic or acidic catalysts , such as crystalline zeolites at about 400- 600 ^{o}C. The molecular weight reduction process involves carbocationic intermediates and the mechanism is based on β-scission of carbenium ions [equation 14]. The main goal of catalytic cracking is to

(14)

upgrade higher boiling oils to yield lower hydrocarbons in the gasoline range (32). The development of highly acidic superacid catalysts in the 1960s focussed attention in acid-catalyzed cracking reactions. HSO_3F+SbF_5 (1:1), trade named Magic Acid derived its name due to its remarkable ability to cleave higher molecular weight hydrocarbons, such as paraffin wax, to lower molecular weight components, preferentially C_4 and other branched isomers.

`   Under strongly acidic conditions, β-scission is not the only pathway by which hydrocarbons are cleaved. The C-C bond can also be cleaved by protolysis. Thus, the protonation of alkanes induces cleavage of molecules by two competing pathways: (a). protolysis of a C-H bond followed by β-scission, (b). direct protolysis of a C-C bond yielding a low molecular weight alkane and a low molecular weight carbenium ion. This reaction which is of enormous economic importance in upgrading higher boiling fractions to gasoline range hydrocarbons, has been shown to work in coal depolymerization and hydroliquefaction processes (33).

Alkylation

The alkylation of alkanes by alkenes, from a mechanistic point of view, must be considered as the alkylation of carbenium ion by the protonation of alkene. The well known acid-catalyzed comercially practised isobutane-isobutylene reaction to produce isooctane demonstrates the mechanism rather well [equation 15]. The last step occurs through the well known Bartlett-Nenitzescu-Schmerling type hydride transfer inspite of considerable steric repulsion between the interacting groups. Such hydride transfer has also been observed intramolecularly, particularly, in medium ring compounds. In fact, under superacidic stable ion conditions, transannular hydrogen bridged carbonium ions have been observed (2). Such systems clearly demonstrate the stability of 3 center two electron bonding.

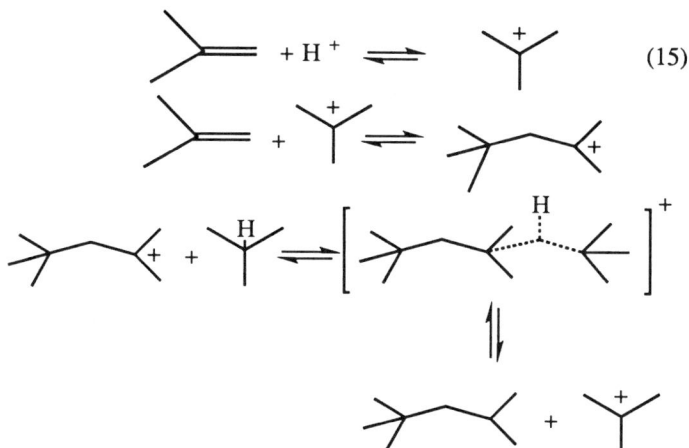

$$(15)$$

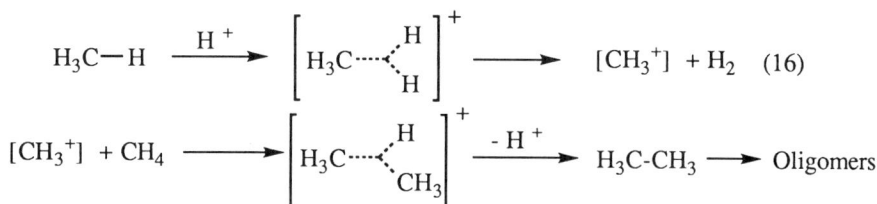

The protolytic oxidative condensation of methane in Magic Acid solution at 60 °C is evidenced by the formation of higher alkyl cations such as *tert*-butyl and *tert*-hexyl cations. The initial reaction process is shown below [equation 16].

$$H_3C-H \xrightarrow{H^+} \left[H_3C\cdots\overset{H}{\underset{H}{\cdots}} \right]^+ \longrightarrow [CH_3^+] + H_2 \quad (16)$$

$$[CH_3^+] + CH_4 \longrightarrow \left[H_3C\cdots\overset{H}{\underset{CH_3}{\cdots}} \right]^+ \xrightarrow{-H^+} H_3C\text{-}CH_3 \longrightarrow \text{Oligomers}$$

Combining two methane molecules to ethane and hydrogen is endothermic by 16 kcal mol^{-1}. Any condensation of methane to ethane and subsequently higher hydrocarbons must overcome the unfavorable thermodynamics. This can be achieved in the condensation process of oxidative nature, where hydrogen is removed by the oxidant (34).

Alkylation of methane , ethane, propane and n-butane by the ethyl cation obtained by the protonation of ethylene in superacid media has been investigated by Siskin, (35) Sommer *et al* (36) and Olah and coworkers (37), respectively. The difficulty lies in generating in a controlled way a very energetic primary carbocation in the presence of alkane and at the same time avoiding oligocondensation of the ethylene itself. Siskin carried out the reaction of methane/ethylene (86:14) gas mixture through a 10:1 HF:TaF$_5$ solution under pressure with strong mixing. (equation 17). Along with recovered starting materials 60% of C$_3$ was found (propane and propylene). Propylene is formed through the isopropyl cation formed by the hydride abstraction of propane by the ethyl cation (equation 18). propane as a degradation product of polyethylene was, however, ruled out because ethylene alone under identical conditions did not give any propane. Under similar conditions under hydrogen pressure, polyethylene reacts quantitatively to form C$_3$ to C$_6$ alkanes, 85%

$$CH_2{=}CH_2 \xrightarrow{\text{HF:TaF}_5} CH_3\text{-}CH_2^+ \xrightarrow{CH_4} \left[CH_3\text{-}CH_2\overset{\overset{\displaystyle H}{\vdots}}{\cdots}\overset{+}{CH_3} \right] \qquad (17)$$

$$\downarrow {-}\,H^+$$

$$CH_3CH_2CH_3$$

$$CH_3CH_2CH_3 \;+\; CH_3\text{-}CH_2^+ \longrightarrow CH_3CH_3 \;+\; CH_3\overset{+}{C}HCH_3 \qquad (18)$$

$$\downarrow {-}\,H^+$$

$$H_3C{-}CH{=}CH_2$$

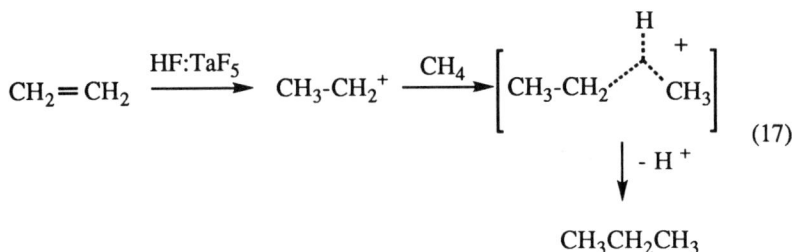

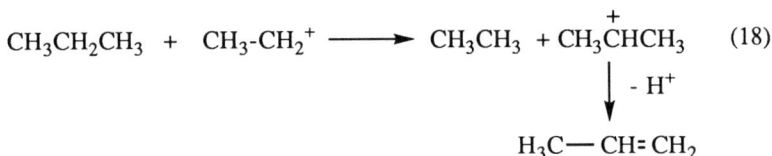

of which are isobutane and isopentane (35). These results further substantiate the direct alkane-alkylation reaction and the intermediacy of the pentacoordinate carbocation (2). Siskin also found (35) that when ethylene reacts with excess ethane in a flow system n-butane resulted as the sole product (equation 19) indicating that the ethyl cation is alkylating the primary C-H bond through the pentacoordinate carbonium ion. If the ethyl cation had reacted with excess ethylene, primary 1-butyl cation would have been formed which irreversibly would rearrange to the more stable *sec*-butyl cation and subsequently to the *tert*-butyl cation giving isobutane as the end product.

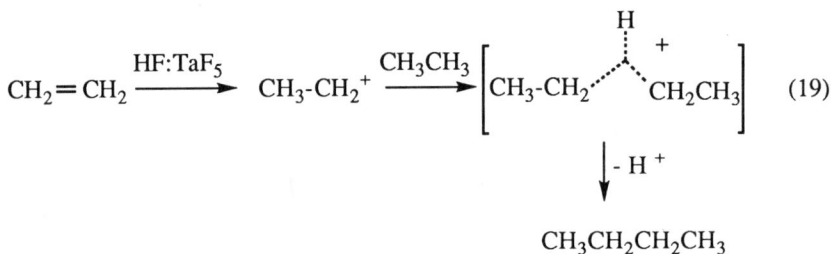

$$CH_2{=}CH_2 \xrightarrow{\text{HF:TaF}_5} CH_3\text{-}CH_2^+ \xrightarrow{CH_3CH_3} \left[CH_3\text{-}CH_2\overset{\overset{\displaystyle H}{\vdots}}{\cdots}\overset{+}{CH_2CH_3} \right] \qquad (19)$$

$$\downarrow {-}\,H^+$$

$$CH_3CH_2CH_2CH_3$$

Despite unfavorable experimental conditions in a batch system for kinetically controlled reactions, Sommer *et al* have found (36) that selectivity of 80% in n-butane was achieved through the ethylation of ethane. The results indicate that to succeed in the direct alkylation the following conditions have to be met. (a) The alkene should be completely protonated to the reactive cation or complex (incomplete protonation leads to polymerization and cracking processes). (b) The alkylation product should be removed from the reaction mixture before it transfers a hydride to the reactive cation. (c) The substrate to cation hydride transfer should not be easy; for this reason the reaction shows the best results with methane and ethane.

Direct evidence for the ethylation of methane with ethylene was provided by Olah *et al* (37) using ^{13}C labled methane (99.9%) over solid superacid catalysts such as TaF$_5$:AlF$_3$, TaF$_5$ and SbF$_5$:graphite. The results show high selectivity in mono ^{13}C labeled propane.

Other Functionalizations of Alkanes

Using a variety of electrophiles selective functionalization of alkanes has been achieved (38) under strong acid catalysis (equation 20). These include halogenation, nitration, nitrosation, amination, formylation, acylation, hydroxylation, sulfuration etc. All these reactions would involve the intermediacy of pentacoordinate carbonium ions.

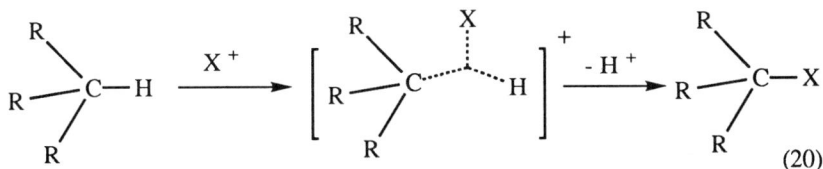

$$R\underset{R}{\overset{R}{>}}C-H \xrightarrow{X^+} \left[R\underset{R}{\overset{R}{>}}C \overset{X}{\underset{\cdots}{\cdots}} H \right]^+ \xrightarrow{-H^+} R\underset{R}{\overset{R}{>}}C-X$$

$$(20)$$

X= Cl, Br, NO_2, NO, HCO, RCO, OH, SH, NH_2, SR etc.

Electrophilic Reactions on Aromatics

Although electrophilic reactions of aromatics (i.e.. Friedel-Crafts reactions) have been well recognized over Lewis and Bronsted acid catalysts certain electrophilic reactions of aromatics occur only under strong acid catalysis (2). For example, ethylation and propylation of benzene with ethylene and propylene to ethylbenzene and cumene, respectively, occur only under strong acid catalysis. Apart from $AlCl_3$: HCl, BF_3:HF and related strong conjugate acids, Nafion-H is a good catalyst for such reactions (2). Solid silicophosphoric acid and related solid acids have also been used for the vapor phase ethylation of benzene. Isomerization and disproportionation of alkylbenzenes are also achieved under the influence of strong acids (2). Many selective functionalizations of deactivated aromatics such as hydroxylations, aminations, nitrations, halogenations can be achieved in strong acid medium. The reactivity of the electrophilic species is enhanced due to superelectrophilic activation (23b). However, these discussions are outside the scope of this review.

Conclusion

This short review demonstrates the role of carbocations, particularly, pentacoordinate electron deficient intermediates in strong acid catalyzed hydrocarbon transformations. Many industrially important processes such as MTG process (methanol to gasoline over H-ZSM-5) developed by Mobil also involve cationic (onium) intermediates (39-41). A wide array of heterogeneous and homogeneous acid catalyzed hydrocarbon reactions involve both trivalent carbenium as well as higher coordinate carbonium ions as distinct intermediates.

Acknowledgements

Partial support of our work by the Loker Hydrocarbon Institute and the National Institutes of Health is gratefully acknowledged. Prof. George A. Olah is thanked for his interest and encouragement.

Literature Cited

1. Asinger, F. *Paraffins, Chemistry and Technology*, Pergamon Press, New York, N.Y., 1965.

2. Olah, G. A.; Prakash, G. K. S.; Sommer, J. *Superacids*, Wiley-Interscience, New York, N. Y, 1985; Olah, G. A.; Prakash, G. K. S.; Williams, R. E.; Field, L.; Wade, K. *Hypercarbon Chemistry* , Wiley Interscience, New York, N.Y., 1987.
3. For an extensive review see, *Mass Spectrometry of Organic Ions,* McLafferty, F. W., Ed, Academic Press, New York, N. Y., 1963.
4. Brown, H. C.; Pearsall, H. W.; Eddy, L. P. *J. Am. Chem. Soc.,* 1950, 42, 5347.
5. Wertyporoch, E.; Firla, T. *Ann. Chim.,* 1933, 500, 287.
6. Olah, G. A.; Kuhn, S. J.; Olah, J. A. *J. Chem. Soc.,* 1957, 2174.
7. Fairbrother, F. *J. Chem. Soc.,* 1945, 503.
8. Rosenbaum, J.; Symons, M. C. R. *Proc. Chem. Soc. (London),* 1959, 92; Mol. Phys., 1960, 3, 205 ; *J. Chem. Soc.,* 1961, 1; Finch, F. C.; Symons, M. C. R. *J. Chem. Soc.,* 1965, 378.
9. Deno, M. C. *Prog. Phys. Org. Chem.,* 1964, 2, 129.
10. Olah, G. A.; Tolgyesi, W. S.; Kuhn, S. J.; Moffatt, M. E.; Bastien I. J.; Baker, E. B. *J. Am. Chem. Soc.,* 1963, 85, 1328.
11. Beacon, J.; Dean, P. A. W.; Gillespie, R. J. *Can. J. Chem.,* 1969, 47, 1655 ; Commeyras, A.; Olah, G. A. *J. Am. Chem. Soc.,* 1968, 90, 2929.
12. Olah, G. A.; White, A. M, *J. Am. Chem. Soc.,* 1969, 91, 5801.
13. (a) Laube, T. *J. Am. Chem. Soc.,* 1993, 115, 7240.
    (b) Saunders, M.; imenz-Vasqquez, H, A. *Chem. Rev.,* 1991, 91 , 375.
14. (a) Lyerla, J. R.; Yannoni, C.; Fyfe, C. A. *Acc. Chem. Res.,* 1982, 15, 208.
    (b) Siehl, H. U. *Adv. Phys. Org. Chem.,* 1987, 23, 63.
15. Bloch, H. S.; Pines, H.; Schmerling, L.*J. Am. Chem. Soc.,* 1946, 68, 153.
16. Olah, G. A.; Lukas, J. *J. Am. Chem. Soc.,* 1967, 89, 2227, 4739.
17. Bickel, A. F.; Gasbeek, G. J.; Hogeveen, H.; Oelderick, J. M.; Platteuw, J. C. *Chem. Commun.,* 1967, 634 ; Hogeveen, H.; Bickel, A.F. *Chem. Commun.,* 1967, 635.
18. Prakash, G. K. S.; Husain, A. F.; Olah, G. A. *Angew. Chem.,* 1983, 95, 51.
19. Olah, G. A.; Lukas, J. *J. Am. Chem. Soc.,* 1968, 90, 933.
20. (a) Hogeveen, H.; Bickel, A.F. *Rec. Trav. Chim. Pays-Bas,* 967, 86, 1313.
    (b) Pines, H.; Hoffman, N. E. in *Friedel-Crafts and Related Reactions,* Olah, G. A., Ed., Vol. II Wiley-Interscience, New York, N. Y., 1964, p.1216.
21. (a) Olah, G. A.; Halpern, Y.; Shen, J. ; Mo, Y. K. *J. Am. Chem. Soc,* 1971, 93, 1251.
    (b) Sommer, J.; Bukala, J.; Rouba, S.; Graff, S.; Ahlberg, P. J. *J. Am. Chem. Soc.,* 1992, 114, 5884.
22. Otvos, J. W.; Stevenson, D. P.; Wagner, C. D.; Beeck, O. *J. Am. Chem. Soc.,* 1951, 73, 5741.
23. (a) Olah, G. A.; Hartz., N.; Rasul, G.; Prakash, G. K. S. *J. Am. Chem. Soc.,* 1993, 115, 6985.
    (b) Olah, G. A. *Angew. Chem. Int. Ed. Engl.,* 1993, 32, 767.
24. Lucas, J.; Kramer, P. A.; Kouwenhoven, A. P. *Recl. Trav. Chim. Pays-Bas.,* 1973, 92, 44.
25. Culman, J-C. and Sommer, J., J. Am. Chem. Soc., 1990, 112, 4057.
26. Sommer, J.; Bukala, J. *Acc. Chem. Res.,* 1993, 26, 370.
27. Haag, W. O.; Dessau, R. H. *International Catalysis Congress (W. Germany),* 1984, II, 105.
28. Nenitzescu, C. D.; Dragan, A. *Ber. Dtsch. Chem. Ges.,* 1933, 66, 1892. .
29. Germain, A.; Ortega, P.; Commeyras, A. *Nouv. J. Chem.,* 1979, 3, 415.
30. Schleyer, P. v. R. *J. Am. Chem. Soc.,* 1957, 79, 3292.
31. Olah, G. A.; Farooq, O.; Husain, A.,; Ding, N.; Trivedi, N. *Catal. Lett.,* 1991, 10, 239.
32. Schuit, G. C.; Hoog, H.; Verhuis, J. *Recl. Trav. Chim. Pays-Bas.,* 1940, 59, 743
33. Olah, G. A.; Bruce, M.; Edelson, E. H. ; Husain, A. F. *Fuels,* 1984, 63, 1432.

34. Olah, G. A., U.S. Patent **1984**, 4,443,192; 4,465,893; 4,467,130 **1985**, 4,513,164.
35. Siskin, M. *J. Am. Chem. Soc.,* **1976**, 98, 5413.
36. Sommer, J.; Muller, M.; Laali, K. *Nouv. J. Chim.,* **1982**, 6, 3.
37. Olah, G. A.; Felberg, J. D.; Lammertsma, K. *J. Am. Chem. Soc.,* **1983**, 105, 6529.
38. Olah, G. A.; Prakash, G. K. S. in *The Chemistry of Alkanes and Cycloalkanes,* Patai. S.; Rappoport, Z., Eds, John Wiley and Sons Ltd., London, U.K., 1992, Ch 13.
39. Chang, C.in *Perspectives in Molecular Sieve Science,* Flank, W. H.; Whyte, T. E., Jr., Eds.; ACS Symposium Series 368; American Chemical Society: Wahington, DC, 1988; Ch 39.
40. Olah, G. A. *Acc. Chem. Res.,* **1987,** 20, 422.
41. Jackson, J. E.; Bertsch, F. M. *J. Am. Chem. Soc.,* **1990**, 112, 9085.

RECEIVED June 17, 1994

Chapter 2

Role of Hydrogen Transfer in Isobutene–Isobutane Selectivities

Gretchen M. Mavrovouniotis[1], Wu-Cheng Cheng, and Alan W. Peters[2]

W. R. Grace and Company—Connecticut, 7379 Route 32, Columbia, MD 21044

A mechanisms for isobutene formation and disappearance is described. The results show that under cracking conditions high hydrogen transfer catalysts such as REY can convert isobutene to isobutane. The hydrogen transfer reaction requires relatively severe conditions or a high conversion level. Some catalysts including beta and ZSM-5 have little hydrogen transfer activity and produce higher amounts of isobutene. Non-zeolitic silica alumina materials also have low hydrogen transfer activity and can be utilized as a component of a cracking catalyst for improved isobutene production.

The Clean Air Act Amendments of 1990 require reformulation of gasoline sold in ozone and CO non-attainment areas. The specifications for this reformulated gasoline include a minimum limit of 2.0 Wt% oxygen in the gasoline. Because of their superior blending properties, methyl t-butyl ether (MTBE), as well as ethyl t-butyl ether (ETBE) and t-amyl methyl ether (TAME), will likely supply much of the oxygenate needed to meet this requirement. These ethers are produced by reacting methanol or ethanol with isobutene or isoamylene. The FCCU, as the primary source of light olefins in the refinery, will be called upon to provide increased yields of isobutene and isoamylene. While one might expect that isobutene would be formed in approximate thermodynamic equilibrium with the other three linear butenes, but under cracking conditions in an FCCU this is not so. The amount of isobutene can be less than half of the amount expected in the case of REY catalysts, and even in the case of USY based catalysts without rare earth the observed isobutene yield is only 60-70% of the amount expected (1). It has been suggested in the literature that isobutene can react with a hydrogen donor to produce isobutane (2). Recent results of cracking over USY catalysts of different unit cell show the characteristic loss of isobutene and increase in isobutane with unit cell size. The degree of both C4 and C5 branching was approximately conserved independently of unit cell size (1). These results are consistent with the suggestions that isobutene is being converted to isobutane. Using test reactions both the zeolites Beta and ZSM-5 have low hydrogen transfer activity compared to rare earth stabilized USY (3). A comparison of Beta and USY for cracking heptane showed a characteristic relative increase in butene and decrease in isobutane selectivities for Beta (4).

Since isobutane is also a valuable product of the FCCU, it is important to be able to control the relative amounts of isobutane and isobutene. In this paper we show that the

[1]Current address: Allied Signal Research and Technology, 50 East Algonquin Road, Des Plaines, IL 60017–5016
[2]Corresponding author

relative amounts of these products can be controlled catalytically. Hydrogen transfer is found to play a key role in determining the overall isobutene and isobutane selectivity. High hydrogen transfer catalysts make relatively more isobutane at the expense of isobutene, and lower hydrogen transfer catalysts make less isobutane and more isobutene.

Experimental

Catalysts tested include a USY catalyst tested by itself and with a 10% blend of a catalyst particle containing ZSM-5. The properties of the feedstock used for the MAT testing (ASTM D-3907) are given in Table 1. Additional catalysts were compared in a circulating pilot plant riser (DCR) (5) using the same feed. These catalysts include a commercial catalyst containing USY as the active component, GO-35™, available from Grace Davison, Baltimore, MD, a silical alumina cogel containing 75 Wt% alumina (no zeolite) and to a cotalyst consisting of the same cogel with a USY component.

Table 1.   Feedstock Inspections for the MAT and DCR Gas Oil Cracking Experiments

API Gravity @ 60°F	22.5	Specific Gravity @ 60°F	0.919
Aniline Point, °F	163	Sulfur, wt.	2.59
Total Nitrogen, Wt%	0.086	Basic Nitrogen	0.034
Conradson Carbon, Wt%	0.25	Ni, ppm	0.8
V, ppm	0.6	Fe, ppm	0.6
Cu, ppm	0.1		

D-1160 Distillation	
Vol. %	°F , 1 atm.
10	615
50	755
90	932

In a second set of experiments isobutene and cis-2-butene feeds were reacted over four different catalysts. The reactions were performed at 500°C using 2.5 Wt% of the olefin feed diluted in nitrogen. Between 0.02 and 0.7 grams of catalyst was used with a hydrocarbon flow rate of 50 cc/min. Activity was controlled by varying the amount of catalyst in each run. The product was sent directly to the GC using a sampling valve. Samples were taken as soon as the first aliquot of product entered the sampling loop. This provides the product yields prior to catalyst deactivation. The four different zeolites studied were REY, USY, ZSM-5, and Beta. The REY and USY zeolites were bound in a matrix, while the ZSM-5 and Beta were tested as a pure zeolite mixed in a bed of quartz chips. All of the catalysts in all of the experiments were steamed at 816 °C for 4 hours in 100% steam prior to testing.

Results and Discussion

Blends of REUSY and ZSM-5 Catalysts. A REUSY (rare earth exchanged USY) catalyst was MAT tested by itself and with a 10% blend of a catalyst particle containing ZSM-5. The properties of the catalysts are given in Table 2.

The blended catalyst was tested both mixed and layered with the ZSM-5 catalyst on the bottom. If isobutene is being transformed to isobutane over the high hydrogen transfer REUSY catalyst, the mixed blend should make more isobutane and less isobutene compared to the layered blend. This is the result observed, Table 3. The mixed blend produced nearly twice the amount of isobutane, while the layered blend produced close to thermodynamic quantities of isobutene and much less isobutane.

Table 2. Properties of REUSY and ZSM-5 catalysts

	REUSY	ZSM-5
Wt% RE_2O_3	2.6	0
unit cell size	2.428 nm	-
Si/Al	-	26
t-plot		
zeolite area, m2/g	107	45
matrix area, m2/g	45	16

Table 3. MAT Yields at 60 Wt% Conversion of REUSY with 10% ZSM-5 Additive as a Physical Blend or Layered Bed (ZSM-5 on the bottom)

		REUSY/ZSM-5		
Catalyst	REUSY	Blend	Layered	
C/O	2.90	3.10	2.96	
H2	0.07	0.06	0.06	
Tot C1+C2	1.54	1.90	1.83	
C3=	3.61	7.82	5.93	
Total C3s	4.27	9.00	6.76	
Isobutylene	0.96	2.04	2.33	
Total C4=	4.00	6.26	5.65	
iC4	2.93	4.65	2.80	
Total C4s	7.51	11.69	9.11	
C5+Gaso	43.64	34.28	39.41	
LCO	23.63	22.61	23.36	
640+Btms	16.40	17.39	16.67	
Coke W% Feed	2.76	2.90	2.62	
C4 Olefin Distribution				**Thermodynamic**
1-Butene/C4=	0.25	0.22	0.19	0.140
Trans-2-butene/C4=	0.26	0.24	0.21	0.245
Cis-2-butene/C4=	0.24	0.21	0.19	0.166
Isobutene/C4=	0.24	0.33	0.41	0.448

C4 Olefin Selectivities from Cis-2- butene and Isobutene Reaction. In another series of experiments isobutene and cis-2-butene were reacted under gas oil cracking conditions over a series of steamed catalysts or zeolites including REY, a low unit cell size (2.423 nm) USY, ZSM-5, and Beta. Beta and the ZSM-5 are known to be low hydrogen transfer zeolites, USY is intermediate, and REY is a high hydrogen transfer zeolite having a higher site density. The properties of the catalysts and the reaction yields interpolated to 15% conversion are shown in Table 4.

Olefin isomerization was observed to be quite rapid. The three unbranched butene species were formed in equilibrium amounts: 41% trans-2-butene, 30% cis-2-butene, and 29% 1-butene from both cis-2-butene and isobutene under our experimental conditions. Thus cis-2-butene conversion is expressed in terms of the total conversion of all unbranched butenes to other species. Isobutene is the major product of the conversion of cis-2-butene, and linear butenes are the major products of isobutene conversion. For both isobutene and cis-2-butene, the pentene species were found to be formed in approximately equilibrium proportions: 41% 2-methyl-2-butene, 22% 2-methyl-1-butene, 16% trans-2-pentene, 9% cis-2-pentene, 6% 3-methyl-1-butene, and 6% 1-pentene. First order rate constants were determined from the data plotted in Figures 1a and 1b. REY, ZSM-5, and Beta had similar activity on a zeolite basis, while the activity of the USY catalyst was considerably less.

Table 4.   Catalyst Composition and Major Products from the Reaction of Cis-2-Butene and Isobutene (moles/100 moles cracked, 500°C)

Catalyst	20% REY	50% USY	ZSM-5	Beta
Properties				
Matrix	Silica	Silica	none	none
SA, m^2/g	115	221	410	640
Wt% Na_2O	0.32	0.42	0.016	0.135
Wt% Al_2O_3	-	-	6.78	5.44
Wt% SiO_2	-	-	92.98	94.23
Si/Al	10	37(2.423 nm uc)	12	15
Cis-2-Butene Cracking				
Mole % conversion	15.0	15.0	15.0	15.0
methane	1.7	2.1	1.2	1.6
ethene	1.1	1.1	1.8	1.2
propene	21.2	10.4	19.5	18.0
isobutane	**2.8**	**1.3**	**0.3**	**0.7**
n-butane	1.6	0.8	0.9	1.3
isobutene	57.0	73.0	65.0	65.0
pentenes	11.8	5.4	8.6	9.5
other C_5+	3.7	5.0	5.0	5.0
Rate constant, hr^{-1}	0.3	0.07	0.8	0.6
Isobutene Cracking				
Mole % conversion	15.0	15.0	15.0	15.0
methane	-	0.8	-	-
ethene	0.5	0.6	0.6	0.3
propene	16.4	9.8	8.5	11.0
isobutane	**5.6**	**4.0**	**0.9**	**1.5**
linear butenes	62.8	77.5	83.2	78.4
pentenes	13.8	7.7	5.6	8.5
Rate constant, hr^{-1}	0.4	0.09	0.6	0.8

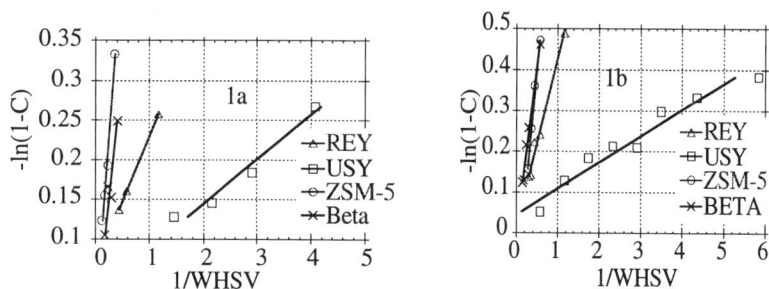

Figure 1.   Kinetic plot for the conversion of a) cis-1-butene and b) isobutene.

The major products from the reaction of isobutene are formed by skeletal isomerization, disproportionation, and hydrogen transfer. Changes in product composition during isobutene cracking are shown in Figures 2a-d. Over the REY catalyst isobutane increases with conversion at the expense of linear butenes. Very little isobutane is formed over the lower hydrogen transfer USY catalyst, and even less isobutane is formed over the ZSM-5 and Beta catalysts.

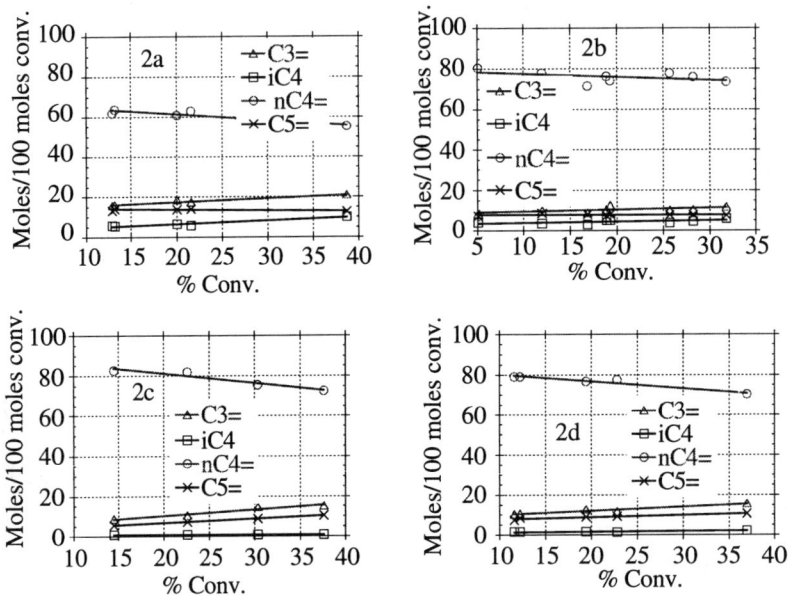

Figure 2.   Product selectivity for the reaction of isobutene over zeolites a) REY, b) USY, c) ZSM-5, and d) beta.

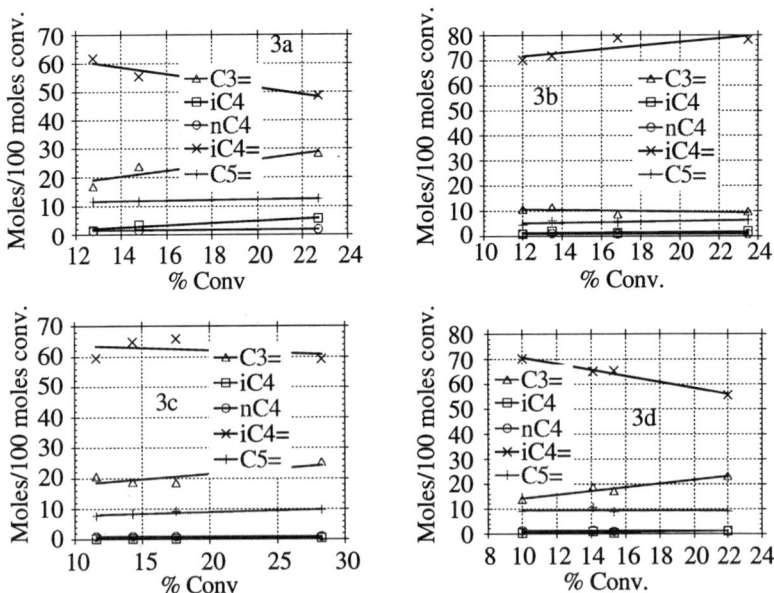

Figure 3.   Product selectivity for the reaction of cis-1-butene over zeolites a) REY, b) USY, c) ZSM-5, and d) beta.

The results of cis-2-butene reaction are similar, Figures 3a-d. The REY catalyst makes increasing amounts of C3 olefins and isobutane with conversion, although less isobutane is produced compared to the case of isobutene cracking. ZSM-5, Beta, and USY do not make isobutane from cis-2-butene under these conditions.

The four zeolites, REY, USY, ZSM-5, and Beta are different in their selectivities for the C5 olefins from the reaction with isobutene, Figures 4a-d. The REY catalyst , Fig. 4a, produces relatively large amounts of C5 olefins decreasing slightly with conversion. The ZSM-5 and Beta catalysts produce smaller amounts of C3 and C5 olefins, but the amounts of both increase with conversion, Figs. 4b,c. The results suggest that these catalysts transfer methyl groups rather than hydrogen under these conditions. ZSM-5 is known to possess transalkylation or methyl transfer activity in aromatic systems such as toluene (6). As is the case with hydrogen transfer effects, the same trends to a lesser extent are observed with the ZSM-5 catalyst in the cis-2-butene experiments, Fig 4d.

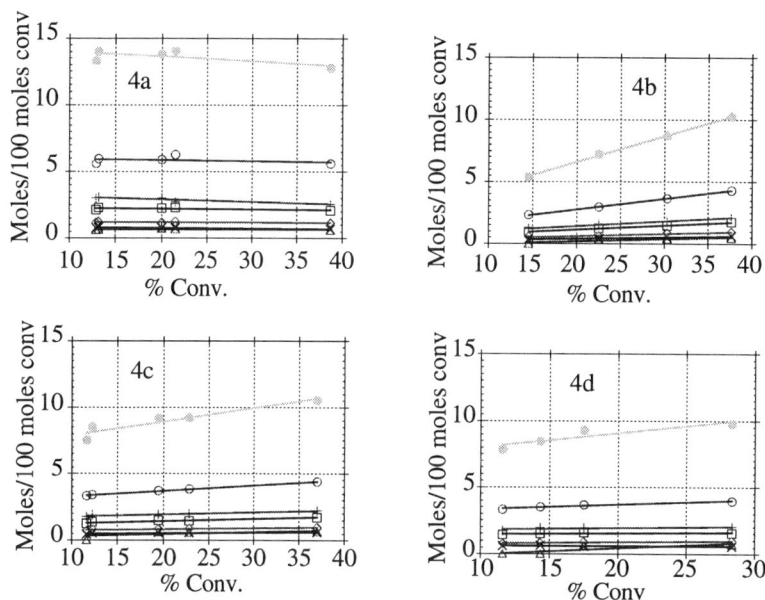

Figure 4. Selectivity for C5 olefins for the reaction of a) isobutene over ZSM-5, b) isobutene over beta, c) isobutene over REY, and d) cis-2-butene over ZSM-5. Legend: • total pentene; o 2-m-2-butene; + 2-m-1-butene; ∆ 3-m-1-butene; ◊ cis-2-pentene; ☐ trans-2-pentene; x 1-pentene.

Silica Alumina Catalysts for the Production of Isobutene.   Silica alumina materials are frequently used as bottoms cracking components of cracking catalysts. Such materials have been observed to produce a high octane olefinic gasoline consistent with low hydrogen transfer activity (7). The gas oil cracking results at constant coke in Table 5 obtained from the DCR unit also show that these materials function as active catalytic cracking materials with low hydrogen transfer activity. The properties of the catalysts of this study are also described in Table 5.

The base catalyst, GO-35™, a commercial catalyst containing a fully rare earth exchanged USY zeolite, makes relatively large amounts of isobutane and reduced amounts of isobutene consistent with the high rare earth content and moderate unit cell size, and makes low coke. The non crystalline silica alumina active matrix material for this study produced enhanced yields of isobutene and correspondingly reduced isobutane yields. The catalyst also produced an olefinic, higher octane gasoline, although at reduced conversion at constant coke and with reduced gasoline selectivity. The catalyst containing USY in the high activity silica alumina matrix produced a gasoline with a RON and an isobutene yield similar to the high values produced by the matrix alone, but with a constant coke conversion and a gasoline yield closer to the GO-35. The selectivies of these mixtures of high and low hydrogen transfer components appear to be synergistic with respect to the desired selectivities for gas oil cracking.

Table 5. Catalyst Properties and Selectivity Comparison in DCR (riser)Testing, at 970°F and Constant Coke using the Table 1 Feedstock

Catalyst	GO-35	Matrix	Matrix / USY
Properties			
Chemical, Wt.			
Al_2O_3	27.4	76.2	70.4
Rare earth	2.79	0	0
Na_2O	0.46	0.52	0.04
Bulk Density, g/cc	0.79	0.38	0.55
DI (attrition index)	3	30	19
Ave. Particle size, μ	69	74	72
Surface area, m^2/g	256	335	336
Activity and Selectivity			
Conversion, Wt.%	71.2	64.6	69.5
Yields, wt.%			
H2	0.05	0.07	0.05
C1 + C2	2.9	3.8	3.4
C3=	4.9	4.7	5.2
Total C3	6.2	5.9	6.5
Total C4=	6.2	7.3	8.1
Isobutene	1.2	2.6	2.7
Isobutene/C4=	0.19	0.37	0.33
Isobutane	3.4	1.7	2.3
C5+ Gasoline	45.6	39.5	42.8
LCO	16.9	19.6	18.2
640°F + HCO	11.9	15.8	12.3
Coke	5.0	5.0	5.0
Gasoline properties			
RON	92.0	95.2	95.3
MON	81.2	81.1	81.4
Isoamylene	4.3	7.9	8.2
Olefins PONA	22.8	45.4	41.8
Aromatics PONA	30.7	24.4	25.5

Conclusions

The relative amounts of isobutene and isobutane present in the product of the catalytic cracking process can be controlled catalytically. Materials with a high hydrogen transfer activity including rare earth exchanged Y or USY make larger amounts of isobutane, while materials with lower hydrogen transfer activity give a product containing larger amounts of isobutene and less isobutane. Materials low in hydrogen transfer activity include non crystalline silica alumina and some high silica zeolites such as ZSM-5 and Beta. Although the zeolites have low hydrogen transfer activity, both Beta and especially ZSM-5 appear to have activity for transferring methyl groups between molecules. Finally, mixtures of materials with high and low hydrogen transfer activities appear to be synergistic with respect to desirable FCCU selectivies including higher than expected gasoline yields, octane number, and isobutene yields, and lower than expected coke yields.

Acknowledgments

We thank Delnora M. Roberts for her technical assistance in the catalytic testing of the samples.

References

1. Cheng, W.-C., Suarez, W., and Young, G. W., AIChE Sym. Ser. 291, 1992, **88**, 38.
2. Parra, C. F., Goldwasser, M. R., Fajula, F, and Figueras, F., Appl. Catal., 1985, **17**, 217.
3. Cheng, W.-C., Peters, A. W., and Rajagopalan, K., in **The Hydrocarbon Chemistry of FCC Naphtha Formation**, H. J. Lovink and L. A. Pine, eds., ´Editions Technip, Paris, 1990, p. 105.
4. Corma, A., Fornés, V., Montón, J. B., and Orchillés, A. V., J. Catal., 1987, **107**, 288.
5. Young, G. W., in **Studies in Surface Science and Catalysis**, 76, *Fluid Catalytic Cracking*, J. S. Magee and M. M. Mitchell, eds., Elsevier, New York, 1993, p. 270.
6. Vinek, H. and Lercher, J. A., J. Mol. Catal., 1991, **64**, 23.
7. Edwards, G. C., Rajagopalan, K., Peters, A. W., Young, G. W., and Creighton, J. E., in **ACS Symposium Series 375**, *Fluid Catalytic Cracking*, M. L. Occelli, ed., Amer. Chem. Soc., Washington D.C., 1988, p. 101.

RECEIVED June 17, 1994

Chapter 3

Benzene Levels in Fluid Catalytic Cracking Gasoline

C. A. Yatsu[1], T. A. Reid[2], D. A. Keyworth[2], R. Jonker[3], and Marcelo A. Torem[4]

[1]806 Devonport, Seabrook, TX 77586
[2]Akzo Chemicals, 13000 Bay Park Road, Pasadena, TX 77507
[3]Akzo Chemicals bv, P.O. Box 26223, Netherlands
[4]Petrobras, Petroleo Brasolieiro S. A. Petrobras Research Center (CENPES), Cidade Universitaria, Quadra 7, Ilha, Brazil

Benzene is present in FCC gasoline typically at a level of 0.5 to 1.5%. The Clean Air Act Amendment has defined benzene as a toxic and will limit its level to 1% maximum in the gasoline pool. In the FCC unit benzene tends to be present in a fixed ratio (binomial distribution) to the methylbenzenes. Based on pilot plant and commercial data, benzene levels in FCC gasoline are very feedstock dependent. CREYs tend to form commercially considerably more benzene (1.0 - 1.3%v) than octane catalysts (0.5 - 0.8%v). Benzene level increases with conversion. Increased reactor temperature, increased RE which increases conversion, and increased oil-vapor residence time all increase benzene levels in FCC gasoline. While the primary factor controlling benzene levels in FCC gasoline is feedstock type, FCC catalyst factors are very important.

Historically the fluid catalytic cracking process (FCC) has been geared towards the production of maximum gasoline. As lead phase out hit the gasoline industry, a need for balancing the maximum gasoline make with octane value arose. One strategy of increasing the octane value was to increase the aromatics content of the gasoline pool by blending in reformate. Tougher legislation is now directing refiners to lower the benzene and total aromatics content in the finished gasoline. Benzene's carcinogenic properties have rendered it an undesirable component of gasoline. The national concern with gasoline aromatics content has been shown in the Clean Air Act Amendment (CAAA) where by 1995, gasoline, which is distributed in ozone non-attainment and opt-in areas, can contain no more than 1%v benzene and no more than 25%v total aromatics. The California Air Resources Board (CARB II) also will put into effect more stringent rulings which state that the benzene content shall not exceed 0.8%v by 1996. Several other states are also adopting the CARB II rulings indicating that refiners will have to find new solutions for their processes. Major

0097–6156/94/0571–0024$08.00/0
© 1994 American Chemical Society

sources of benzene in finished gasoline are reformate product, FCC gasoline and hydrocrackate material. Although in the present formulation of the gasoline pool FCC contributes only about 10%v of the total benzene, the contribution of benzene from the FCC process must be monitored (1). Control of the benzene in the pool from FCC gasoline requires an understanding of the chemistry and conditions involved in the formation of benzene in the FCC process.

Influence of FCC Unit Variables on Benzene

Benzene levels in the FCC process are most greatly affected by 1.) feedstock characteristics, 2.) operating conditions, and 3.) catalytic factors.

Feedstock Effects

A feedstock's characteristics play a large role in the product slate. In Table I three different feedstocks have been examined on a pilot riser using the same catalyst and run conditions.

The higher the aromatic character of a hydrocarbon mixture, the lower the aniline point (2). Hence, in the study shown above, the feedstock with the higher amount of aromatics made the higher amount of benzene and total aromatics in the gasoline product. Since aniline point shows only total aromatic character, it is not possible to conclude from this data whether or not the higher benzene production is a result of certain aromatic species such as only alkylbezenes or benzocycloparaffins. It is known that polynuclear aromatics tend not to crack in the FCC unit but remain in the 430 °F fractions (3). In studies comparing benzene, KVGO which is highly aromatic, and a Gulf Coast feed highly naphthenic but less aromatic, the feed with the lower aromatic character made more benzene at constant coke (4).

In the FCC unit benzene tends to be present in a fixed ratio to the other members of the methylbenzene series (5). The ratio is however dependent as shown (Figure 1).

Operating Parameters

Although feedstock effects are very important in determining aromatic and benzene levels, operating conditions also affect the benzene production. Increasing reactor temperatures causes an increase in the production of benzene and aromatics in general (4). High conversion levels also contribute to high benzene production. Below is a table showing three different feedstocks used in a study on a microsimulation unit (MST). Two of the feeds are typical FCC feeds with slightly different aromatics content.

TABLE I. BENZENE IN FCC GASOLINE AT 75% CONVERSION

	HVGO BRAZILIAN-1	HVGO BRAZILIAN-2	HVGO COLUMBIAN
FEED			
DENSITY 20/4 C	0.875	0.937	0.922
ANILINE POINT C	113.0	86.0	75.0
NAPHTHENES vol%	41.5	53.3	38.3
GASOLINE PRODUCT			
BENZENE wt%	0.4	0.9	1.3
TOTAL AROMATICS wt%	23.6	43.6	47.1

PILOT RISER CONDITIONS: RST=520 C
NAPHTHENES CALCULATED BY MASS SPECTROSCOPY
CORRELATIONS

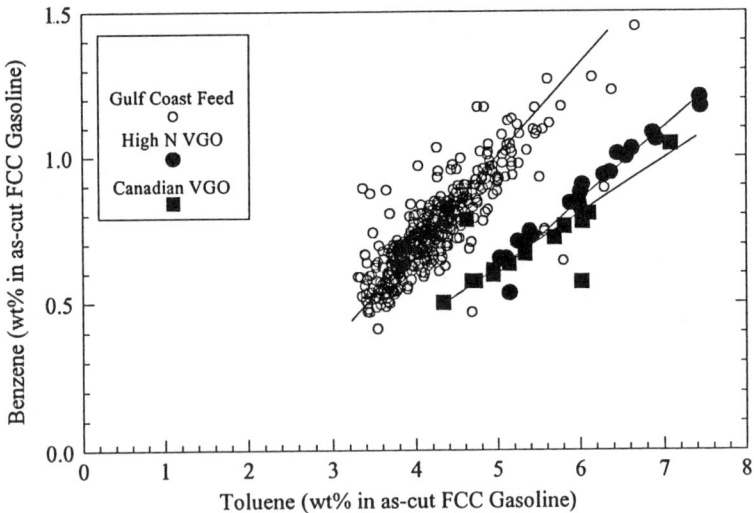

FIGURE 1. Feedstock effect on Benzene to Toluene ratio

The C_A, C_N and C_P values are calculated parameters indicating the amount of aromatic carbon, naphthenic carbon and paraffinic carbon (6). In this case the aromatic character of the feeds is much lower than those used in the pilot riser study cited above as seen by the aniline point.

Results of the MST test are shown in figure 2. Here, the increase in benzene level with conversion can be noted for all feeds. Note in the figure below that a feedstock completely devoid of aromatic character can produce benzene at equivalent levels as the feeds containing aromatics. The conversion of mineral oil feed is much higher than that of the other two feeds because there are no aromatic clusters which are difficult to crack and are coke formers (7). At a constant conversion, there is not a significant difference in benzene production between the KVGO and Gulf Coast feed even though a difference in the total aromatics of the feeds is seen in Table II.

Temperature has a big effect on benzene production in the FCC unit. Shorter contact times have been shown to reduce benzene levels (4).

Catalytic Factors
One of the important variables in the FCC process is the type of catalyst used. In general CREY (calcined rare earth Y type zeolites) catalysts tend to make the greatest amount of benzene (1.0 - 1.3%v), where octane catalysts produce the least (0.5 - 0.8%v) amount of benzene (4). In looking at 90 commercial FCC gasolines from nine different refineries with different operating conditions and catalysts the data summarized in Table III were found.

Chemistry

The mechanisms forming alkylbenzenes and cycloparaffins are operative in the formation of benzene in FCC gasoline. As observed in the MST study using mineral oil, aromatics can be formed from linear paraffins. From previous work it was shown that the amount of benzene relative to toluene formed in the FCC process using paraffin wax feed is smaller (Table IV) than that observed with a commercial FCC feedstock (5). These relative ratios show that about half of the benzene must come from mechanisms other than the cracking of straight chain paraffins.

The hydrogen transfer mechanism in the FCC process transforms olefins to paraffins. It is well accepted that olefins cyclize to form cycloparaffins with subsequent hydrogen transfer to aromatics. In the case of benzene formation, the relationship

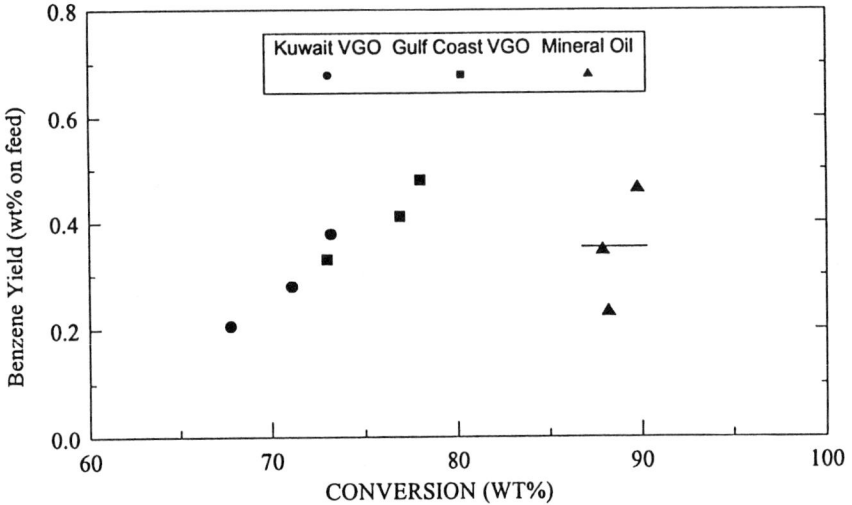

FIGURE 2. Feedstock Effect on Benzene Yield

TABLE II. FEEDS USED IN MST TESTING

FEED	Ca	Cn	Cp	ANILINE PT.
KVGO	23.5	11.4	65.1	81
GULF COAST	13.3	22.4	64.3	98
MINERAL OIL	0.0	0.0	100.0	NA

TABLE III. BENZENE IN COMMERCIAL FCC GASOLINES

WT% IN GASOLINE	RANGE	AVERAGE
BENZENE	0.30-1.04	0.65
TOTAL AROMATICS	25.7-48.5	33.60

between decreasing 1-methylcyclopentene (1-MCP) with increasing benzene (4) suggests that first a 1-MCP ring forms by cyclization of hexene, and that the 1-MCP then ring expands to form a cyclohexene. The cyclohexene then, by hydrogen transfer, forms benzene. Other aromatics such as toluene form by cyclization of heptene to methylcyclohexene. By increasing the reactor temperature the ratio of benzene to toluene is increased (Figure 3).

Both benzene and 1-MCP increase with reactor temperature, and depending on the catalyst the ratio of 1-MCP to benzene either decreases or remains constant (Figure 4). These phenomena imply that either the increased benzene yield is a result of increased 1-MCP being transformed by cyclization into benzene or some benzene is being made by alternative routes such as dealkylation of alkyl cycloparaffins and cycloolefins.

Once formed, the aromatics undergo transalkylation to form a series (5). The ratio of aromatics in this series will vary slightly with different feedstocks as shown in figure 1. However, the amount of benzene remains very low in comparison to the other members of the methyl benzene series.

Determination of Aromatics in the Feed

Several methods exist for measuring total aromatics in feeds.

        1.) ASTM D-3238 n-d-M Method
        2.) ASTM D-611 Aniline Point Method
        3.) ASTM D-1319 Fluorescent Indicator Absorption
        4.) Total aromatics by HPLC using Dielectric Constant
Detector (8)

Measuring Benzene in FCC Product

The determination of benzene in FCC gasoline is difficult because of the relatively low level of benzene in the presence of interfering compounds like (1-MCP) and other mono and di-olefins. The following methods of analysis can be applied to the analysis of benzene in FCC gasolines.

Gas Chromatography

Both single and multi-column systems can be used with gas chromatography to analyze for benzene.

Single column GC methods are either performed by high efficiency, nonpolar columns or medium to low efficiency columns with a highly polar stationary phase.

TABLE IV. Paraffin Wax Study
Ratio of Benzene: Toluene: Ethylbenzne: Xylene

FEED	BENZENE+1MCP	TOLUENE	ETHYLBENZENE	XYLENES
PARAFFIN	0.07	0.20	0.05	0.74
VGO	0.14	0.32	0.07	0.47

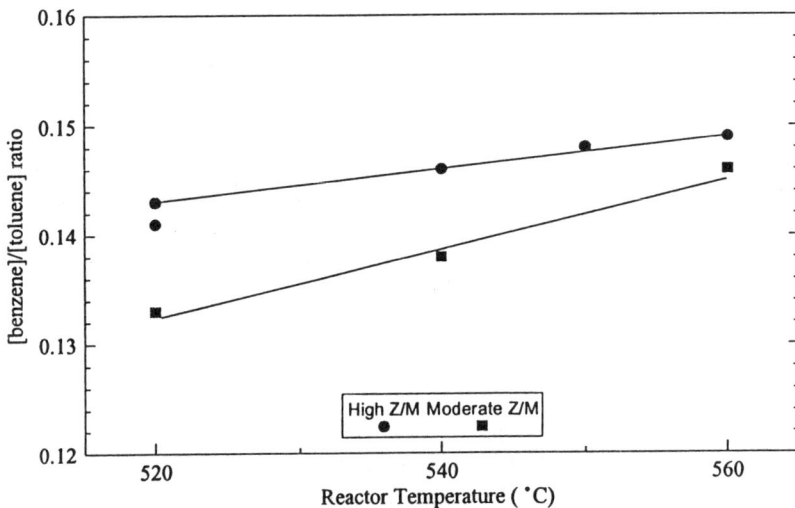

FIGURE 3. Effect of reactor temperature and Zeolite
to Matrix ratio on Benzene:Toluene

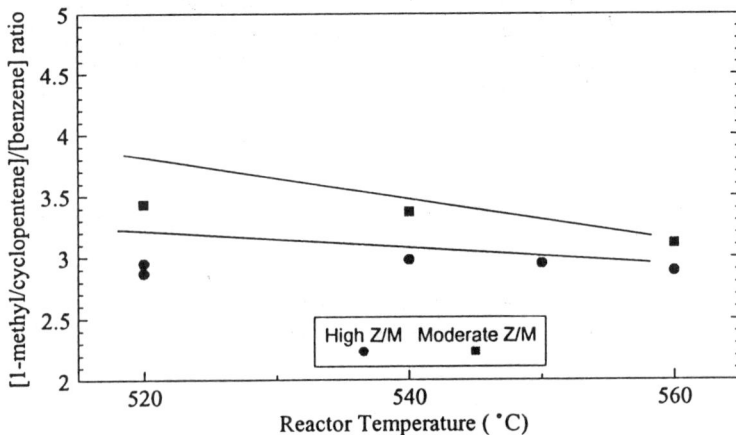

FIGURE 4. Effect of reactor temperature on Benzene:1-Methycyclopenten

The high polarity stationary phase columns (e.g. OV 275, carbowax), in which the benzene elutes after the non-aromatics from the column, cannot be applied to full range gasolines. In this instance, benzene most likely elutes with higher di-olefins. However, this approach may be appropriate for a light naphtha.

Benzene determination using a highly efficient, nonpolar column (100 meter, Supelco, Petrocol DH) can suffer from the coelution of benzene with 1-MCP and other olefins. The other olefins are normally in low concentration in FCC gasoline and in any case are largely separated through GC optimization. When 1-MCP is present at a threefold concentration higher than benzene special attention to GC optimization is necessary. Such optimization includes controlling the temperature and use of a 100 meter column (Figure 5). Separation of the benzene 1-MCP pair is very sensitive to changes in the polarity of the stationary phase and the temperature of the run. Different column suppliers, differences in the temperature program, changes in the carrier gas and in GC dead time can lead to significantly different elution patterns.

With multicolumn GC methods the coelution of the benzene peak with other peaks can be avoided. The transferring of compounds, which would coelute on one column, to two separate columns with stationary phases of different polarity eliminates the coelution problem. The ASTM D-3606 method (9) and the PNA analyzer (10) are examples of the use of this approached based on packed columns.

Column switching techniques can also be applied. Two gas chromatographs (for example, HP 5890s), can be equipped with capillary columns of opposite polarity, where column switching is controlled by changing the pneumatic conditions in the system (Gerstel, Mulheim a/d Rurh, FRD) is an example. The instrument can be coupled to a MAS spectrometer for peak identification. In Gerstel's experimentation two compounds other than 1-MCP were found to coelute with benzene. The mass spectra of those peaks showed that they were branched mono-olefins.

Conclusions

Benzene is a product of the FCC process as are other members of the methylbenzene series. Operating parameters can be changed to optimally reduce the amount of benzene in the product. In particular, lower run temperatures, use of octane catalysts, short vapor contact times and lower conversions will result in lower benzene production. Feedstock effects are important, but not easy to predict.

Due to the low levels of benzene found in the FCC product, analysis of benzene is best accomplished by gas chromatography. Careful optimization of run conditions can result in good separation power on a single column GC. Multicolumn GC has advantages over single column GC in that less chance of coelution can result. Successful work using a dual column capillary system has shown the possibility of significant interference of olefinic compounds in the separation of benzene. Such findings indicate that the actual levels of benzene are smaller than previously reported.

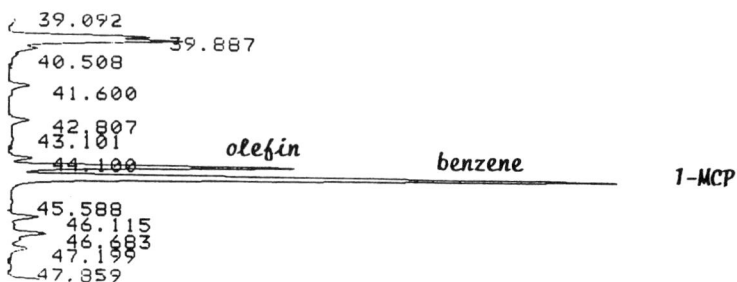

```
  ⟨  39.092
   =============39.887
   ⌐ 40.508
   ⌐  41.600
   ⌐ 42.807
   ⌐ 43.101         olefin              benzene            1-MCP
   ╾──44.100══════════════════════════════════════════
   ⌐──45.588
   ⌐  46.115
   ⌐  46.683
   ⌐  47.199
   ⌐──47.859
```

FIGURE 5. 100 Meter Separation of Benzene/1-MCP

Literature Cited

(1) Keesom, W. H., Kelly, A. P., Raman, Bozzano, U. G., National Petroleum Refiners Association, AM-91-36, "Benzene Reduction Alternatives".

(2) ASTM Standard D611-82, "Standard Test Methods for Aniline Point and Mixed Aniline Point of Petroleum Products and Hydrocarbon Solvents, " 1982, Annual Book of ASTM Standards, 1990.

(3) Keyworth, D. A., Reid, T. A., and Wilson, J., "Legislative and Market Trend's Effect on Octane Catalyst Selection," National Petroleum Refiners Association, AM-88-69.

(4) Keyworth, D. A., Reid, T. A. Kreider, K. R., Yatsu, C. A., Zoller, J. R., "Controlling Benzene Yield from the FCCU", National Petroleum Refiners Association, AM-93-49.

(5) Yatsu, C. A., Keyworth, D. A., "Aromatics Formation in FCC Catalysis"; Editor, Lovink, H. J. and Pine, L. A., ACS Symposium, Division of Petroleum Chemistry,Miami, Florida, 1989, Editions Technip, pp. 147-158.

(6) ASTM Standard D3238-82, "Calculation of Carbon Distribution and Structural Group Analysis of Petroleum Oils by the n-d-M Method", 1982, American Petroleum Institute, Washington D. C., Second Edition, 1970.

(7) Wojciechowski, B. W., Corma, A., Catalytic Cracking, Marcel Dekker, Inc., pp. 179-180.

(8) Hayes, P. C. Jr., Anderson, S. D., "Hydrocarbon Group Type Analyzer System for the Rapid Determination of Saturates, Olefins and Aromatics in Hydrocarbon Distillate Products", Anal Chem, 58 (1986), pp. 2384-2388.

(9) ASTM Standard D3606, "Standard Test Method for Benzene and Toluene in Finished Motor and Aviation Gasoline by Gas Chromatography", 1987, Manual on Hydrocarbon Analysis, Fourth Edition, 1987.

(10) Arkel, P. van, Beens, J. Spaans, H., Grutterink, D., Verbeek, R., J Chromatographic Science, 25, 141, (1987), p. 14.

RECEIVED June 17, 1994

Chapter 4

Deep Catalytic Cracking Process for Light-Olefins Production

Zaiting Li, Wenyuan Shi, Xieqing Wang, and Fuking Jiang

Research Institute of Petroleum Processing, China Petrochemical Corporation (SINOPEC), Beijing 100083, People's Republic of China

The newly developed process, Deep Catalytic Cracking (DCC) is a petrochemical extension of FCC for propylene and light isoolefins production. For maximum propylene operation, the propylene yield can be as high as 23 wt% (feed basis). In maximized isoolefin mode, both isobutylene and isoamylene yield can be over 6 wt%. The ratio of isobutylene to total butenes reached 0.42 which was close to the thermodynamic equilibrium value 0.44 at operation temperature. The isoamylene ratio was also near equilibrium. A demonstration unit was started up in 1990 and the results verified the pilot plant data. The effects of principal variables, such as feedstock, catalyst, reaction temperature, severity and hydrocarbon partial pressure are discussed.

The light olefins are increasingly useful for petrochemicals. The conventional technique for light olefins production is the well-known tubular furnace pyrolysis with light hydrocarbons such as natural gas, naphtha or light gas oil as feedstocks. In recent years, great attention has been paid to the development of catalytic processes by using heavy feedstocks to produce the light olefins. DCC (Deep Catalytic Cracking Process) is a novel technology developed by RIPP to convert heavy feedstocks into light olefin products.

DCC is a petrochemical extension of FCC which aims at producing maximum propylene as well as butenes, especially isobutylene from VGO or VGO blended with DAO and resid feedstocks. In maximum propylene mode operation, the propylene yields from various VGO feedstocks are in the range of 13-23 wt% (feed basis) with 4-7 wt% ethylene and 10-16 wt% butenes. It can be attractive as a supplement to an ethylene plant to achieve a desired propylene/ethylene balance. DCC can also be operated in maximum isoolefin mode in which isobutylene and isoamylene yield can both be as high as 6 wt%, still with significant amount of propylene. These olefins are the building blocks of many clean fuels components such as MTBE, TAME and alkylates. This enables the refiner to maximize the reformulated gasoline production from an existing crude slate.

0097–6156/94/0571–0033$08.00/0

Process Description

The process scheme is similar to that of FCC, the main facilities of which consist of a reaction-regeneration zone and a fractionation zone (1). Feedstock dispersed with steam is fed to the system, then contacted with the hot regenerated catalyst either in a riser plus fluidized bed (maximum propylene operation) or a riser (maximum isoolefin operation) and is catalytically cracked. Reactor effluent proceeds to a fractionator for separation. The coke-deposited catalyst is stripped with steam and transferred to a regenerator where air is introduced and the coke on the catalyst is removed by combustion. The hot regenerated catalyst is returned to the reactor at a controlled circulation rate to achieve the heat balance of the system. The innovations involved are the tailor-made catalyst, the variation of operating parameters and certain coking-prevention means.

Maximum Propylene Operation

Catalyst. Catalyst is always the key factor of a catalytic process. A DCC catalyst needs to be tailored to light olefins production as well as heavy oil conversion. CHP-1 is a specially designed catalyst for propylene production from heavy oil via primary and secondary reactions with low hydrogen transfer ability. The physical and chemical properties of the catalyst are given in Table I. These physical properties, such as attrition index and particle size distribution, all meet the conventional FCC catalyst specification.

Table I.   Properties of CHP-1 Catalyst

Catalyst		CHP-1
Chemical composition	wt%	
Aluminum oxide		47.9
Sodium oxide		0.10
Ferric oxide		0.46
Physical properties		
Pore volume	ml/g	0.22
Surface area	sq.m/g	154
Apparent bulk density	g/ml	0.84
Attrition index	%	2.1
Particle size distribution	wt%	
0-20 micron		2.1
20-40 micron		16.0
40-80 micron		66.0
>80 micron		15.9

Light Olefin Yields from Various Feedstocks. DCC process is applicable to various feedstocks for propylene production. Table II lists the pilot plant test results of six representative VGOs and one VGO blended with atmospheric resid which mixed CCR (Conradson Carbon Residue) is 2.8%. The No. 1 VGO from

Indonesia is a high paraffinic feedstock which propylene yield can be as high as 23.73 wt% (feed basis). The No. 2 Chinese Daqing VGO is also paraffinic which propylene yield is 21.03%. No. 3 and 4 VGOs are from Chinese intermediate-base crudes of which the propylene yields are somewhat less than the paraffinic ones but still near 17-18 wt%. No. 5 is a VGO of a naphthene-base Iranian crude, whose property is similar to No. 6 Chinese Liaohe VGO. The propylene yields are around 13 wt%. The Daqing VGO blended with 50% ATB still exhibits good crackability, and its propylene yield can be reached 17.86 wt%.

**Table II. DCC Light Olefins Yields from Various Feedstocks
(maximum propylene mode)**

Number	1	2	3	4	5	6	7
Feedstock	Indonesia VGO	Daqing VGO	Linsheng VGO	Shengli VGO	Iranian VGO	Liaohe VGO	Daqing VGO/ATB
Density wt%							
	0.8449	0.8579	0.8808	0.8868	0.9004	0.9100	0.8815
API	36.0	33.1	29.1	28.1	25.6	24.0	29.0
UOP K	12.7	12.4	12.0	11.9	11.7	11.6	12.2
Light olefin yield wt%							
Ethylene	5.79	6.10	4.93	4.29	3.51	3.58	4.06
Propylene	23.73	21.03	18.39	16.71	13.57	13.16	17.86
Butylene	17.78	14.30	14.16	12.73	10.11	10.60	14.89

Commercial Results.   An idle FCC unit with 120,000t/yr original designed capacity was revamped for DCC operation. For minimizing reconstruction and cost, the throughput was half of the FCC, which was constrained by the capacity limitation of the wet gas compressor. The reactor-regenerator section was revamped to have the flexibility of either riser plus fluidized bed or riser with zero level operation. This 60,000 t/yr unit was started up in Jinan Refinery, SINOPEC, in November (2, 3). It is under revamping to enlarge the capacity to 150,000 t/yr with a compressor replacement and is planned to start up in March 1994.

The feedstocks used in the commercial unit were similar to the No. 3 feedstock listed in Table II, all the UOP K factors were 12.0 from the same crude. The commercial feedstocks were blended with some deasphalted oil. The CCRs of the feedstocks were in the range of 1.18-1.41 wt% for self-heat balance operation. The commercial feedstock analysis is shown in Table III. The yield structure and gas composition in comparison with the 2 b/d pilot plant results are listed in Table IV and V.

The product yields and light olefin selectivities obtained from this commercial run almost duplicated those obtained from the pilot plant. Total gas yield was around 50 wt% in which propylene reached about 19 wt% as a major product. The butylene content was 14 wt% as an important co-product. More promising was the isobutylene content which accounted for more than 40 wt% of total butylene isomers. Ethylene yield was in the range of 4-5 wt% which was much less than propylene and butylene, but its concentration in dry gas was two times higher than that in FCC. From the standpoint of using the diluted ethylene to produce ethylbenzene, DCC gas is more valuable than that of FCC.

Table III. Commercial Feedstock Analysis

Feedstock		1	2	Pilot plant
Density (20°C)	g/ml	0.8946	0.8862	0.8808
API		26.0	28.2	29.1
CCR	wt%	1.18	1.41	0.24
Pour point	°C	45	34	44
Viscosity	mm²/s			
100°C		7.85	6.79	5.23
Distillation	°C			
10%		367	315	328
50%		451	440	405
90%		550	541	499
EBP		-	(80%)	545
Sulfur	wt%	0.48	0.33	0.33
UOP K		12.0	12.0	12.0

Table IV. DCC Commercial Yield Structure in Comparison with Pilot Plant

Unit		Commercial		Pilot plant
		1	2	
Yield structure	wt%			
C_4- gas		50.58	50.33	51.53
where in				
Hydrogen		0.25	0.26	0.56
Methane		3.56	4.39	4.00
Ethane		2.17	2.32	2.47
Ethylene		4.43	5.36	4.93
Propane		3.33	2.91	3.30
Propylene		19.01	19.25	18.39
Butane		3.15	2.68	3.72
Butenes		14.68	13.16	14.16
C_5+ Naphtha		22.72	20.94	24.77
LCO		16.10	18.37	15.08
Coke		10.60	9.39	8.10
Loss		0	0.97	0.52
Total		100.00	100.00	100.00

The trace diolefins, alkynes and impurities in LPG and dry gas are shown in Table VI. The impurities are far less than that in pyrolysis gas. The naphtha is a high-octane blending component in which RONC and MONC was 96.5 and 83, respectively. The naphtha can be selectively hydrotreated to saturate the diolefin. The BTEX amounted to about 30 wt% in naphtha. A further hydrotreating of naphtha for extracting of BTEX is also reasonable.

Special means have been considered in the revamping, so although the DCC reaction temperature was higher than FCC, all the possible coking places — such

Table V. DCC Commerical Gas Composition in Comparison With Pilot Plant

Unit		Commercial		Pilot plant
		1	2	
Gas composition wt%				
Hydrogen sulfide		-	-	0.33
Hydrogen		0.49	0.52	0.76
Methane		7.04	8.72	7.76
Ethane		4.29	4.61	4.79
Ethylene		8.76	10.65	9.76
Propane		6.58	5.78	6.40
Propylene		37.59	38.25	35.68
Isobutane		4.45	3.81	5.30
n-Butane		1.78	1.51	1.92
1-Butylene		5.30	4.55	4.93
Isobutylene		12.12	10.73	11.20
Trans-2-butene		6.60	6.20	6.37
Cis-2-butene		5.00	4.67	4.72
Butadiene		-	-	0.08
Total		100.00	100.00	100.00

as the top of the disengager, cyclones, dip legs, reactor effluent transfer line, the inlet and the bottom of the main fractionator and the slurry system — were virtually free of coking and fouling after a four-month operation.

Table VI.   Trace Hydrocarbon and Impurities in DCC Gas

Gas		LPG	Dry Gas
Trace hydrocarbons vppm			
Propadiene		22	<1
Ethyne		128	50
Propyne		28	<1
Total nitrogen vppm		32	5.0
Basic nitrogen		26	-
Nitrile & other nitrides		6	-
Total sulfur vppm		674	1861
H_2S		664	1860
Organic sulfur		10	0.8
Mercaptan sulfur		8.6	<0.05
COS & other sulfides		0.6	<0.05
Sulfur ether		0.8	<0.05

Maximized Isoolefin Operation

Experiments were conducted in a 2 b/d riser pilot plant with various feedstocks. The feedstock properties are listed in Table VII. Feed A, B and C was VGO from paraffin-base, intermediate-base and naphthene-base crude, respectively. Feed D was a blended feedstock of feed A with 40% vacuum residue, which CCR was 3.3 wt%. Three catalysts — CS-1, CZ-1 and CZ-2, with different zeolite types developed by RIPP — were used in the studies.

The preheated feedstock was atomized by steam and contacted with the regenerated catalyst at the bottom of the riser reactor. The catalytic cracking reaction can be conducted either once through or in recycle mode.

Table VII. Properties of Various Feedstocks

Feedstock		A	B	C	D
Gravity (20°C)	g/ml	0.8788	0.8781	0.9249	0.8880
API		29.5	29.7	21.5	29.3
CCR	wt%	0.10	0.44	0.20	3.30
Aniline point	°C	101.5	96.0	79.9	>105
Pour point	°C	45	46	30	40
UOP K		12.4	12.0	11.5	12.6

DCC Product Slate and Olefin Yield Structure for Maximizing Isobutylene and Isoamylene. Table VIII presents the DCC product slate and light olefin yield structure for maximizing isoolefin production in comparison with DCC maximum

Table VIII. Comparison of DCC product Slate and Olefin Yield Structure (Feed A, Pilot plant data)

Process	I DCC Maximum Propylene mode	II DCC Maximum Isoolefin mode	FCC
Material balance wt%			
C_2-	11.91	5.59	3.47
C_3-C_4	42.22	34.49	17.62
C_5+ Naphtha	26.60	39.00	54.81
LCO	6.60	9.77	10.20
HCO	6.07	5.84	9.30
Coke	6.00	4.31	4.30
Loss	0.60	1.00	0.30
Total	100.00	100.00	100.00
Light olefin yield wt% (feed basis)			
Ethylene	6.10	2.26	0.75
Propylene	21.03	14.29	4.88
Butenes	14.30	14.65	6.10
Isobutylene	5.13	6.13	1.67
Petenes	-	9.77	-
Isoamylene	-	6.77	-

propylene mode and FCC. For maximum isoolefin mode operation, the reaction temperature was milder than maximum propylene mode but still slightly higher than that of conventional FCC. From these data, one can see that DCC produces much more light olefins at the expense of naphtha. It is clear that gasoline range fraction further cracks under stringent reaction conditions. In maximum isoolefin operation, isobutylene and isoamylene yield was 6.13 and 6.77 wt% (feed basis), respectively, with 14.29 wt% propylene and 39.0 wt% naphtha for feedstock A.

The butylene and amylene compositions are listed in Table IX and X. The ratio of isobutylene to total butenes reached 0.42 which was close to the thermodynamic equilibrium value (*4*). The isoamylene ratio was also near equilibrium. The double bond-shift and chain branching of pentenes are even faster than that of butenes in this operation condition. The limited isobutylene and isoamylene concentrations in conventional FCC are due to the high hydrogen transfer activity of the catalysts. The high olefin concentration in cracked gas and high isoolefin ratios clearly indicate that less hydrogen transfer reaction occurs in the DCC process.

Table IX. Distribution of Butylene Isomers
(feed A)

	Equilibrium Value		*I* *DCC maximum*	*II* *DCC maximum*	*FCC*
	I Temp.	*II Temp.*	*propylene*	*isoolefin*	
1-Butene	0.156	0.147	0.176	0.128	0.193
Trans-2-butene	0.246	0.245	0.250	0.267	0.301
Cis-2-butene	0.168	0.167	0.213	0.186	0.231
Isobutylene	0.430	0.441	0.361	0.419	0.274

Table X. Distribution of Amylene Isomers
(feed A)

	Equilibrium Value *at II Temperature*	*II* *DCC maximum* *Isoolefin Mode*
1-Petene	0.052	0.052
Trans-2-petene	0.122	0.176
Cis-2-petene	0.120	0.079
Isoamylene	0.706	0.693

Light Olefin and Isoolefin Yields from Different Feedstocks.   Three types of VGOs and one VGO blended with residue oil were tested for maximizing isoolefin production. Table XI shows the olefin yields from A, B, C VGOs and the resid feedstock D. The same kind of catalyst with 3000 ppm contaminated nickel was used for the residue feed test. Paraffinic feed A was the best for producing the light olefins. Naphthenic feed C seems to be the worst, but the isobutylene and isoamylene yield reductions were only around 1 wt% in comparison with the best.

The propylene and isobutylene yields of the residual feedstock D were less compared to VGO; however, the isoamylene yield was at the same level.

Table XI. DCC Isoolefin Yields of Various Feedstocks
(In comparison with A feedstock)
(CS-1 catalyst)

Feedstoock		A	B	C	D
Light olefin yield	wt%				
Ethylene		2.26	+0.16	-0.71	-0.33
Propylene		14.29	-0.31	-2.34	-1.66
Butenes		14.65	-1.14	-3.69	-1.08
Isoolefin yield	wt%				
Isobutylene		6.13	-0.67	-1.14	-1.21
Isoamylene		6.77	+0.07	-1.20	-0.22

Catalyst Selection.   Catalyst test results obtained at constant conditions with A feed are shown in Table XII. Propylene selectivity and the ratios of isobutylene/total butenes and isoamylene/total pentenes are used as the selectivity criterions for DCC catalyst selection. The propylene selectivity is defined as propylene/total C_3. Propylene selectivities of all three test catalysts were extremely high, which were around 90%. The isobutylene/total butenes and isoamylene/total pentenes of CS-1 catalyst was over 0.4 and close to 0.7, respectively. CS-1, CZ-1 and CZ-2 were ranked first, second and last in all above selectivities.

Table XII. Isoolefin Selectivities of Various Catalysts

Catalyst	CS-1	CZ-1	CZ-2
Selectivity			
Propylene/total C_3	0.904	0.890	0.882
Isobutylene/butenes	0.419	0.368	0.361
Isoamylene/pentenes	0.693	0.666	0.618

The Effects of Operating Variables

The effects of operating variables have been studied both in laboratory and in a 2 b/d pilot plant. The DCC light olefin yields depend on the following principal variables:
1. Feedstock composition
2. Catalyst activity and selectivity
3. Temperature
4. Residence time and catalyst-to-oil ratio
5. Hydrocarbon partial pressure.

The constituents normally present in DCC feedstocks fall into four principal classes: normal paraffins, branched paraffins, cycloparaffins or naphthenes, and aromatics. The desired olefins are cracked from the first three classes. The aromatic side-chain scission can   occur in the DCC condition; however, unsubstituted aromatics undergo relatively slow cracking because of the stability of the aromatic ring. The pilot plant result shows that the paraffins are the preferable feedstocks. Aromatic feedstocks are highly refractory and are not suitable as DCC feed. The composition of the DCC effluent is not only determined by the primary decomposition of the saturates but also by the secondary reaction. The secondary reaction occurring after the initial cracking steps has the most important influence on the DCC product composition.

The catalyst used in DCC should consist of a large molecular conversion component and a specified pore diameter component for the selective secondary reactions. Cracking reactions catalyzed by these acidic surfaces proceed via surface carbonium ion intermediates produce the desired olefins. Catalytic reaction is predominant in the DCC process, which results in the effluent that is rich in propylene, butenes and isoolefins. Hydrogen transfer reaction is also a secondary reaction which needs to be controlled for maximizing olefin production.

In comparison with the selectivity, the activity of a DCC catalyst is less important. It is very difficult to design a catalyst with both high activity and light olefin selectivity. The deficient catalyst activity can be compensated by the increased severity of the process.

The overall reactions are endothermic in the DCC process in which heat of reaction is 2 to 3 times that of FCC for the high conversion and the large quantity of the gas produced. The secondary cracking reaction occurring in DCC needs more reaction heat and higher temperature than primary reaction; however, the exothermic hydrogen transfer reaction will be less at high temperature. The optimum DCC temperature is normally 30° to $80^\circ C$ higher than that of FCC. Further increasing the temperature will increase the thermal reaction and result in an unexpectedly high dry gas yield.

DCC uses longer residence time and higher catalyst-to-oil ratio than that of FCC. Longer residence time favors the secondary reaction for deep conversion. The high catalyst-to-oil ratio increases the dynamic activity of the catalyst. Both variables increase the process severity to compensate for the lower test activity of the catalyst.

The partial pressure of the hydrocarbons affects the chemical equilibria and reaction rates, and thereby influences the product distribution. High yields of the desired olefins are favored by low hydrocarbon partial pressure. Reducing hydrocarbon partial pressure also decreases the velocity of the bimolecular reaction such as hydrogen transfer and increases the product olefin concentration.

Steam dilution is normally used in the DCC process to decrease the hydrocarbon partial pressure. It encourages higher selectivity to the desired olefinic ·products and also decreases the tendency to form coke in the reaction zone and the downstream system. DCC uses more steam than FCC but still much less than in steam cracking. Steam dilution can be varied in the DCC process. One should achieve a balance between the olefin gained and the energy consumption.

Acknowledgments

The authors wish to thank all the people involved in the DCC experiment and development works.

Literature Cited

(*1*) Li Zaiting, Jiang Fukang, Min Enze and Wang Xieqing, *NPRA Annual Meeting*, AM-90-40, March 25-27 (1990)
(*2*) Wang Xieqing, Li Zaiting, Jiang Fukang and Yu Bende, *AIChE Meeting*, April 8 (1991)
(*3*) Li Zaiting, Jiang Fukang, Yang Qiye and Tang Qinlin, *Proceedings of the international conference on petroleum refining and petrochemical processing, Beijing*, **1991**, 345
(*4*) Alberty, R.A. and Gehrig, C.A., *J. Phys. Chem. Ref. Data*, **1985**, *14*(3)

RECEIVED June 17, 1994

Chapter 5

Comparison of Laboratory and Commercial Results of ZSM-5 Additives in a Fluid Catalytic Cracking Unit

L. Nalbandian[1], I. A. Vasalos[1], I. Dimaratos[2], and K. Vassilakis[2]

[1]Chemical Process Engineering Research Institute and Department of Chemical Engineering, P.O. Box 1517, 54006 Thessaloniki, Greece
[2]Motor Oil Hellas, Corinth, Greece

A laboratory evaluation of a ZSM-5 containing commercial additive has been performed simultaneously with a commercial test of the same additive at the FCCU of a Greek refinery. Incremental yields and RON gain, derived independently from both tests, are in very good agreement to each other. Significant gasoline loss and LPG yield increase, accompanied by a moderate RON gain have been observed when ZSM-5 was added to a unit running with a low unit cell size USY zeolitic catalyst and a highly aromatic feed.

In order to overcome pollution problems, most of the countries have removed or are considering reducing the lead content of gasolines. Thus, the need for alternative means of gasoline octane enhancement has emerged.

A major source of gasoline octane in a modern petroleum refinery is the Fluid Catalytic Cracking (FCC) unit, in which heavy feeds are broken down to smaller molecules, such as gasoline and C_{4-} gases.

Because of the need for higher gasoline octane, refiners are looking for economical ways of improving octane in their FCC units, either by changing operating conditions or applying new cracking catalyst. Two catalytic routes have been commercially successful. First the use of catalyst containing low unit cell size, ultrastable Y zeolite (USY); and second the use of ZSM-5.

ZSM-5 is introduced to the reactor either as a separate additive which commonly comprises about 25% ZSM-5 in a relatively inert matrix, or as a composite i.e. a catalyst containing ZSM-5 and a faujasitic cracking component in the same particle.

Since the first commercial test of ZSM-5, in 1983, this additive has been used in a large number of units (1, 2). Its performance has varied slightly in these evaluations depending largely on feedstock quality, catalyst type and unit operating conditions; nontheless, the octane enhancing capability of ZSM-5 in catalytic cracking has always been observed.

0097–6156/94/0571–0043$08.00/0

The octane gain achieved by ZSM-5 addition to an FCC unit is usually accompanied by a gasoline yield loss of 1-2 vol % per RON gain (*1 - 8*). Most of the material lost from the gasoline appears as C_3-C_4 olefins which can be used to make alkylate or ethers. ZSM-5 has no significant effect on the dry gas yields (H_2, C_2) (*1, 4*).

The MOTOR OIL HELLAS refinery, in its effort to enhance the octane of the gasoline produced from its FCC unit has decided to test the performance of ZSM-5. Two kinds of tests have been performed, a laboratory test at an MAT unit and a commercial test at the refinery FCC unit. "Delta yields" i.e. differences between yields before and after ZSM-5 addition have been derived independently from both tests and compared to each other. Based on the above results, economic calculations have been made on the profitability of ZSM-5 addition to the commercial unit. In addition the capability of the MAT unit to predict " delta yields" of the commercial unit has been examined.

Experimental

Description of the MAT Unit. The Microactivity Test (MAT) Unit used for the experiments has been designed according to the ASTM D-3907 method, with minor modifications. A schematic diagram of the MAT unit is shown in Figure 1.

It is made of a pyrex, fixed bed reactor, heated by a three-zone furnace. Feed is injected, through an independently controlled preheat device, just above the catalyst bed. Feed rate is controlled by a syringe-pump while the duration of all experiments is constant (50 sec).

The vapor products are cooled to $0°C$ at the exit of the reactor where part of it is condensed and collected in the specially designed liquid receiver. The remaining uncondensed gas products are driven to a gas collecting apparatus where the volume of the gas is measured at atmospheric pressure and room temperature.

Three N_2 flows are used during MAT experiments in order to drive the feed and products along the reactor and purge the remaining traces of feed from the injection system. Two of the above mentioned N_2 flows are continued for long time (~30 min) after the experiment, in order to strip the liquid products collected in the liquid receiver. The three N_2 flows are controlled accurately by mass-flow controllers (Brooks 5850E).

The reaction temperature is measured by a type J thermocouple just above (~1 cm) the catalyst bed. The pressure in the reactor is measured with a pressure transmitter (Validyne).

The gaseous products are analysed at a specially designed GC called "Refinery Gas Analyser" (Varian model 3400). It is equipped with four columns and two detectors (TCD and FID) and is able to detect all gaseous products of the catalytic cracking reaction.

Conversion of liquid products is measured by a Gas Chromatograph equipped with a Megabore OV-101 Column (Varian model 3400).

The gasoline produced from each MAT experiment is analysed by Capillary Gas Chromatography (HP 5880 A). The Research Octane Number of the gasoline is predicted by the use of a well known model (*9*).

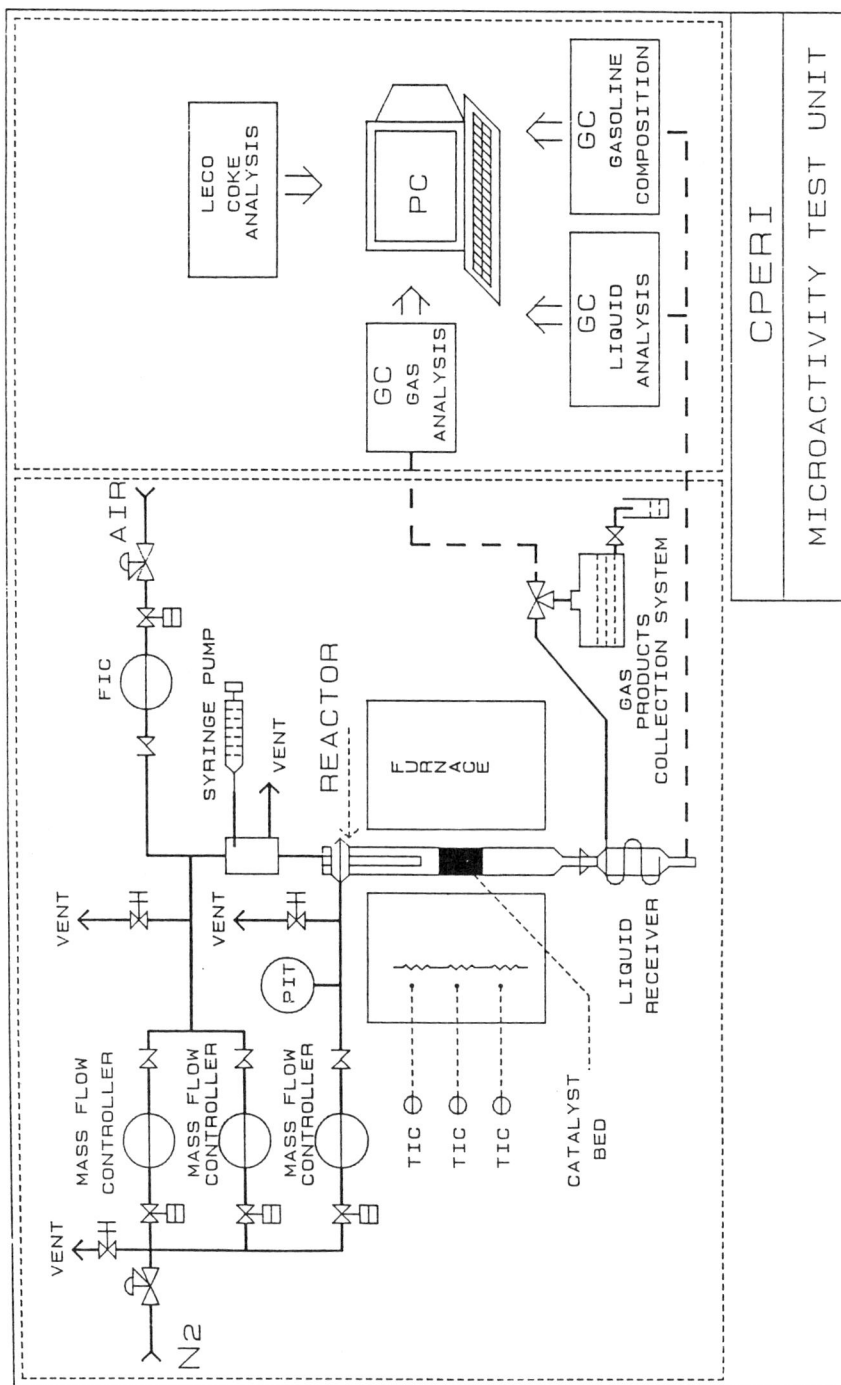

Figure 1. Schematic diagram of the MAT unit.

The weight of coke deposited on the catalyst is measured by an "Elemental Analyser" (Leco CHN-800).

Sample preparation. An equilibrium USY catalyst, obtained from the FCC unit of MOTOR OIL HELLAS was used in this study. As ZSM-5 additive, the product commercialized by Engelhard under the name Z-100 (20-25% ZSM-5 crystal) was used.

Equilibrium catalyst samples were taken from the FCCU inventory just before the ZSM-5 addition started (base case), as well as during the test. Prior to test at the MAT unit the equilibrium catalyst samples were calcined at 1000° F for 10 hrs.

Catalyst samples containing ZSM-5 were either taken directly from the FCCU as equilibrium catalyst or prepared in the laboratory by mixing preweighted amounts of Z-100 with the "base case" catalyst (equilibrium catalyst obtained from the FCCU just before the ZSM-5 addition started).

The Z-100 additive was added to the catalyst either fresh or after severe hydrothermal treatment (1450° F, 9 hrs, 100% steam, 1 atm).

Physicochemical characterization of catalysts. Unit Cell Size (UCS), total surface area and zeolite area measurements have been performed for the catalyst samples used in this study. Unit Cell Size is measured at a Siemens D500 X-ray diffractometer according to the standard ASTM D-3942-85 method. Total surface area of catalysts is measured by nitrogen adsorption at liquid nitrogen temperature and is calculated by the BET method. Zeolite area and micropore volume are derived by the t-plot method (ASTM D-4365-85) from nitrogen adsorption at liquid nitrogen temperature measurements. Total surface area and zeolite area measurements are performed at a Micromeritics Accusorb 2100E instrument.

Table I presents the total surface area and zeolite area measurements of all the catalyst and ZSM-5 samples tested in this work.

Experimental details. A typical feedstock taken from the FCC unit of MOTOR OIL HELLAS during the test period was used in the MAT experiments. Representative feedstock properties have been given elsewere (10). All MAT tests were performed at 970° F for 50 secs. Catalyst to oil ratio was varied in the range 2-7 while the catalyst weight was held constant (3 gr).

Results and discussion

Effect of ZSM-5 addition on the product yields (MAT study)

Effect of the additive concentration. Figure 2 shows representative selectivity vs. conversion curves obtained by testing at the MAT unit the "base case" catalyst and mixtures of "base case" with fresh Z-100 at three different concentrations. The above selectivity curves were derived from experiments conducted at different conversions by varing the cat to oil ratio.

The plots indicate that an increase in the additive concentration results in decrease in the gasoline production, an increase in the yields of C_3's and C_4's

(LPG), increase in the olefinicity ratios of C_2, C_3 and C_4 and increased RON of the produced gasoline.

Table I. Surface area and zeolite area of catalysts

CATALYST	STATE	TOTAL SURFACE AREA (m^2/gr)	ZEOLITE AREA (m^2/gr)	MICROPORE VOLUME (cm^3/gr)
Motor Oil Equilibrium Cat.	Calcined	180 (± 2)	102 (± 3)	0.0416
Z-100	Fresh	46 (± 1)	20 (± 2)	0.0087
	Steamed 1450 °F for 9 hours	47 (± 1)	21 (± 2)	0.0093
	Steamed 1450 °F for 5 hours	46 (± 1)	24 (± 2)	0.0103

On the other hand ZSM-5 has no influence on conversion and dry gas (H_2, C_2_) production.

Figures 3-6 have been derived from the selectivity curves (Figure 2) and show the effect of ZSM-5 addition on product selectivity at constant conversion (65% wt). The same results are also shown in Table II.

Figure 3 shows the gasoline yield drop occuring simultaneously with RON increase, as a result of increase in the additive concentration in equilibrium catalyst. Based on the above plots, a RON gain of 0.3 has been derived for every unit % wt of gasoline loss. The obtained value is quite small compared to previous studies (*1 - 8*) where RON gain ranging from 0.5 to 2 have been reported. For the interpretation of the above observation, the type of the base catalyst, as well as the properties of the gas-oil feed should be taken into account. In the present study a low Re_2O_3 USY (equilibrium UCS: 24.27 A) base catalyst was used in combination with an aromatic feed. The base RON value was, as a result, relatively high (92.4, Table II). Complete compositional analysis of the gasoline samples was performed in order to interpret the above data. The results of this study, that have been published elswere (*10*), as well as the conclusions of previous workers (*3, 6 - 8, 10*) have indicated that ZSM-5 increases FCC gasoline octane by decreasing the yield of low octane gasoline components, mainly C_6-C_9 parafins and olefins. In a relatively high RON gasoline, such as that obtained in the base case of this study, the concentration of those components is already low, thus it becomes diffucult to decrease it further.

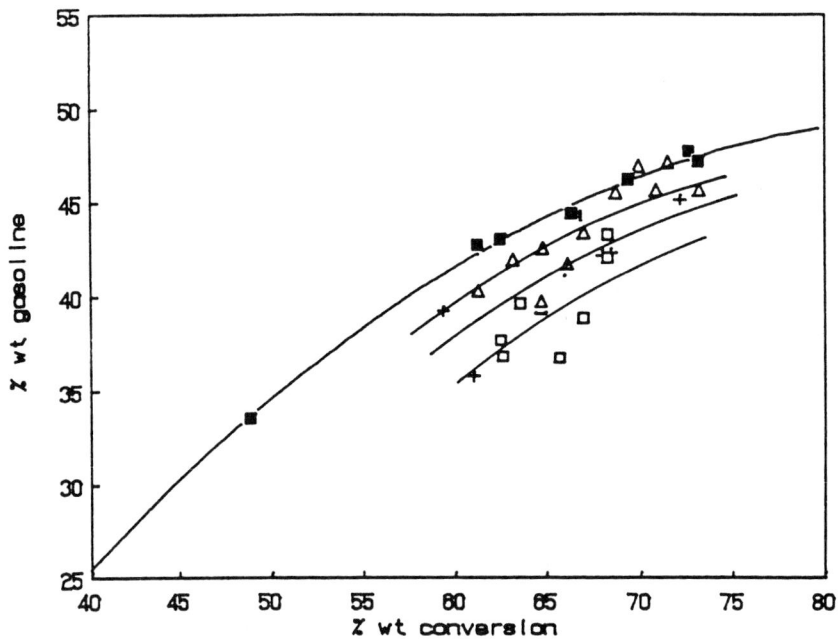

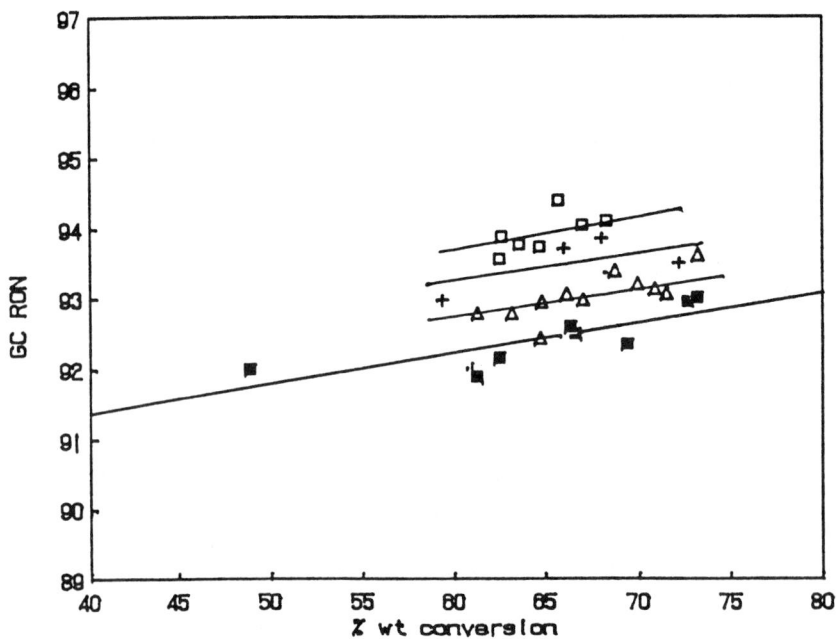

Figure 2. Selectivity vs. conversion curves. ■: base, △: base+4% wt Z-100 fresh, +: base+6% wt Z-100 fresh, □: base+8% wt Z-100 fresh.

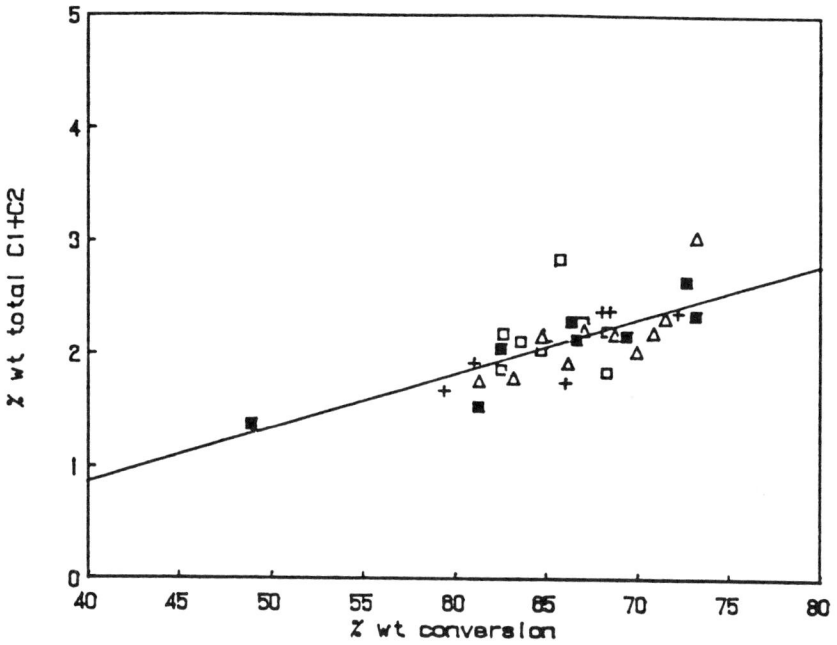

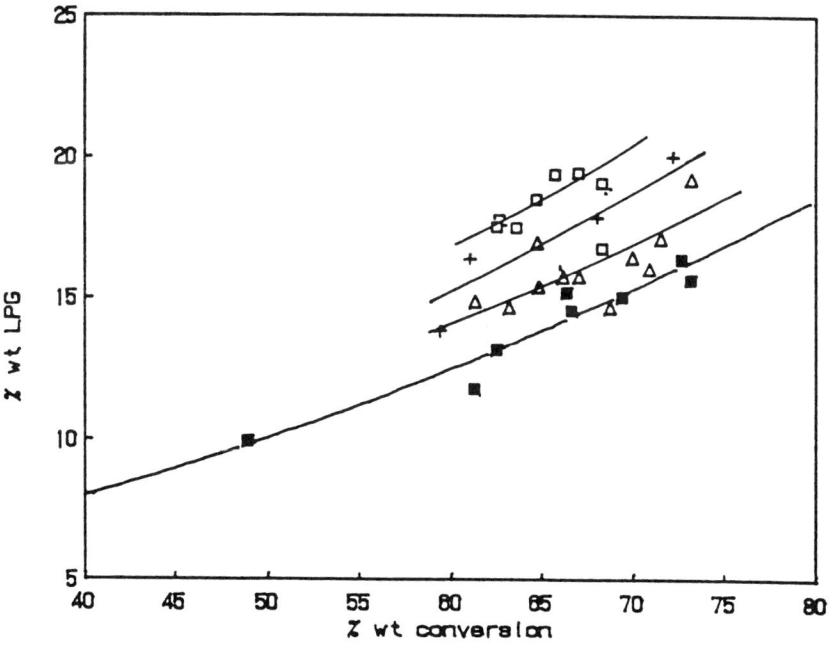

Figure 2. Continued.

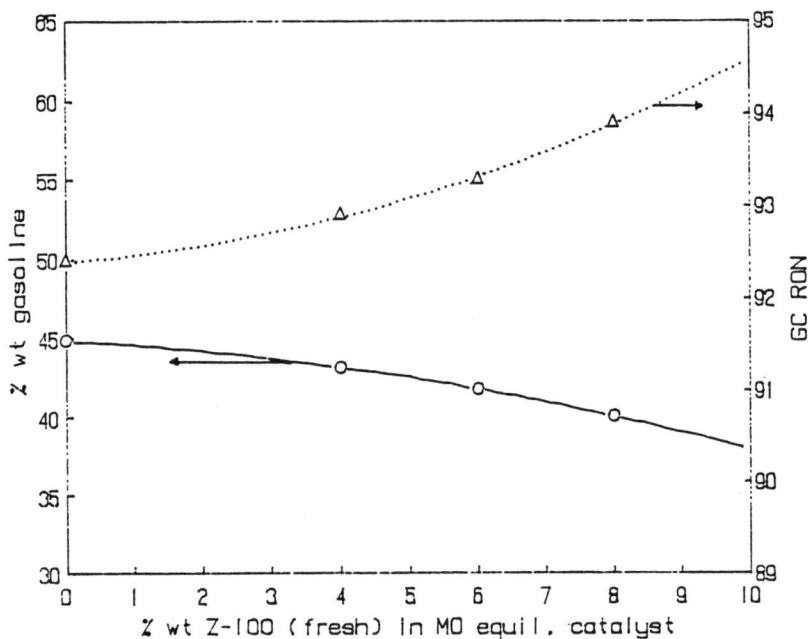

Figure 3. Effect of ZSM-5 addition on gasoline selectivity and RON. O: gasoline,
Δ: RON.

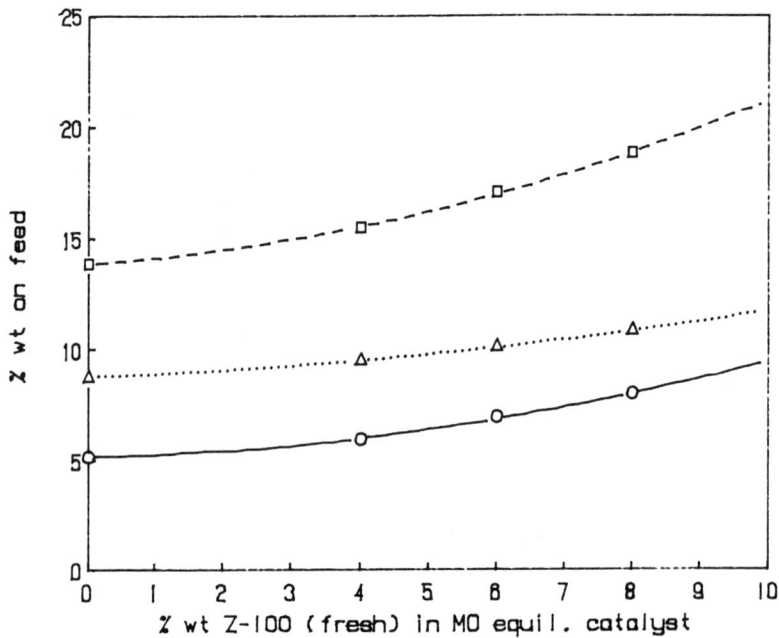

Figure 4. Effect of ZSM-5 on LPG yields. O: total C_3, Δ: total C_4, □: total LPG.

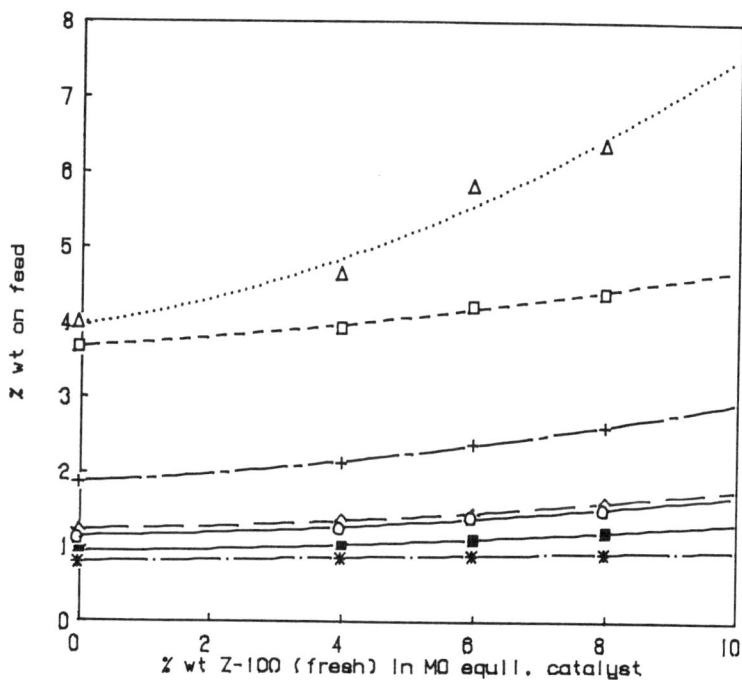

Figure 5. Detailed analysis of the effect of ZSM-5 addition on the yields of C_3 and C_4 gases. O: n-propane, $\triangle$: propylene, *: n-butane, $\square$: iso-butane, +: butene-1, $\Diamond$: trans-butene, ■: cis-butene.

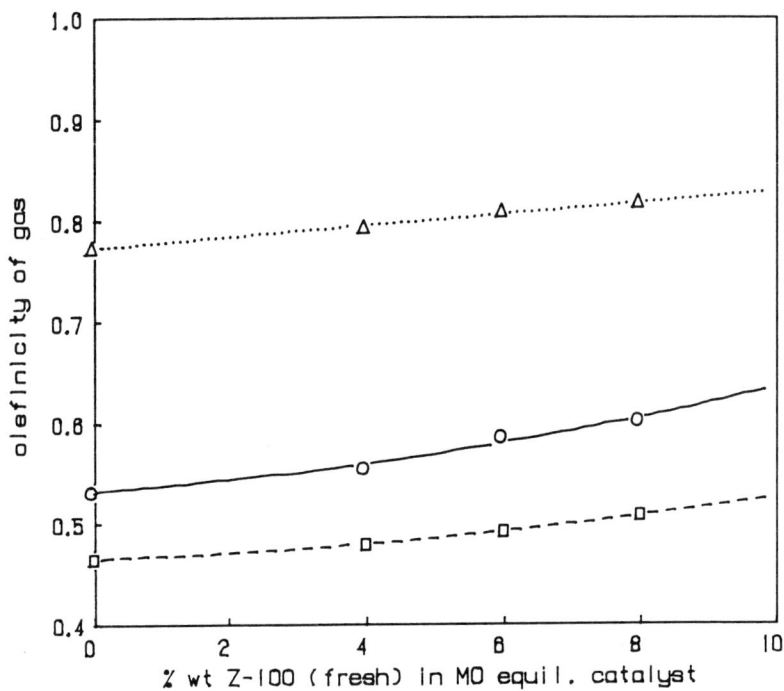

Figure 6. Effect of ZSM-5 addition on the olefinicity of product gas.
o: olefinicity C_2, Δ: olefinicity C_3, $\square$: olefinicity C_4.

Table II. Equilibrium Catalyst Yields and "Delta Yields" after ZSM-5
addition, at 65% wt conversion, MAT data
(MAT conditions: 970° F, cat to oil 4-8)

	Equilibrium + 0% Z-100	Equilibrium + 4% Z-100	Equilibrium + 6% Z-100	Equilibrium + 8% Z-100
cat to oil	5.0	0	0	0
gasoline (% wt)	44.8	-1.66	-3.17	-4.9
H_2 (% wt)	0.16	0	0	0
$C_1 + C_2$ (% wt)	2.07	0	0	0
Total C_3's (%wt)	5.13	+0.87	+1.80	+2.82
Total C_4's (%wt)	8.74	+0.79	+1.37	+2.08
Total LPG (%wt)	13.87	+1.66	+3.17	+4.90
Olefinicity C2	0.533	+0.023	+0.053	+0.070
Olefinicity C3	0.773	+0.021	+0.036	+0.044
Olefinicity C4	0.466	+0.014	+0.027	+0.042
Olefinicity LPG	0.580	+0.021	+0.044	+0.058
GC RON	92.4	+0.5	+0.9	+1.5

Figure 4, showing the increase in LPG yields, indicates that ZSM-5 selectively boosts the production of C_3's while a moderate increase in C_4's yield is observed. Figure 5 shows in more detail the distribution of total LPG among the various C_3 and C_4 components. Propylene yield undergoes the highest increase by about 60% wt, followed by butene-1 which increases by 40% at the highest additive concentration tested, 8% wt. Trans- and cis-butene as well as n-propane undergo a moderate increase of about 30% at the same Z-100 concentration while iso-butane and n-butane increases only by 20 and 15% wt respectively. This ZSM-5 selectivity towards C_3 olefins has been also observed in commercial operation with ZSM-5. Incremental propylene yield has been reported to exceed butylene yield by 1.5-3 to one, depending on ZSM-5 concentration, feed quality and FCCU operational conditions (*12*).

All the above compositional changes are also reflected on the olefinicity ratios calculated for the gas products and shown in Figure 6. It is worth mentioning at this point that, although the total $C_1 + C_2$ yield remains unaffected on ZSM-5 addition, the ethylene to ethane ratio increases, in the same manner as do the C_3 and C_4 olefinicity ratios.

Effect of the additive pretreatment. Laboratory prepared mixtures of the "base case" catalyst with steam pretreated (1450° F, 9 hrs, 1 atm, 100% steam) additive at three different concentrations (4, 6, 8% wt) were tested for selectivity with the MAT unit.

Figures 7-9 show representative product yields obtained with the above "steamed" mixtures compared with the corresponding yields with the "fresh" mixtures, mentioned in paragraph 3.1.1, and with the "base case" catalyst. From Figures 7-9 it is evident that no systematic variation can be detected in the selectivity of ZSM-5 mixtures when the additive Z-100 is fresh or severely steam deactivated. From Table I, where the total surface area and the zeolite area of Z-100 samples, fresh and steamed at two conditions, are compared, it can be seen that the above pretreatment does not affect also the total surface area and the zeolite area of the samples.

It has been proved, however, from commercial experience, as well as from the work of previous investigators (5, 11), that ZSM-5 deactivates, modifying its initial catalytic action, after equilibration in a commercial unit or after laboratory steaming. As ZSM-5 ages the ratio (Delta gasoline) to (Delta RON) decreases. The fact that, to the additive tested in this work, nothing seems to change after steaming, might suggest that the steaming conditions used are not severe enough to deactivate it. However, fresh FCC catalysts deactivate properly when subjected to the same steaming. Furthermore, previous investigators have used equal (16) or even milder (3, 7, 11) steaming conditions and still they observed changes in the additives performance.

Thus the only explanation that could fit the experimental observations is that, since a commercial additive was used, it can be speculated that it has already undergone some aging pretreatment, at the manufacturing stage.

Performance of equilibrium catalyst samples. Equilibrium catalyst samples, obtained from the FCC unit of the industrial partner, 2, 14 and 19 days after the Z-100 addition started, were tested in the MAT unit.

Based on the addition program of the fresh catalyst and additive to the FCCU, the Z-100 concentration in the above samples was calculated to be 0.77, 2.63 and 2.92% wt respectively.

Figure 10 shows representative MAT yields obtained with the above samples. Although the ZSM-5 concentration was fairly low (Z-100 ≤3% wt → ZSM-5 ≤0.75% wt) the obtained product yields show that the effect of ZSM-5 on the equilibrium catalyst performance is similar to that observed with the laboratory prepared mixtures.

Further elaboration of the data of Figure 10 was considered not to be justified because of the uncertainty concerning the concentration of Z-100 in the equilibrium catalyst samples. The above results however confirm that no basic qualitative differences exist between the performance of laboratory prepared ZSM-5 containing mixtures and commercially equilibrated samples.

Effect of the ZSM-5 addition on the FCCU performance

ZSM-5 addition policy. The FCC unit of the industrial partner has an approximately 70 metric tons catalyst inventory. It can handle up to 180 ton/h of fresh feed. Feed to the unit consists of vacuum gas oil and residue. The quality of

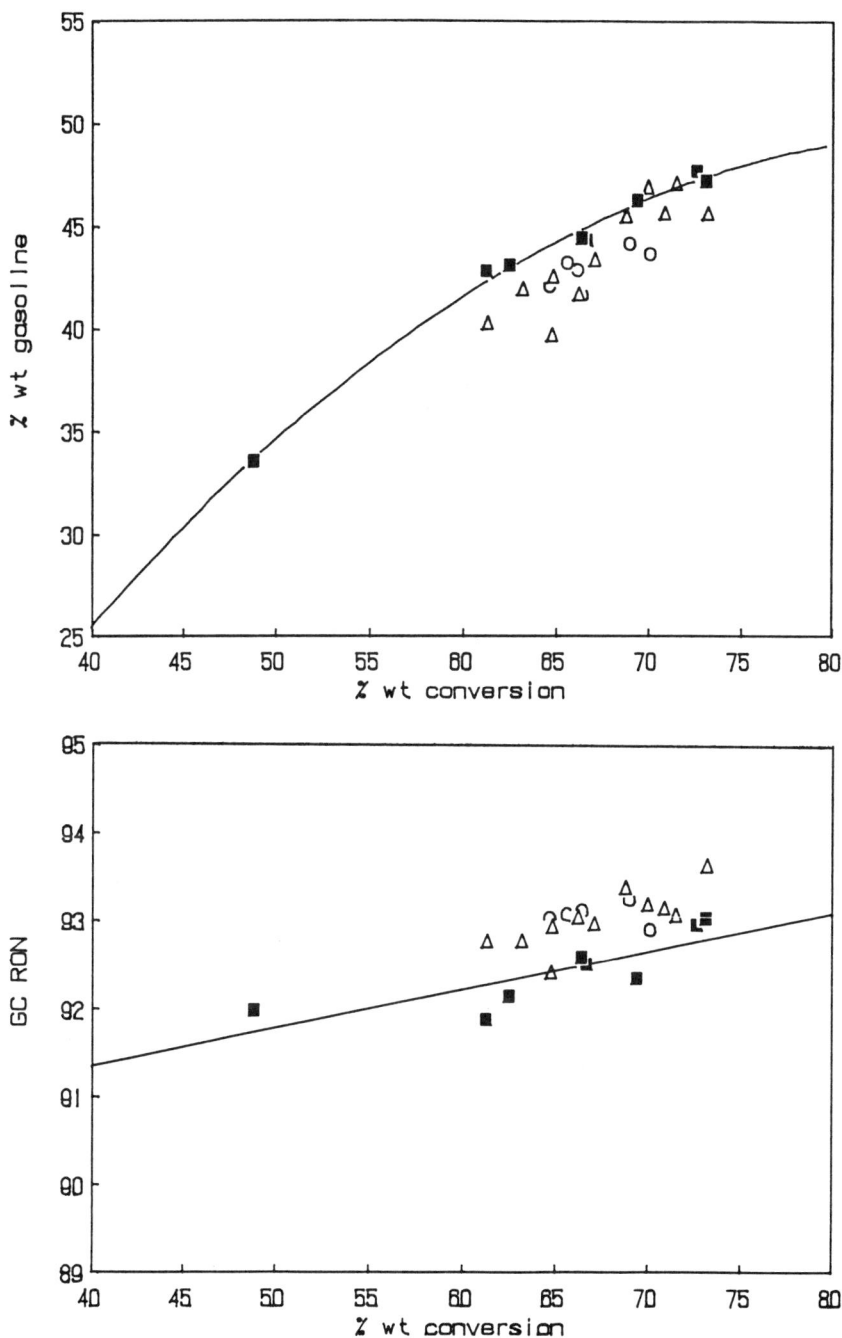

Figure 7. Gasoline selectivity and RON vs. conversion. ■: base, Δ: base+4% wt Z-100 fresh, ○: base+4% wt Z-100 steamed.

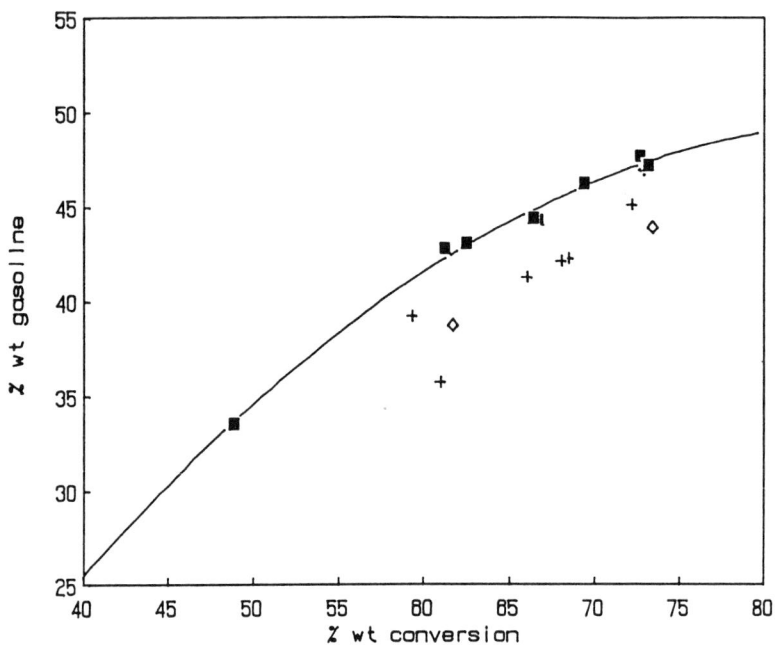

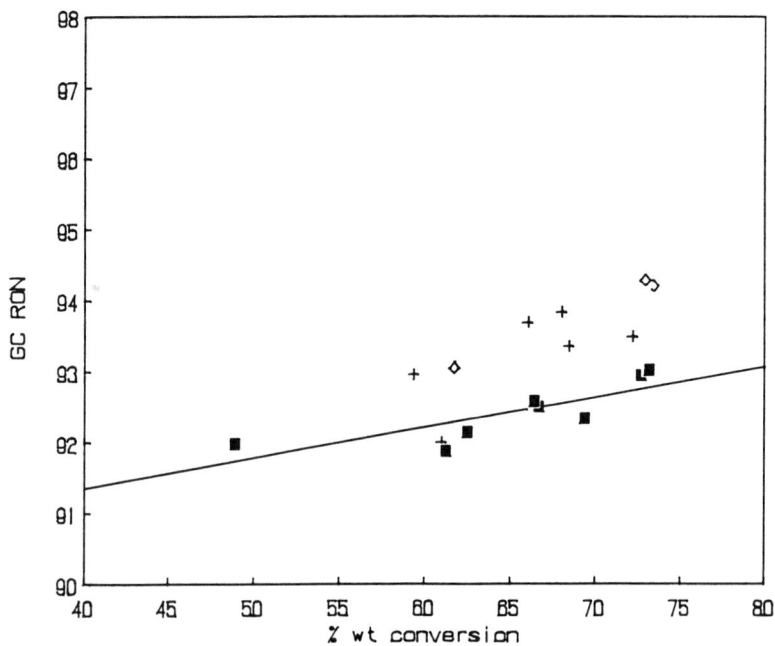

Figure 8. Gasoline selectivity and RON vs. conversion. ■: base, +: base+6% wt Z-100 fresh, ◇: base+6% wt Z-100 steamed.

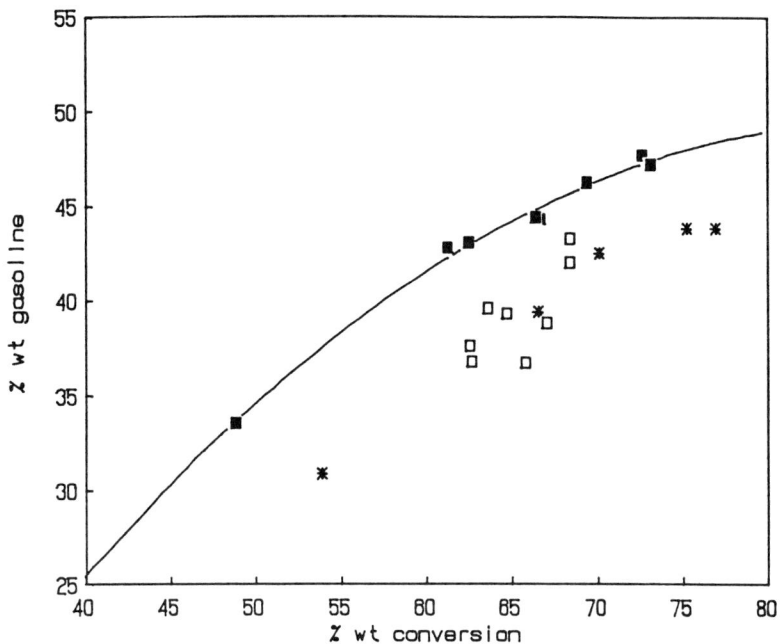

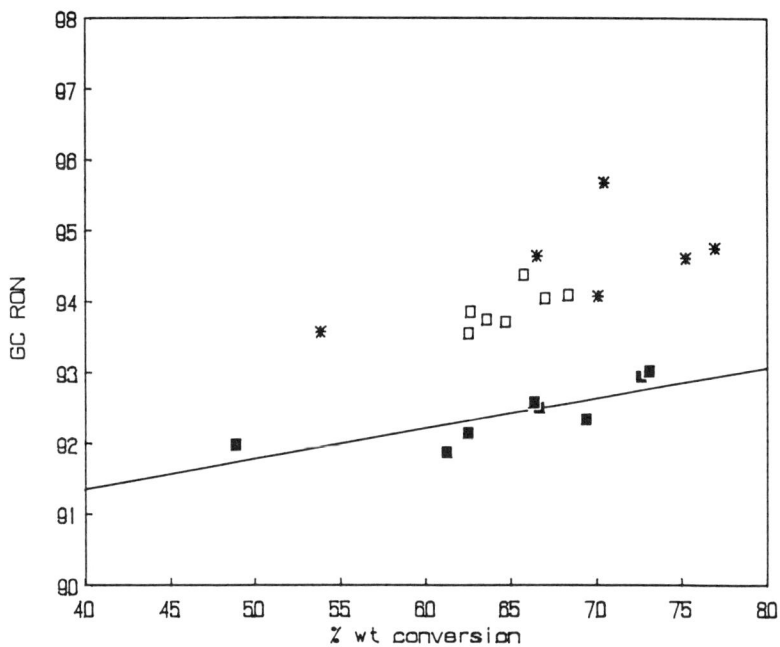

Figure 9. Gasoline selectivity and RON vs. conversion. ■: base, □: base+8% wt Z-100 fresh, *: base+8% wt Z-100 steamed.

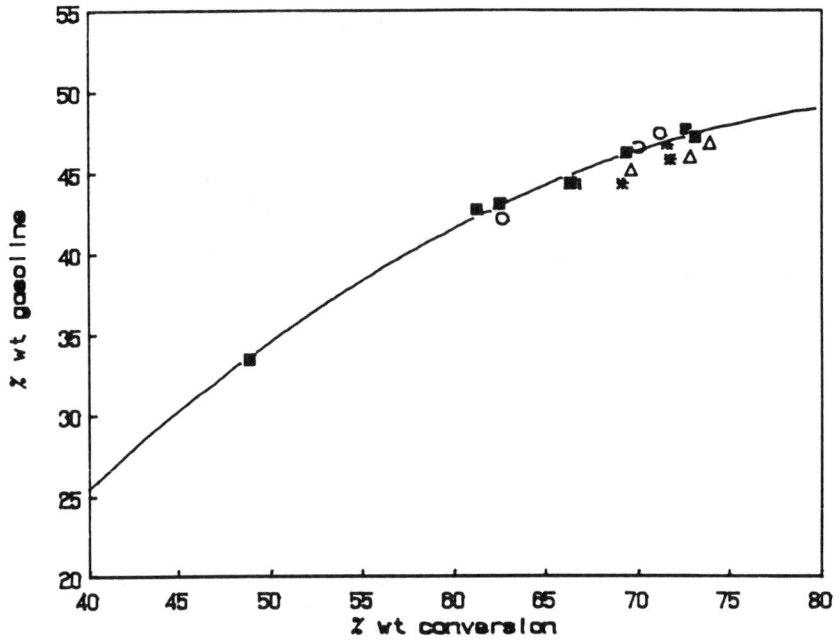

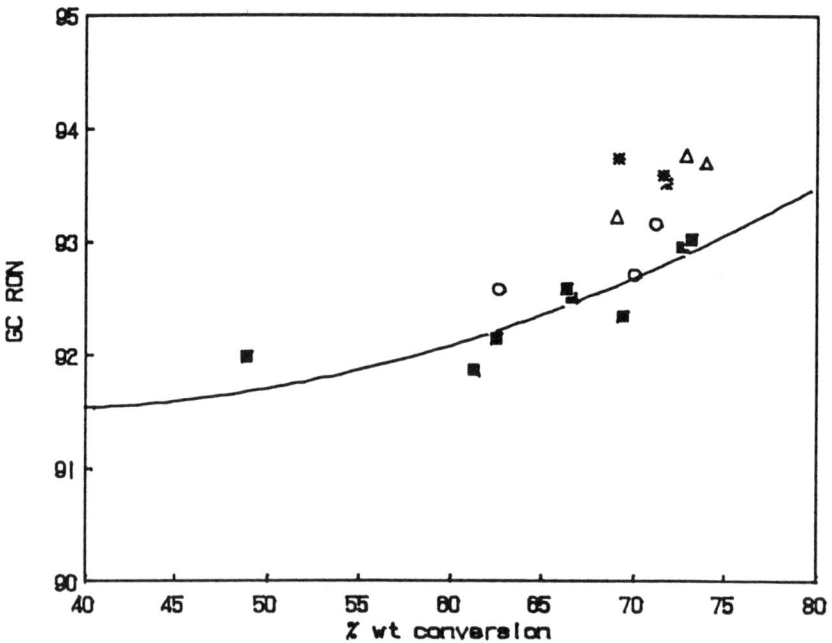

Figure 10. Selectivity vs. conversion curves with equilibrium catalyst samples.
■: 0 day (base), ○: 2nd day from Z-100 introduction, Δ: 14th day, *: 19th day

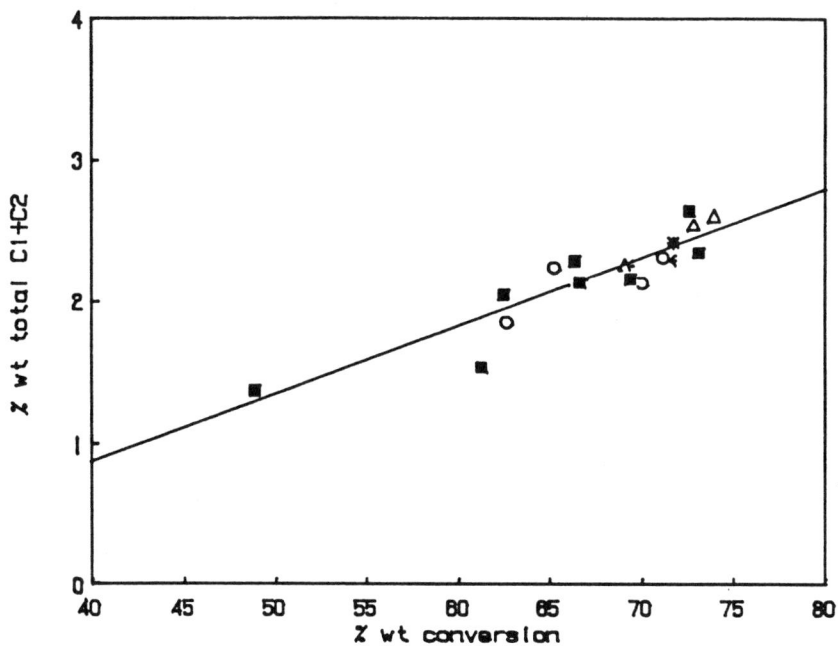

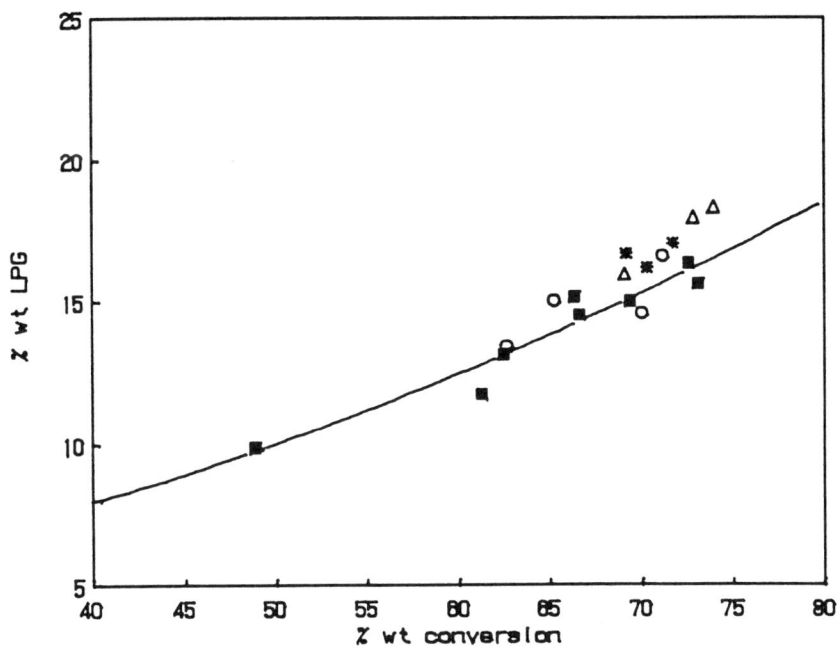

Figure 10. Continued.

feed is generally good with relatively low metals. During the whole period of the commercial test of ZSM-5, a highly aromatic feed was fed to the unit (*10*). A low Re_2O_3 containing USY was the base catalyst in the FCC unit at the time of the test.

The property of the ZSM-5 of most interest was its reported ability to increase the octane of the FCC gasoline.

The duration of the test run with ZSM-5 was one month. The commercial additive Z-100, produced by Engelhard was used. The addition rate of Z-100 was initially accelerated in order to achieve the desired concentration in less than 10 days. It was then reduced to maintain the desired ZSM-5 level in inventory. The base FCC catalyst addition policy remained as it was prior to the introduction of ZSM-5.

In order to estimate the additive percentage in the unit inventory, catalyst samples obtained at the period of constant Z-100 concentration were analysed by the manufacturing company and found to contain ca. 4% wt Z-100.

ZSM-5 effects in FCCU yields. The estimation of the effects of ZSM-5 addition in the performance of the FCC unit is not straightforword since the operational parameters as well as the quality of the feed were not constant during the entire period of the test run, but changed according to the daily schedule of the Refinery.

In order to arrive at some conclusions concerning the effects in question, two periods of the FCC unit performance were compared, namely the month just before the introduction of ZSM-5 and the period during which the Z-100 concentration in the inventory was considered constant. The base catalyst in those two periods was the same. Specific days from the two periods were isolated, when the operational parameters of the FCC unit as well as the quality of the feed were very close. This method was considered to be more realistic than trying to refer all data to the same operating conditions and feed properties, because of the uncertainty that this referring would introduce.

The obtained results are shown in Table III. Gasoline yield has been decreased by 1.6% wt while LPG has increased by 1.4% wt. However the RON gain by 0.4 is quite small. The above fact has been attributed to the already high cat gasoline RON (93.0) attained by the FCC unit at normal operation, due to the highly aromatic feed used (*10*) and the type of the base catalyst.

Comparison of MAT data to commercial data. Table IV presents a direct comparison of the "delta yields" caused by the addition of 4% wt Z-100 on the base catalyst, obtained from MAT experiments and the commercial test. Although the absolute yields between MAT and FCCU are, as expected, quite different (Tables II, III), it is evident fron Table IV that the effect of ZSM-5 addition on the FCCU yields can be predicted accurately based on MAT data.

The most important effects, namely gasoline decrease, LPG increase and RON gain as derived from the two different tests, are very close to each other. It is very important to note that MAT data predict accurately the unexpectedly small RON gain, which was observed in the FCC gasoline and was attributed to the specific combination of feed and catalyst type.

Table III. FCC Unit Yields before and after the introduction of ZSM-5

	0% Z-100 in FCCU inventory	4% Z-100 in FCCU inventory	"Delta Yields"
Product Yields			
gasoline (% wt)	44.9	43.3	-1.6
LPG (% wt)	14.1	15.5	+1.4
Fuel gas (% wt)	6.2	6.4	+0.2
Coke (%wt)	5.0	5.0	0
LCO (%wt)	20.6	20.0	-0.6
Cl.O (%wt)	9.2	9.8	+0.6
LPG olefinicities			
C_3	0.797	0.808	+0.011
C_4	0.574	0.583	+0.009
Gasoline octanes			
RON	93.0	93.4	+0.4
MON	81.3	82.1	+0.8

Furthermore, the fact that the commercial results are very close to the MAT data obtained with samples containing either steamed or fresh additive shows that ZSM-5 maintains its initial catalytic activity after equilibration in an FCC unit for at least 30 days and confirms the laboratory observation concerning its superior hydrothermal stability.

Conclusions

Addition of ZSM-5 to an ultra-stabilised fluid catalytic cracking (FCC) catalyst leads to significant losses in gasoline yield, the gasoline being cracked preferentially to propylene and butylene. The octane boost resulting from ZSM-5 addition, when this type of catalyst is used in combination with an aromatic feed, is quite a bit smaller than expected.

The commercial additive tested in this work has very high hydrothermal stability, not undergoing any substantial change after severe steaming. Commercialy equilibrated additive containing catalyst samples perform in the same way as laboratory prepared mixtures.

By working with a bench scale MAT unit, it has been possible to access the effect of ZSM-5 addition to a USY equilibrium catalyst in a commercial FCC unit.

Table IV."Delta Yields" after ZSM-5 addition,MAT data vs. Commercial Test

	Equilibrium Cat.+ 4% Z-100, MAT data	Equilibrium Cat.+ 4% Z-100, Commercial test
Gasoline (% wt)	-1.66	-1.6
Fuel gas (% wt)	0	+0.2
Total LPG (%wt)	+1.66	+1.4
LCO (% wt)	-	-0.6
Cl.O (% wt)	-	+0.6
Olefinicity C3	+0.021	+0.011
Olefinicity C4	+0.014	+0.009
GC RON	+0.5	+0.4

Acknowledgments

The authors gratefully acknowledge the financial assistance granted by the General Secreteriat for Science and Technology under the auspices of the STRIDE-HELLAS program. The permission of MOTOR OIL HELLAS to publish the commercial results is also appreciated.
    The authors would like also to thank Ms Ioanna Dourou for her contribution in operating the MAT unit.

References

1. Dwyer, F.G.; Schipper, P.H.; Gorra, F. NPRA Annual Meeting, March 29-31, San Antonio, Texas
2. Yanik, S.J.; Demmel, E.J.; Humphries, A.P.;Campagna, R.J. *Oil and Gas J.,* **1985,** May 13, 108
3. Biswas, J; Maxwell, I.E. *Appl. Catal.,* **1990,** *58,* 1
4. Madon, R.J. *J. Catal.,* **1991,** *129,* 275
5. Rajagopalan, K.; Young, G.W. Am. Chem. Soc., Div. of Petrol. Chem., *Preprints,* **1987,** *32,* 627
6. Pappal, D.A.; Schipper, P.H. Am. Chem. Soc., Div. of Petrol. Chem., *Preprints,* Washington Meeting, 26-31 August, **1990,** 678
7. Miller, S.J.; Hsieh, C.R. Am. Chem. Soc., Div. of Petrol. Chem., *Preprints,* Washington Meeting, 26-31 August, **1990,** 685
8. Elia, M.F.; Inglesias, E.; Martinez, A.; Perez Pascual, M.A. *Appl. Catal.,* **1991,** *73,* 195
9. Anderson, P.C.; Sharkey, J.M.; Walsh, R.P., *J. Inst. Pet.,* **1972,** *58,* 83
10. Nalbandian, L; Lemonidou, A.A.; Vasalos, I.A. *Appl. Catal.,* **1993,** *105,* 107
11. Buchanan, J.S. *Appl. Catal.,* **1991,** *74,* 83
12. NPRA Q & A - 1, *Oil and Gas J.,* **1992,** March 16, 37

RECEIVED July 11, 1994

Chapter 6

Effect of Secondary Porosity on Gas-Oil Cracking Activity

Pei-Shing E. Dai[1], L. D. Neff[1], and J. C. Edwards[2]

[1]Texaco Research and Development, Port Arthur, TX 77641
[2]Texaco Research and Development, Beacon, NY 12508

Studying the effect of secondary pores on the cracking activity of heavy gas oils we conclude that for Y zeolites having unit cell sizes in the range of 24.25 to 24.35 Å the cracking activity of heavy gas oil increases with increasing secondary porosity and secondary pore mode. We have been able to generate a range of total acid site densities, secondary porosities and pore modes in ultrastable Y zeolites using a combination of hydrothermal and acid post treatments. In contrast with the secondary pore effects, the heavy oil cracking activity did not correlate with the total acid site density.

In catalytic cracking processes the important properties required for Y type zeolites are high hydrothermal stability, strong acid strength, and a large number of accessible active sites. Dealumination of zeolite Y has been extensively investigated in the past with an aim of achieving optimum acidity for catalytic cracking of hydrocarbons. Hydrothermal treatment of NH_4Y zeolite causes the dealumination of framework aluminum and the formation of secondary mesopores (1-4). In our previous work (5-6), systematic approaches were adapted to generate varied amounts of secondary mesoporosity and different sizes of secondary pore mode by employing hydrothermal, acid, and the combined hydrothermal/acid treatment of ultrastable Y zeolites. Very recently, high resolution electron microscope studies (7) of steam/acid treated neat ultrastable Y zeolite (USY) gave clear evidence for an inhomogeneous formation and distribution of the 50-500 Å defect regions attributed to mesopores which occurs concomitantly with a further dealumination. The study reported a deficiency in aluminum within the mesopore. In regions with high defect concentration, mesopores coalesced to form channels and cracks, which ultimately defined the boundaries of fractured crystallite fragments. The aluminum appeared to be enriched at these boundaries. Arribas et al., (8) investigated the influence of framework aluminum gradients on the gas oil cracking activity of Y zeolites dealuminated by different procedures. Their results supported the idea that gas oil does not penetrate the

0097–6156/94/0571–0063$08.00/0
© 1994 American Chemical Society

zeolite deeply, and therefore the activity of the catalyst is controlled by the framework Si/Al ratio near the external surface, instead of bulk Si/Al ratio of the dealuminated Y zeolites. Very little effort has been put on the influence of transport properties of secondary mesopores on the cracking activities of heavy oils. The objective of our present study is to characerize the dealuminated Y zeolites and make an attempt to segregate the effect of mesporosity from the effect of total acid sites and acid strength of dealuminated Y zeolites on heavy oil cracking activity. Of particular importance here is to examine the effect of secondary pore mode on catalytic performance for dealuminated Y zeolites with unit cell sizes less than 24.40Å.

Experimental

Preparation of Dealuminated Y Zeolites. The starting zeolites used in the hydrothermal dealumination experiments were Linde's LZ-Y82 and PQ's CP300-56 USY. The starting zeolite used in the acid treatment was CP300-35 SUSY zeolite supplied by PQ Corporation. In each case, a 20 gram sample of the starting zeolite was treated at temperatures ranging from 534-800°C for 2-48 hours with 100% steam in a 1" ID fix-bed reactor (one atmospheric pressure.) Both the steamed USY samples and CP300-35 SUSY zeolite were treated with 0.5-2 N nitric acid at 70-100°C for 2-6 hours using a weight ratio of zeolite to acid in the range of 0.02-0.10. The acid-leached sample was filtered and washed with deionized water and then dried in air at 120°C overnite.

Characterization of Dealuminated Y Zeolites. The secondary mesoporosity, defined as pore volume of pores having diameters in the range of 100-600Å, was measured by nitrogen adsorption using a Micromeritics ASAP 2400 instrument. X-ray diffraction (Scintag PAD-V) was used to determine the unit cell size, crystallinity and bulk Si/Al ratio. The bulk Si/Al ratio was calculated using the equation:

$$\frac{Si}{Al} = [\frac{1.6704}{(A_o - 24.19)}] - 1$$

The surface Si/Al ratio was determined by XPS using a VG surface analyzer. Solid-state ^{29}Si and ^{27}Al NMR spectra were obtained using a Varian VXR-300 spectrometer at resonance frequencies of 59.6 and 78.17 MHz, respectively. The probe used was a Chemagnetics 7mm CP/MAS probe. Magic angle spinning (MAS) was utilized with spinning speeds of 3-4 kHz for ^{29}Si and 6-7 kHz for ^{27}Al. Quantitative ^{29}Si single pulse magic angle spinning (SP/MAS) spectra were obtained using a 45° pulse angle and a 20 s recycle delay; ^{1}H-^{29}Si cross polarization (CP)MAS spectra were obtained using a 6 ms contact time and a 3 s recycle delay. The ^{27}Al SP/MAS spectra were obtained with a 5° pulse angle and a 0.5 s recycle delay, while the ^{1}H-^{27}Al CP-MAS spectra were acquired with a 0.5 ms contact time and a 3 s recycle delay. Si/Al ratios were calculated from the ^{29}Si SP/MAS signal intensities using the equation:

$$\frac{Si}{Al} = \frac{\sum\limits_{n=0}^{4} I_{Si(nAl)}}{\sum\limits_{n=0}^{4} (\frac{n}{4})\, I_{Si(nAl)}}$$

The total acidity was measured using ammonia temperature programmed desorption (TPD) over the temperature range of 175-500°C. The cracking activity was measured by charging a high nitrogen-containing feedstock (a blend of vacuum gas oil and atmospheric resid) over a mixture of 1 gram zeolite and 6 grams of Ottawa sand in a fixed-bed reactor at 515°C, WHSV of 20, and catalyst to oil ratio of about 5.

Results and Discussion

Effect of Steaming Time. The changes in pore mode and surface area as a function of steaming time at 770°C for LZ-Y82 zeolites is shown in Figure 1. Both the pore mode of the secondary pores created during hydrothermal treatment and surface area of the treated zeolites appeared to reach the constant values of 125Å and 410 m^2/g. The total pore volume of the steamed zeolites was increased to 0.19 cc/g compared to 0.10 cc/g for the starting material. The pore volume of pores having pore diameters in the 100-600Å region (PV 100-600 Å), shown in Figure 2, was increased from about 0.02 to 0.12 cc/g upon steaming.

Figure 3 shows the influence of steaming time on the unit cell size and crystallinity of the steamed Y zeolites. The unit cell size appeared to reach a stable value at about 24.30Å. Similarly, the crystallinity of the steamed zeolites seemed to line out at about 60% relative to the starting zeolite in about 15 to 24 hours of steaming at 770°C. Figure 4 shows that the surface Si/Al atomic ratio measured by XPS and the framework Si/Al atomic ratio measured by XRD also level off in the same time period at about 0.8 and 15, respectively. The steamed zeolites showed a pronounced enrichment in Al on the outer surface of the crystallites, which is like the results reported by Fleisch et al. (9).

Similar behavior has been found using CP300-56. These results imply that hydrothermally treated zeolites reach a steady-state composition after about 15 hours using these steaming conditions. Physical properties which decreased with steaming time were: zeolite crystallinity, unit cell size and surface area. By contrast, properties which increased with time were: pore volume of pores with pore diameters in the region of 100-600Å, the pore mode of secondary pores, and the bulk Si/Al ratio of the crystalline zeolite.

Effect of Steaming Temperature. Table I summarizes the pore volume distributions of steamed LZ-Y82 and CP300-56 zeolites as a function of steaming temperature (24 hour steaming.) For the steamed LZ-Y82 zeolite samples, the pore volumes of secondary pores having diameters in the region of 100-260Å (PV 100-260Å) show a maximum at 788°C, whereas, the pore volume in the region of 260-600Å (PV 260-600Å) increases continuously with increasing steaming temperatures from 732 to

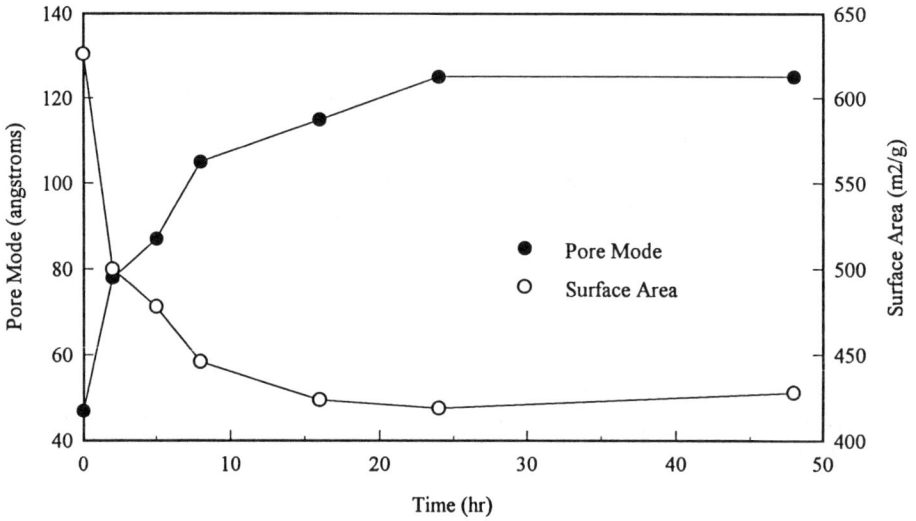

Figure 1. Effect of steaming time on zeolite porosity.

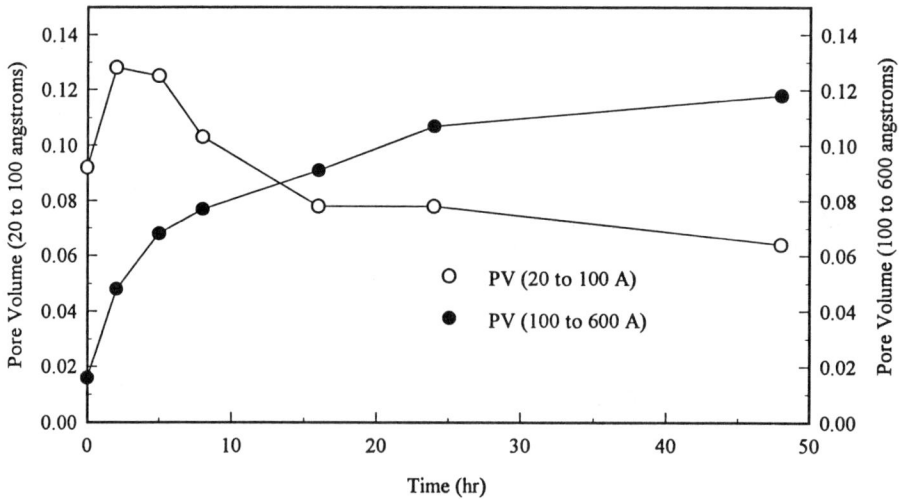

Figure 2. Effect of steaming time on pore volume distribution.

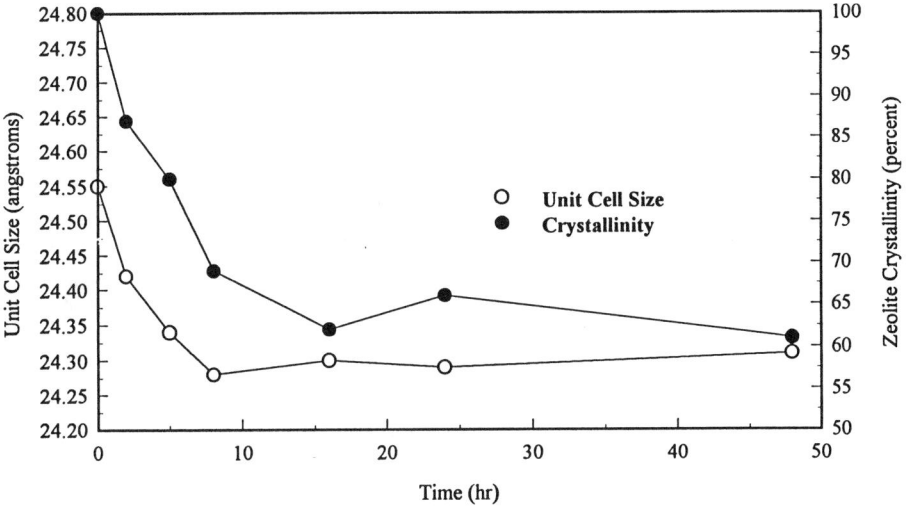

Figure 3. Effect of steaming time on zeolite properties.

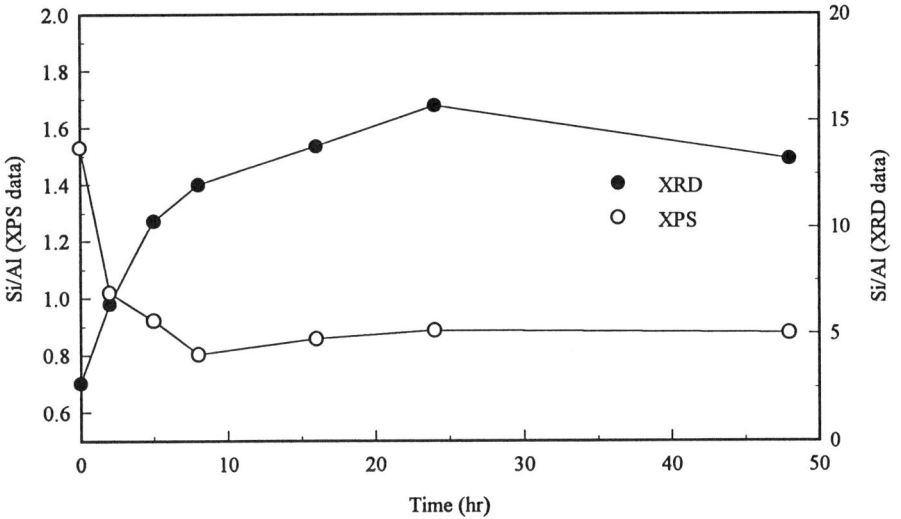

Figure 4. Effect of steaming time on zeolite Si/Al ratio

Table I. Effect of Steaming Temperature on Secondary Porosity

Zeolite	Steam Temp (°F)	Pore Volume 100-260 Å (cc/g)	Pore Volume 260-600 Å (cc/g)	Secondary Pore Mode (Å)	Surface Area (m²/g)
LZY-82	---	0.011	0.005	47	626
	732	0.064	0.008	105	457
	771	0.083	0.020	125	411
	788	0.093	0.024	145	392
	800	0.062	0.031	170	376
CP300-56	---	0.013	0.007	43	687
	593	0.065	0.017	115	559
	677	0.086	0.029	120	498
	732	0.096	0.031	125	473
	760	0.110	0.034	135	464
	771	0.108	0.036	145	454
	788	0.107	0.041	165	439
	800	0.111	0.050	170	408

800°C. This change in secondary pore volume is accompanied by a similar shift in the secondary pore mode from 47Å (fresh LZ-Y82) to 170Å.

For the CP300-56 USY zeolite steamed at temperatures ranging from 593 to 800°C, the 100-260Å pore volume appears to increase linearly with temperature from 593 to 760°C and reaching a plateau at about 0.11 cc/g. Over this temperature range, the 260-600Å pore volume also increases from 0.02 to 0.05 cc/g. The secondary pore mode shifts to 170Å after steaming at 800°C.

The enrichment of non-framework aluminum species and degree of dealumination as a function of steaming temperature is presented in Figure 5 for CP300-56. It is noted that the surface Si/Al atomic ratio decreased from 2.04 to 0.95 with increasing temperatures up 732°C with little change for higher temperatures. In contrast, the framework Si/Al ratio increases with steaming temperature from 3.5 for the untreated zeolite to 32 for the sample treated at 800°C.

The total acidity as measured by ammonia TPD exhibits a general decrease with increasing steam severity, as shown in Figure 6. Upon steaming at 534°C for 5 and 24 hours, the total acidities are 11.5 and 6.6 cc NH_3/g, STP compared to 28.9 cc NH_3/g, STP for the untreated CP300-56 USY zeolite. For samples steamed at high temperatures (>700°C) for longer than 5 hours, the total acidity is in the range of 1-5 cc NH_3/g, STP.

Effect of Acid Treatment. Post-treatment of a series of CP300-56 USY samples (steamed 24 hours at temperatures from 593 to 800°C) with 0.5 N aqueous nitric acid solution at 70°C for 6 hours causes a further increase in the secondary pore volume and an increase in the secondary pore mode. For instance, post-acid treatment boosts the PV 100-260Å of the sample steamed at 760°C from 0.11 to 0.14 cc/g, and the pore mode from 135 to 190Å. These results suggest that some of the secondary pores are blocked (narrowed) by deposited non-framework aluminum species resulting from steam treatment. Subsequent acid leaching partially removes this material resulting in an increase in secondary pore volume and pore mode of the dealuminated Y zeolites.

A correlation between zeolite crystallinity and framework Si/Al ratio for these materials is shown in Figure 7 for both the steamed and steamed/acid-leached samples. For steamed samples with higher crystallinities (>70%), a further dealumination takes place during acid treatment as indicated by the increase in the framework Si/Al ratio. However, for steamed samples with lower crystallinities (55-70%), acid treatment results in little or no additional dealumination of the framework. All of these steamed/acid-leached zeolites had similar unit cell sizes (24.25-24.28Å,) and consequently they had similar framework Si/Al ratios of about 18-27 (see Figure 8)as compared to 8-32 (Figure 5) for the steamed zeolites. It is also noted that the acid-leached samples had similar surface Si/Al ratios (30 to 35, except for the sample that had been steamed at 800°C as can be seen in Figure 8), which is an order of magnitude greater than the starting zeolite (2.04.) Therefore, the post-acid treatment converts the outer surface of the hydrothermally dealuminated Y zeolites from an Al-enriched to an Al-deficient state.

^{29}Si and ^{27}Al MAS-NMR Studies of Dealuminated Y Zeolites. The ^{29}Si MAS-NMR spectrum shows that in the untreated USY sample the peak intensities for

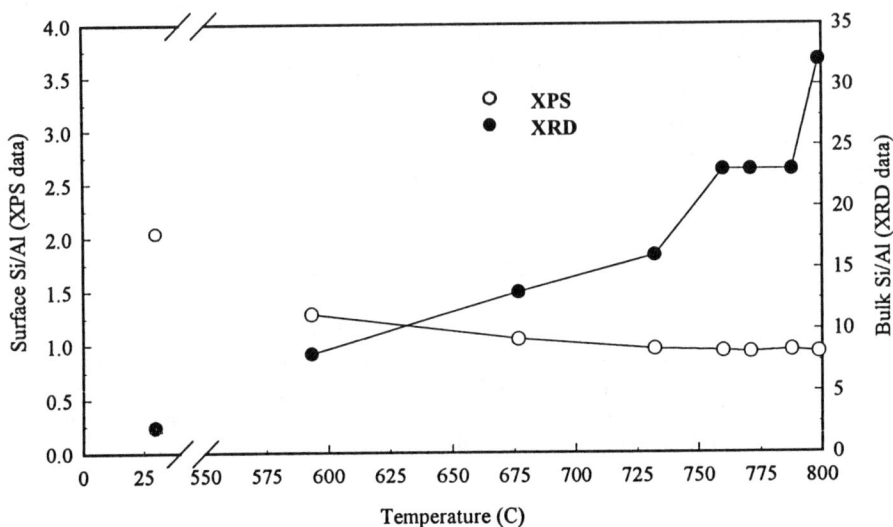

Figure 5. Effect of steaming temperature on framework and surface Si/Al ratio.

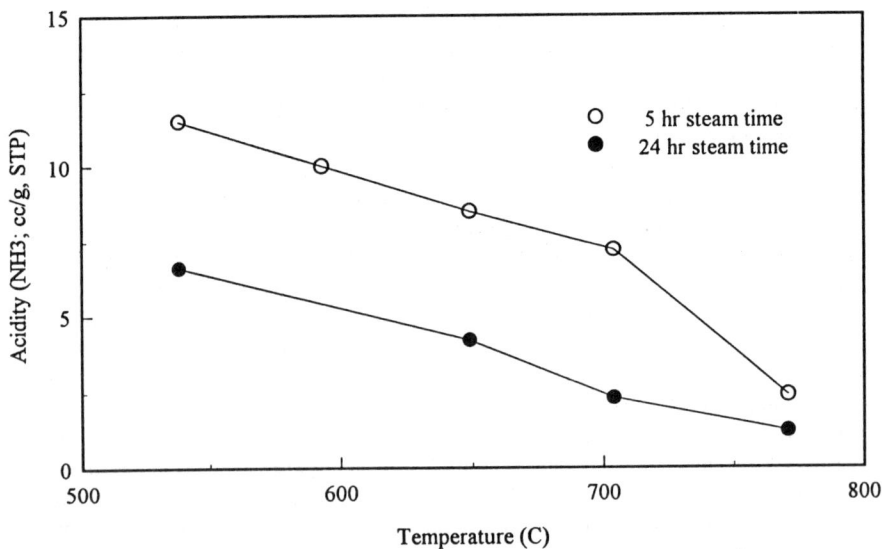

Figure 6. Effect of steaming temperature on acidity of zeolites.

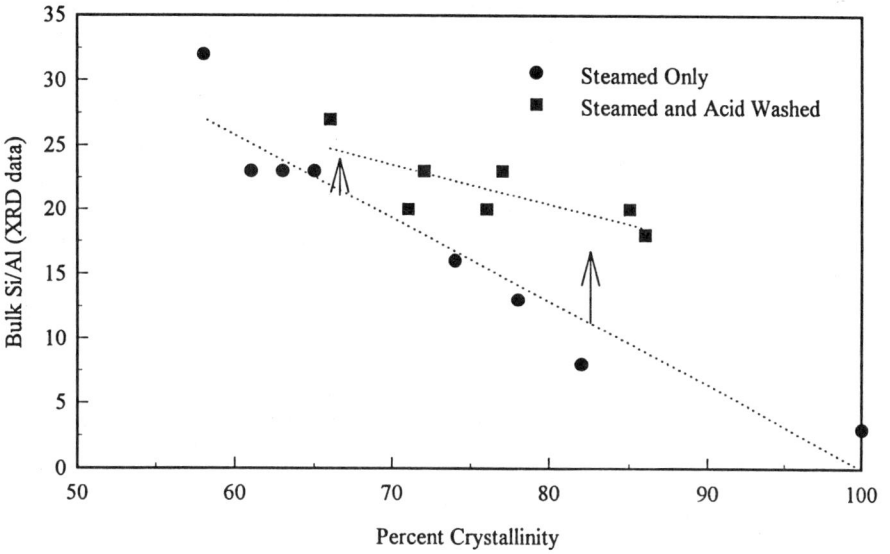

Figure 7. Correlation between zeolite crystallinity and framework Si/Al ratio.

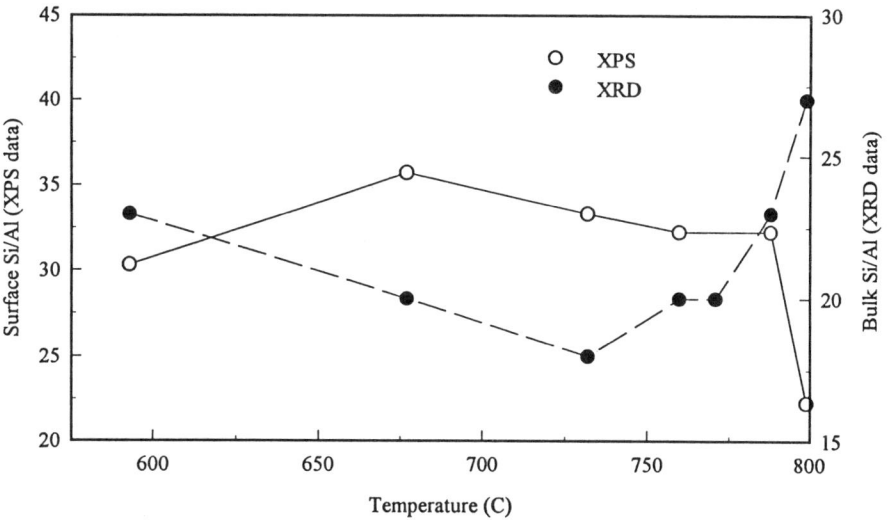

Figure 8. Effect of steaming temperature and acid treatment on framework and surface Si/Al ratio.

Si(3Al),Si(2Al), Si(1Al), and Si(0Al) were 8:14:35:43 having Si(0Al) as the strongest peak. This result is in good agreement with the result reported by Ray et al. (10) for a Davison Z-14 USY zeolite having a Si/Al of 4.8. For the samples having unit cell sizes of 24.26-24.29Å obtained by steaming and combined steaming/acid washing, the ^{29}Si MAS-NMR spectra have only two peaks and the relative intensity ratios of Si(1Al) to Si(0Al) in the range of 1:5 to 1:10. The presence of amorphous oligomeric silicon species is observed particularly in the samples with unit cell sizes less than 24.21Å, a large amount especially when unit cell size is less than 24.15Å. This is more pronounced in the spectrum for sample F in Figures 9 and 11 by the broadening of the peak near -115ppm.

The ^{29}Si MAS-NMR and ^{27}Al MAS-NMR analysis of three acid-washed samples of CP300-35 SUSY zeolite (denoted as samples E, F, and G in Table II) are shown in Figures 9 and 10, respectively. The ^{27}Al MAS-NMR can be used to obtain some idea of the degree of dealumination that occurs upon steaming and acid washing. In hydrated zeolite framework, tetrahedral aluminum appears at a chemical shift of 50-70 ppm, while octahedral, extra-framework aluminum (present as amorphous material on the outside of the zeolite crystallites) appears in the 10 to -20 ppm range. Assuming that all of the aluminum present was observed in the [27]Al MAS experiments, a semi-quantitative analysis of the spectra in Figure 10 shows that as the unit cell size decreased from 24.33 to 24.28Å there was an increase in the ratio of $Al(T_d)/Al(O_h)$ from 2.1 to 4.9, and then a sharp decline from 4.9 to 0.2 as the unit cell size changed from 24.28 to 24.26Å. The assumption that [27]Al data can be treated as semi-quantitative appears to be valid as much of the extra framework aluminum is present in a highly octahedral form - probably $[Al(H_2O)_6]^{3+}$. These results suggest that mild acid treatment removes the amorphous aluminum species, thereby increasing the ratio of $Al(T_d)/Al(O_h)$, while severe acid washing causes extensive extraction of framework aluminum, with heavy deposition of aluminum on the crystallite surface, leading to a decrease in the ratio of $Al(T_d)/Al(O_h)$. The ^{27}Al MAS-NMR spectrum of sample M, with a unit cell size of 24.27Å, which was prepared by mild steam/acid treatment of CP300-56 USY zeolite, indicates that a high concentration of both framework and non-framework aluminum is present in the stream/acid washed zeolite, with the $Al(T_d)/Al(O_h)$ being 3.6. However, none of this aluminum is amenable to cross polarization experiments which transfer magnetization from ^{1}H to [27]Al. This may be due to rapid dynamics of the water coordinated to [27]Al.

The ^{29}Si MAS-NMR and ^{27}Al MAS-NMR analysis of three zeolites with unit cell size of 24.27 to 24.28Å (denoted as samples D, F, and H in Table II) are shown in Figures 11 and 12, respectively. The [27]Al spectrum reveals the formation of amorphous material in tetrahedral (61 ppm), octahedral (1 ppm), and pentacoordinate (34 ppm) environments. This [27]Al spectrum of steamed USY zeolite sample D compares very well with that reported by Pellet et al. (11) for the steamed LZ-Y82 and LZ-210 zeolites. They suggested a minimum of four Al species present in the steamed samples, namely, tetrahedral framework Al, intermediate (probably five-coordinate-trigonal bipyramidal) Al, extra-framework octahedral Al, and a distorted geometry phase Al leading to the broad component.

The results of ^{29}Si MAS-NMR and ^{27}Al MAS-NMR analysis of dealuminated Y zeolites for their framework Si/Al ratios are summarized in Table II along with the

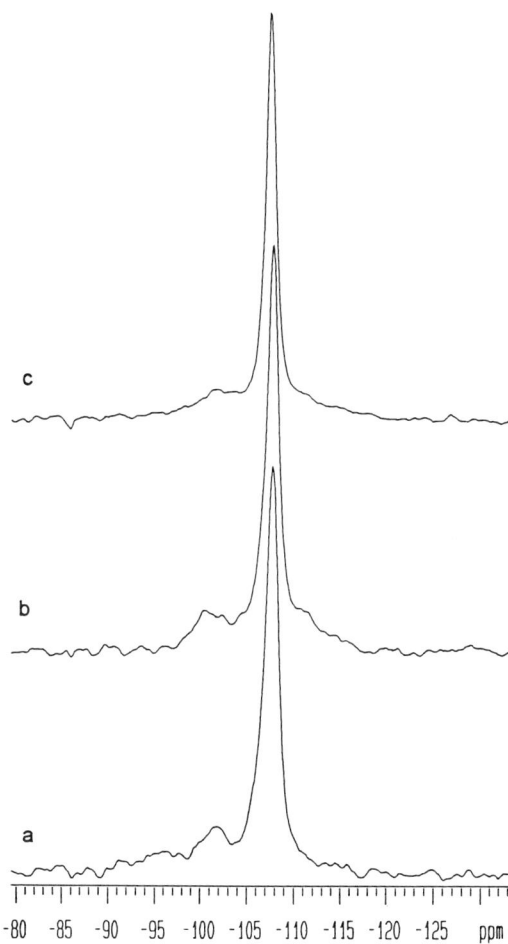

Figure 9. ^{29}Si SP/MAS NMR spectra of dealuminated Y zeolites; (a) sample E, (b) sample F and (c) sample G.

TABLE II. ^{29}Si and ^{27}Al MAS-NMR Analysis of Zeolites

Sample Code	Treat-ment	Si/Al Ratio (XRD)	Lattice Const. (Å)	Si/Al Ratio (XPS)	Si/Al Ratio (NMR)	Al(Td)/Al(Oh) Ratio (NMR)
A	CP300-56	3.5	24.56	2.0	4.6	5.3
B	CP300-35	9.4	24.35	1.6	11.6	2.0
C	CP304-37	8.3	24.37	2.9	9.7	2.7
D	A-St	19.9	24.27	1.1	41.5	N. A.
E	C-AW	10.9	24.33	7.0	10.8	2.1
F	C-AW	17.6	24.28	29.1	23.3	4.9
G	C-AW	22.9	24.26	37.0	22.4	0.2
H	A-St-AW	19.9	24.27	28.0	28.0	2.5
I	A-St-AW	N. A.	24.12	30.1	12.8	1.8
J	A-St-AW	N. A.	24.18	25.7	14.4	3.3
K	A-St-AW	82.5	24.21	22.7	20.5	3.2
L	A-St-AW	40.8	24.23	32.0	21.8	3.8
M	A-St	19.9	24.27	N. A.	19.4	3.6
N	A-St	15.7	24.29	3.9	N. A.	N. A.
O	A-St	7	24.39	5.1	N. A.	N. A.

St - Steaming
AW - Acid washing
St-AW - Steaming followed by acid washing
Si/Al (XRD) = [1.6704/(a-24.19)]-1,   where a = lattice constant in Å

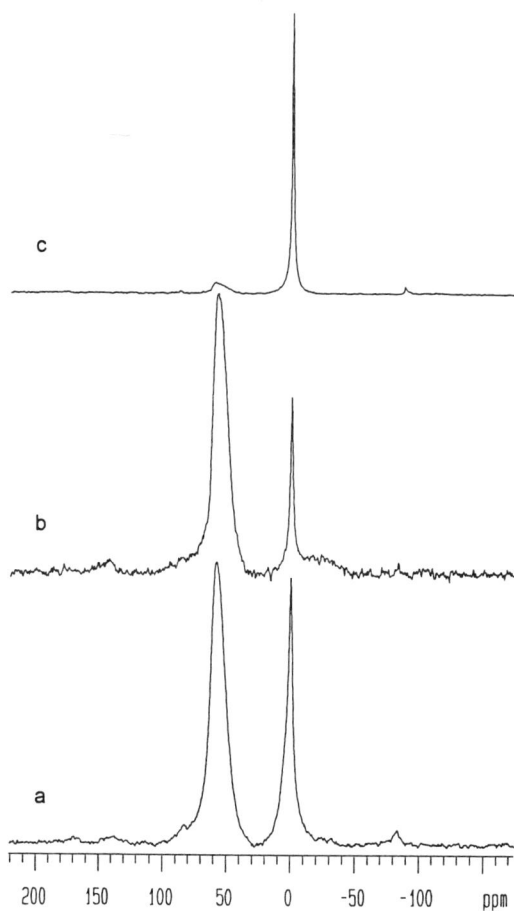

Figure 10. ^{27}Al SP/MAS NMR spectra of dealuminated Y zeolites; (a) sample E, (b) sample F and (c) sample G.

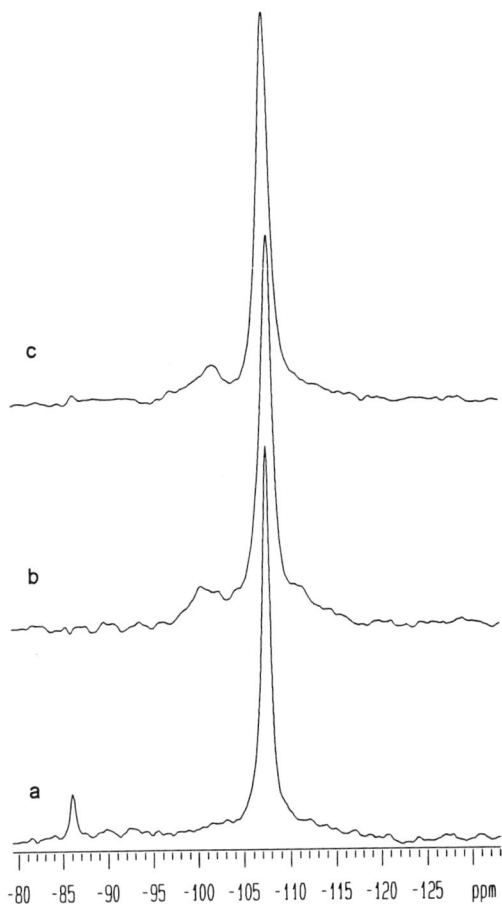

Figure 11. ^{29}Si SP/MAS NMR spectra of dealuminated Y zeolites; (a) sample D, (b) sample F and (c) sample H.

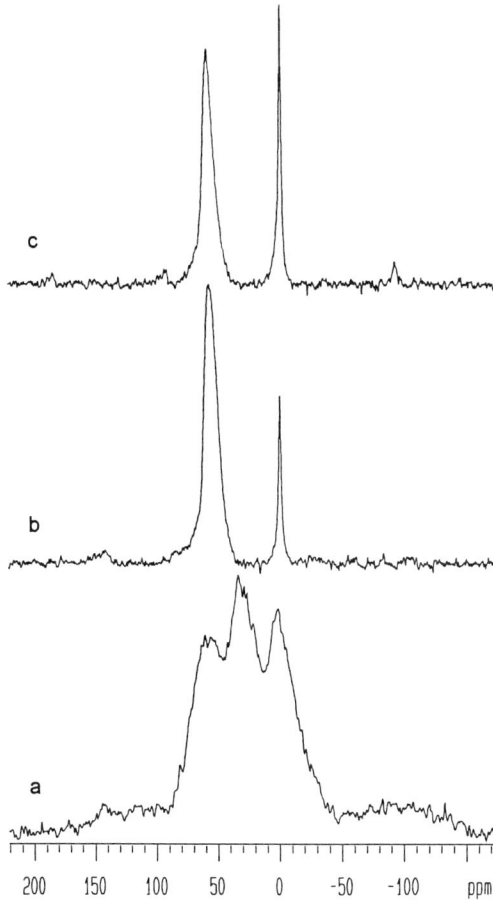

Figure 12. ^{27}Al SP/MAS NMR spectra of dealuminated Y zeolites; (a) sample D, (b) sample F and (c) sample H.

Si/Al ratios determined by XRD and XPS. Figure 13 shows the correlation between Si/Al ratios obtained using NMR and XRD analyses. The Si/Al ratio obtained by ^{29}Si MAS-NMR for the untreated zeolite (sample A) is consistent with that obtained by XRD. For the dealuminated Y zeolite samples having unit cell sizes greater than 24.25Å, there is a fairly good agreement between the Si/Al ratios determined by the two techniques.

Gas Oil Cracking Activity of Dealuminated Y Zeolites. The gas oil cracking activity is plotted against the unit cell size in Figure 14. Samples for these data included untreated CP300-56 USY, steamed CP300-56 USY, the acid-leached CP300-35 SUSY, and the steamed/acid-leached CP300-56 USY samples. For the steamed and steamed/acid-leached samples, the maximum occurs in the range of 24.33-24.35Å. This result is in good agreement with the results reported by Arribas et al. (8) for steamed and SiCl$_4$-treated Y zeolites.

It is noted that there is no correlation between the gas oil cracking activity and the total acidity as measured by TPD of ammonia for these samples. This is thought to be due to the different procedures used to prepare the samples. For example, samples D, F, and N (Table II) have similar unit cell sizes (24.27-24.29Å). Nevertheless, sample D exhibits much lower activity (30%) than the acid-leached sample F (63%) and the steam/acid-leached sample N (58%). The total acidity for samples D, F, and N follows the order of N (7.9) > F (2.8) > D (1.4 cc NH$_3$/g, STP). The secondary mesopore volume, PV 100-600Å, for these samples D, F, and N are 0.14, 0.20, and 0.17 cc/g, respectively. It appears that the conversion of heavy oil cracking generally follows the same order as the secondary mesopore volumes for this set of samples. The results can be rationalized on the basis that upon steaming, the amorphous alumina and silica-alumina debris may block the diffusion of heavy oil and adversely affect the cracking activity. The difference in total acidity perhaps plays little role in the perfomance of these zeolites since mesopores facilitate the transport of heavy oil molecules.

Two steamed/acid-leached samples (E and O) and one steamed sample (C) with similar unit cell sizes in the range of 24.33-24.39Å are compared with fresh CP300-56 (sample A) for their Si/Al ratios, total acidity, secondary porosity and gas oil cracking activity in Table III. The steamed sample is Al-enriched on the surface of the crystallites, whereas, the Al distribution is more uniform for the steamed/acid-leached samples. The total acidity decreased in the order of A (28.9) > O (16.5) > E (10.2) > O (6.8). However, sample E is greater than sample O in both the secondary pore volume (0.15 vs. 0.10) and pore mode (155 vs. 125Å). The gas oil cracking activity follows the order of E (73%) > O (61%) > C (52%) > A (41%.) It should be noted that sample E gave the lowest yield of slurry oil (bp. >343°C) of the four zeolites listed in the table. The activity of sample E may be attributed to the largest secondary pore volume (PV 100-600Å) and the pore mode (155Å.)

Conclusion

Secondary pores with substantial pore volume in the range of 100-600Å were created by steaming, acid leaching, and the combined steaming with acid leaching of ultrastable Y zeolites. The treated zeolites have pore modes of these secondary pores

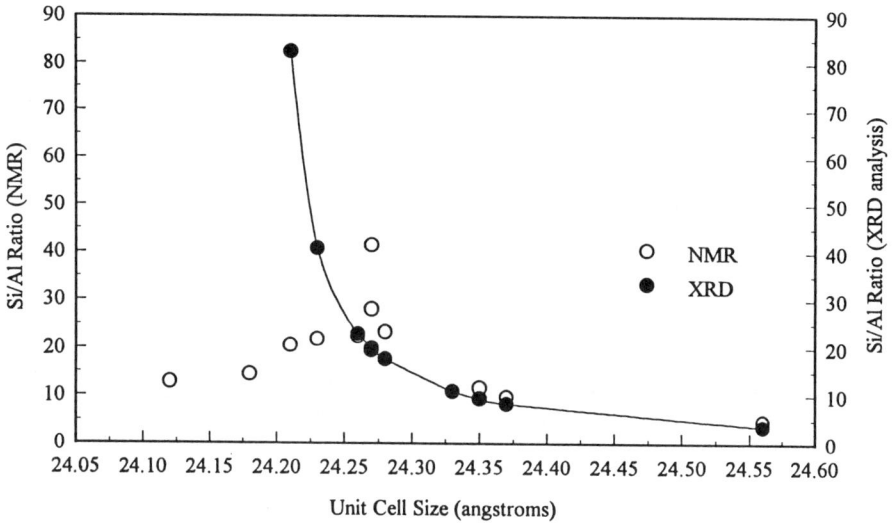

Figure 13. Correlation between Si/Al ratios obtained using NMR and XRD analyses.

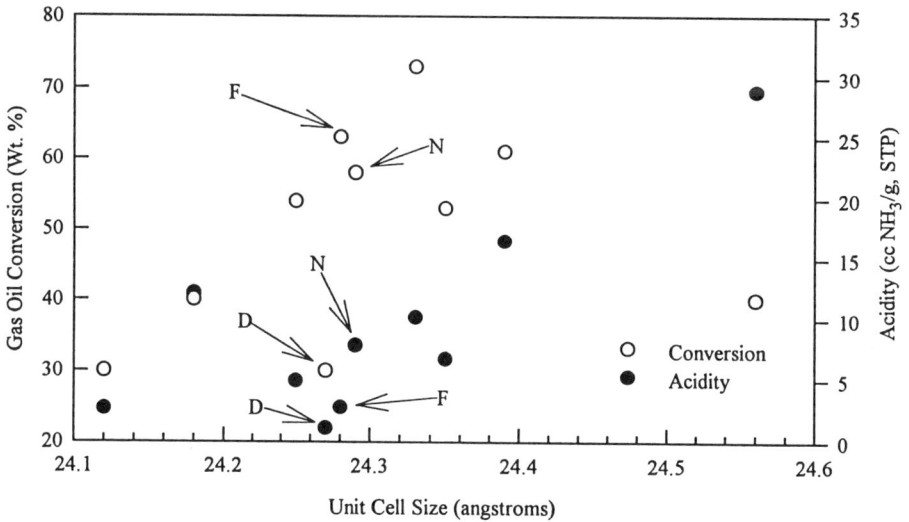

Figure 14. Effect of crystal unit cell size on zeolite acidity and gas oil cracking activity.

TABLE III. Heavy Gas Oil Cracking Activities of Dealuminated Y Zeolites

Sample Code	A	O	C	E
Total Acidity, cc NH$_3$/g, STP	28.9	16.5	6.8	10.2
Secondary Pore Vol., cc/g	0.02	0.10	0.12	0.15
Pore Mode, Å	47	125	87	155
Gas Oil Conv., %	41	61	52	73
LCO, %	23	16	23	17
Slurry Oil, %	23	22	26	9

in the range of 115-170Å. The maximum gas oil cracking activity appears at the unit cell size of 24.33Å for the dealuminated Y zeolites. Over the range of 24.29-24.35Å, the cracking activity was increased with increased secondary porosity and pore mode. In contrast, the cracking activity did not correlate with the total acidity. Thus, secondary pores appear to facilitate the transport of heavy oil molecules into the acid sites and enhance the cracking activity of dealuminated Y zeolites.

Literature Cited

1.   Lynch, J., Raatz, F., and Dufrensne, P. *Zeolites* **1987**, *7*, 333.
2.   Lohse, U. and Milderbrath, M. *Z. Anorg. Allg. Chem.* **1981**, *476*, 126.
3.   Stach, H., Lohse, U., Thamm, H. and Shrimer, W. *Zeolites* **1986**, *6*, 74.
4.   Patzelova, V and Jaeger, N. J. *Zeolites* **1987**, *7*, 240.
5.   Dai, P. E., Sherwood, D. E., and Martin, B. R. U.S. Patent Nos. 5,069,890 (1991) and 5, 087,348 (1992).
6.   Dai, P. E., and Sherwood, D. E., U.S. Patent Nos. 5,112,473 (1992) and 5,143,878 (1992).
7.   Choi-Feng C., Hall, J. B., Huggins, B. J., and Beyerlein, R. A., *J. Catal.*, **1991**, *140*, 395.
8.   Arribas, J., Corma, A., Fornes, V., and Melo, F. *J. Catal.*, **1987**, *108*, 135.
9.   Fleisch T. H., Meyers, B. L., Ray. G. J., Hall, J. B. and Marshall, C. L. *J. Catal.*, **1986**, *99*, 117.
10.   Ray, G. J., Meyers, B. L. and Marshall, C. L. *Zeolites* **1987**, *7*, 307.
11.   Pellet R. J., Blackwell, C. S. and Rabo, J. A. *J. Catal.*, **1988**, *114*,71.

RECEIVED June 20, 1994

Chapter 7

Investigation of Mesopore Formation

Evolution of Nonframework Aluminum Species During Hydrothermal Dealumination of Ultrastable Y Catalysts

R. A. Beyerlein, C. Choi-Feng, J. B. Hall, B. J. Huggins, and G. J. Ray

Amoco Research Center, P.O. Box 3011, Naperville, IL 60566

Combined high resolution electron microscope (HREM) and analytical electron microscope (AEM) investigations have been applied to the formation and evolution of mesopores in hydrothermally dealuminated ultrastable Y (USY) materials. The associated development and evolution of nonframework aluminum species was investigated by high resolution solid state ^{27}Al MASNMR. In regions with high defect concentration, mesopores coalesce to form channels and cracks, which ultimately define the boundaries of fractured crystallite fragments. At these boundaries, a dark band is observed which is highly enriched in aluminum. These dark bands are observed both in a neat USY material subjected to several cycles of steam/acid treatment and in a high temperature steam deactivated USY fluid catalytic cracking (FCC) catalyst. Parallel ^{27}Al MASNMR investigations on the neat USY material show the evolution of several different aluminum species.

The catalytic properties of ultrastable Y (USY) zeolite are directly influenced by the zeolite destruction which occurs during formation of USY and during subsequent hydrothermal treatment. For example, it has been suggested that the internal channels associated with mesopore formation provide greater accessibility to the micropore system defined by the crystalline zeolite. Yet, little information is available regarding mechanisms of mesopore formation and evolution. The nature and role of nonframework Al that is associated with mesopore formation are also poorly understood.

Previous transmission electron microscope (TEM) studies of hydrothermal aging of neat USY materials and also of USY cracking catalysts have shown 5-50 nm defect domains, which were attributed to mesopores (1-9). Such features, which become more pronounced in the presence of vanadium, are characteristic of extended hydrothermal treatment (6,7). Scherzer and Maugé, et al have suggested that these regions of zeolite destruction comprise the silica source for "healing" the tetrahedral site vacancies left by hydrothermal dealumination (5, 10-13). Sorption studies are

0097–6156/94/0571–0081$08.00/0

consistent with the picture of entire sodalite units being destroyed concomitant with the rebuilding or healing of the USY framework (14,15). A schematic rendering of this mechanism for healing is given in Fig. 1. Typical porosity analyses of mildly steamed USY materials show a distribution of mesopore dimensions in the range 5-50 nm, that is skewed towards the smaller sizes, further supporting the association of the light "amorphous" zones observed by TEM with the secondary pore system characteristic of USY materials (2,4).

Previous studies of USY materials subjected to extended hydrothermal treatment have suggested a homogeneous or near homogeneous distribution of mesopores (2-4). In contrast, recent investigations on a hydrothermally aged La-Y cracking catalyst demonstrate significant inhomogeneity both in the extent of dealumination and in defect formation (8).

In the present study, both high resolution electron microscopy (HREM) and analytical electron microscopy (AEM) are utilized to track the formation and evolution of mesopores, and the ultimate fate of the associated extralattice Al in hydrothermally treated USY materials. The associated development and evolution of aluminum species was investigated by high resolution solid state NMR (^{27}Al MASNMR). Results for a laboratory steamed neat USY material and for a high temperature steam deactivated USY fluid catalytic cracking (FCC) catalyst are compared with each other, and with results from a previous study on age-separated equilibrium USY cracking catalyst from the fluid cracking unit (FCU) (9).

Experimental

A highly dealuminated zeolite Y material was prepared by subjecting a sample of commercial high silica Y, LZ-Y82 from Union Carbide, to three cycles of steam/acid treatment in which each aqueous treatment was followed by rinsing and drying steps. The first cycle consisted of 600°C steam for 2 hours, ammonium exchange using 1.8 M NH$_4$Cl, 65°C, 1.5 hours, and then a mild acid treatment using 100 volumes of 0.033 N HCl, for 4 hours at room temperature. The product material was given a second steam treatment at 650°C for 3 hours, followed by two successive acid treatments using 50 volumes of 0.1 N HCl. A third steam treatment was carried out at 650°C for 4 hours, followed by an acid treatment with 0.1 N HCl, and finally calcining at 538°C for 3 hours. All steam treatments were carried out using 100% steam. The extensively dealuminated product of this series of steam/acid treatments on the neat USY material is hereafter referred to as ED-USY. A second highly dealuminated ultrastable Y material was prepared by steam treatment of a commercial fluid catalytic cracking catalyst, KOB-619-1A, obtained from Akzo (Ketjen Catalysts), at 816°C for 24 hours, using 100% steam in a fluidized bed. The highly dealuminated USY catalyst is hereafter referred to as ED-YCAT. Properties of these two dealuminated USY materials after steam treatments are given in Table I.

The as-prepared samples were ground to a fine powder, embedded in LR-White acrylic resin, and then cut with a diamond knife on a Reichert-Jung Ultracut E ultramicrotome to obtain thin sections approximately 60-80 nm thick. The sections were supported on Cu grids and coated lightly with C to gain conductivity. High resolution electron microscopy (HREM) and analytical electron microscopy (AEM)

Table I. Properties of Dealuminated USY Materials

	Steam/Acid Treated Neat USY Materials [a] (ED-USY)	Steam Treated USY Cracking Catalyst [b] (ED-YCAT)
Surface area (m²/g)		
Total	529	169
Pores < 2 nm diameter	436	117
% Crystal, relative to fresh	75	61 [c]
Unit Cell Size (nm)	2.426	2.424
Framework Al/Unit Cell[d]	4.3 ± 1.1	2.0 ± 1.1

[a] The fresh USY parent material (LZ-Y82) showed a unit cell size of 2.456 nm.

[b] The fresh parent USY FCC cracking catalyst (Ketjen KOB-619-1A) showed a unit cell size of 2.440 nm.

[c] The crystallinity of the parent (fresh) catalyst relative to LZ-Y82, determined by X-ray diffraction, was 35%.

[d] Calculated from unit cell size correlation by Kerr, G. T.; *Zeolites*, **1989**, *9*, pp 350.

were used to characterize the prepared samples. For HREM, two microscopes, a JEOL JEM 2000EX and a JEOL JEM 2010, were used. The AEM work was done in-house using a Noran TN5500 energy dispersive X-ray spectrometer (EDXS) attached to a Philips 420T scanning transmission electron microscope (STEM). The ^{27}Al MASNMR spectra were obtained on a VXR-400 at a resonance frequency of 104 MHz. Spectra were taken with a 1 μs excitation pulse (8 μs = 90°), a relaxation delay of 0.1s, and 10,000 scans. The MAS rotor was spun at 8 kHz.

Results

Steam/Acid Treated Neat USY Material.

27Al MASNMR Investigations. The series of ^{27}Al MASNMR spectra shown in Figure 2 track the evolution of the extensively dealuminated USY material, ED-USY (Fig. 2d), used in the electron microscope investigations. All spectra contain peaks at 60, 30, and 0 ppm where the 60 and 0 ppm peaks are due to tetrahedral and octahedral species. The starting USY (Fig. 2a) has sharp resonances at 60 and 0 ppm, which are, respectively, attributed to framework and nonframework tetrahedral species. A 30 ppm "peak" was also found. For the starting USY, this "peak" has been shown to be only a portion of the second-order, quadrupolar broadened line shape caused by an aluminum in a distorted tetrahedral environment (*16*). The subsequent steam treatment (Fig. 2b) causes a 25% reduction of the 60 ppm peak due to framework aluminum, and simultaneously generates a true resonance at 30 ppm,

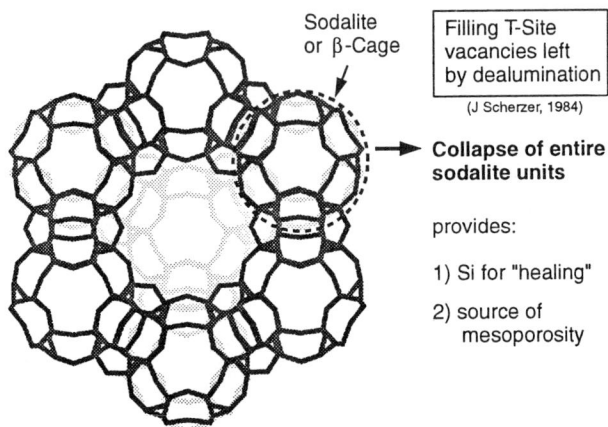

Figure 1. Schematic model for mechanisms of dealumination in FAU. Collapse of entire sodalite units provides 1) Si for "healing" of tetrahedral site vacancies left by dealumination, and 2) the source of mesoporosity. Reproduced with permission from reference 22.

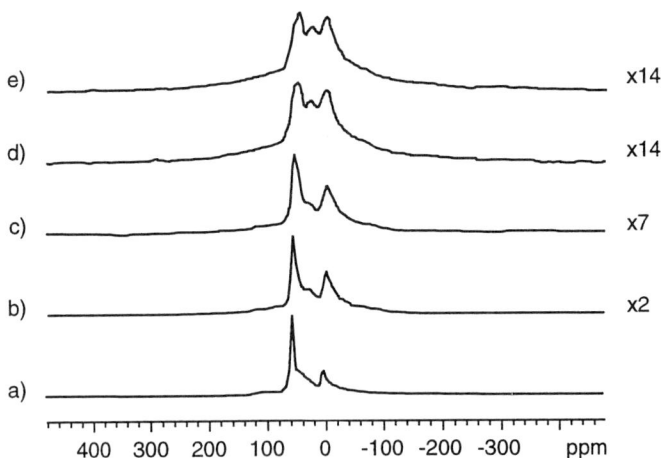

Figure 2. ^{27}Al MASNMR spectra of (a) starting USY zeolite; (b) after steam treatment of (a), 600 °C, 2hr; (c) after steam treatment of (b), 650 °C, 3 hr, with NH_4^+ exchange and acid wash before and after steaming; (d) after acid wash and steam treatment, 650 °C, 4hr, followed by calcination, 538 °C, 3 hr; (e) after steam treatment of (d), 700 °C, 3hr, followed by acid wash. Spectrum (d) is of ED-USY. Reproduced with permission from reference 22.

which has been shown to be due to an aluminum in a penta-coordinate environment (*16*). This "peak" persists as dealumination proceeds. Additional steam treatments (Figs. 2c,d) lead to a continued loss of framework aluminum, and both the tetrahedral and octahedral resonances which remain are considerably broadened. The distribution of aluminum sites and the peak widths were not affected by the final treatment (Fig. 2e).

The tetrahedral resonance of the most dealuminated samples (Figs. 2c - e) shows a substantial broadening. The assignment of this broadened tetrahedral resonance is not clear. Only a small portion of this resonance can be attributed to framework aluminum because the lattice parameter (2.246 nm) of ED-USY indicates a framework aluminum content of only 4 Al/unit cell (Table 1). This resonance is due either to nonframework aluminum species or, more likely to a silica alumina species formed from detrital silicon and aluminum atoms removed from the framework by the steaming process.

TEM Investigations. A TEM image showing the overall structure of the ED-USY, is shown in Figure 3. An inhomogeneous distribution of mesopores is seen among different USY grains and within individual grains. Grain fracture boundaries show random orientations. Large cracks due to ultramicrotoming are seen as fracture lines that are roughly parallel to each other. A TEM micrograph showing only a few ED-USY grains is shown in Figure 4a. The mesopores appear as light pseudospherical domains, approximately 15 to 50 nm in dimension, corresponding to lower density. Although these domains have lost much of their crystallinity, the connecting regions remain crystalline (Fig. 4b). These light, low crystallinity domains have been observed previously by many researchers and are referred to as the secondary pore system or mesopores (*1-5*). The mesopores, evident in most of the USY grains, are inhomogeneously distributed. Some grains appear to contain more mesopores than others. Within individual grains, mesopores are often highly concentrated in localized regions. In regions of high concentration, mesopores appear to coalesce to form channels as seen in Figure 4a. Such channels can apparently evolve into cracks, leading to the fracturing of the USY grains into several smaller fragments as shown in Figure 5. In extreme cases, the original large crystals have broken into crystallites as small as 20 nm. These small crystallites, each with a different crystallographic orientation, are shown in Figure 6.

A STEM image of several USY grains, along with EDXS spectra of the entire imaged area (overall) and of two USY grains (PF1 and PF2), are shown in Figure 7. Numerous mesopores are readily seen in grain 1 (PF1) while grain 2 (PF2) is largely free of mesopores. The average Si/Al ratio of the overall imaged area is lower than that of PF1, but higher than that of PF2. These results indicate that the formation of mesopores in the USY zeolite occurs concomitantly with its dealumination.

HREM investigation of the extensively dealuminated USY material shows characteristic "dark bands" within the channels or cracks that evolve from coalesced mesopores (Figure 5). A STEM image of several similar USY grains along with three EDXS spectra are shown in Figure 8. One spectrum was collected while scanning the partial field, PF1, a second with a stationary electron probe on a dark band (S-1), and a third with a stationary electron probe within a mesopore (S-2). From the spectra, it is evident that the dark band (S-1) is highly enriched in Al, while

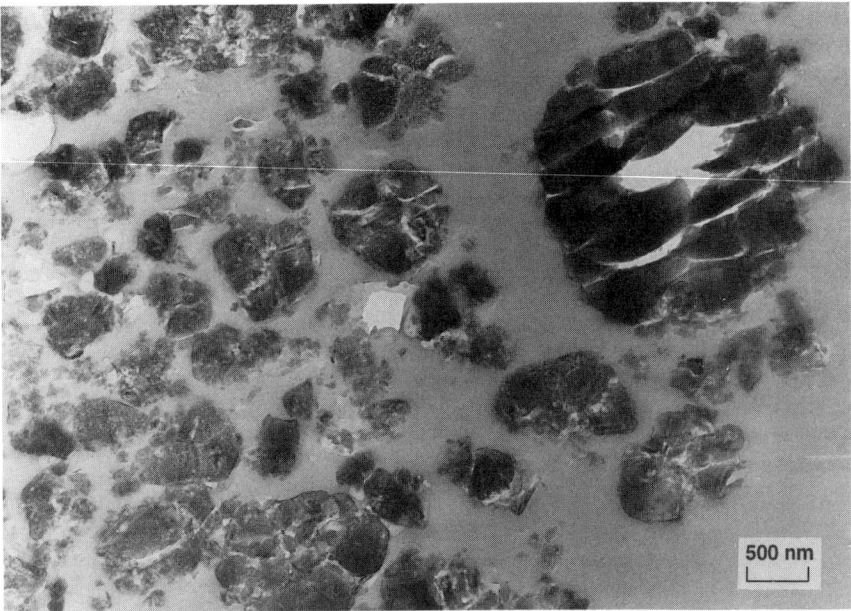

Figure 3.   TEM image showing the overall structure of the extensively dealuminated USY material, ED-USY. An inhomogeneous distribution of mesopores is seen among different USY grains and within individual grains. Grain fracture boundaries show random orientations. Large cracks due to ultramicrotoming are seen as fracture lines that are roughly parallel to each other. Reproduced with permission from reference 22.

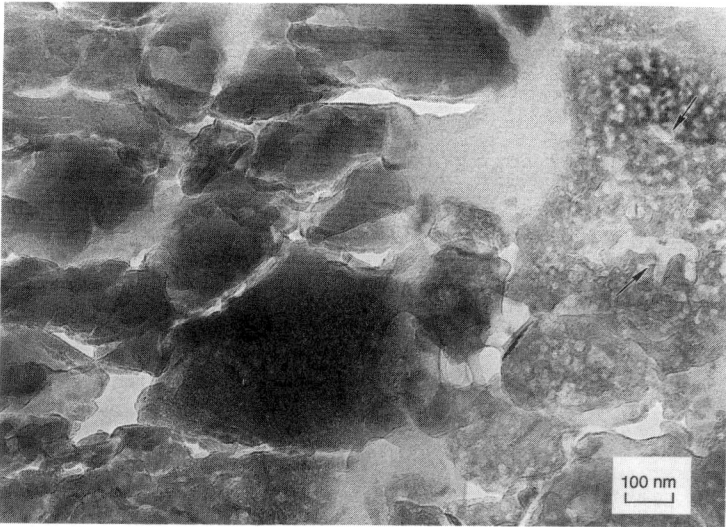

Figure 4a. TEM image of a few ED-USY grains. An inhomogeneous distribution of mesopores is seen within individual grains; some grains contain more mesopores than others. In regions with high mesopore concentration, the pores coalesce to form channels as indicated by arrows. Reproduced with permission from reference 22.

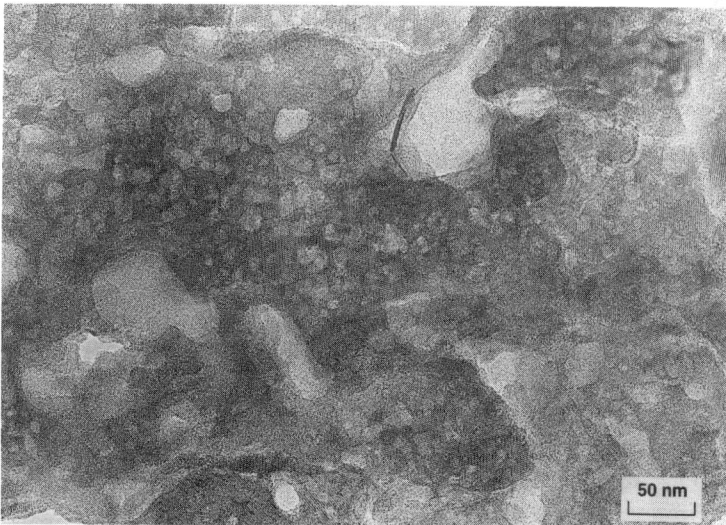

Figure 4b. HREM image of ED-USY grains. Many mesopores are formed. Although localized disorder is observed within the pores, the connecting regions remain crystalline. Reproduced with permission from reference 22.

Figure 5. HREM image of a ED-USY grain. Internal grain fracture boundaries or cracks as indicated by arrows are formed from the evolution of the coalesced mesopores. An example of a "precursor" crack is seen at the upper left. Dark bands are observed along these cracks just outside fracture boundaries. Reproduced with permission from reference 22.

Figure 6. HREM image showing a region where a ED-USY grain has broken into many small crystallites. Each crystallite has a different orientation. Reproduced with permission from reference 22.

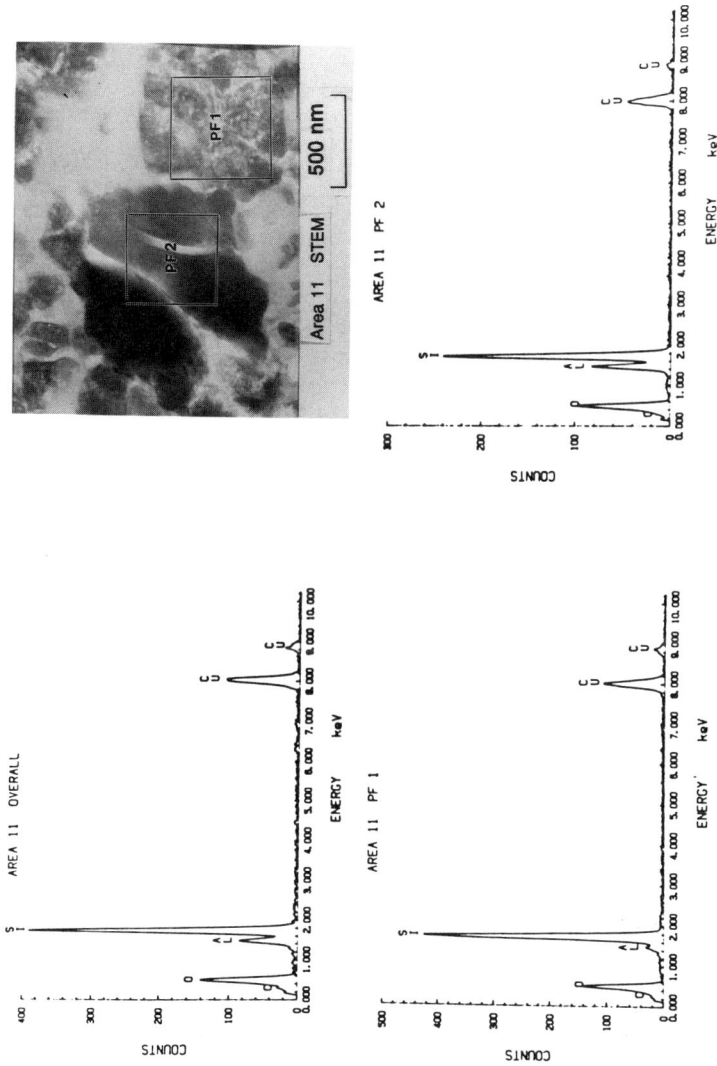

Figure 7. STEM image of a few ED-USY grains along with EDXS spectra of the entire imaged area (overall) and two individual grains (PF1 and PF2). The grain shown in PF1 contains a high density of mesopores and has a higher Si/Al ratio than the average (overall). The grain shown in PF2 is largely free of mesopores and has a lower Si/Al ratio than the average value. Reproduced with permission from reference 22.

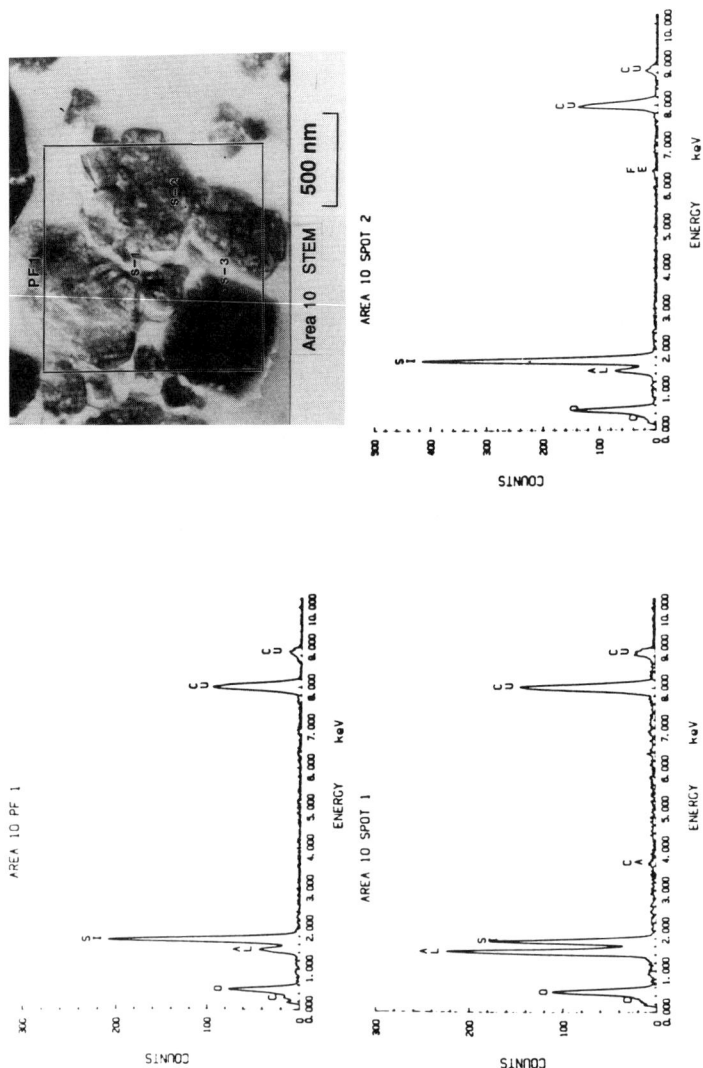

Figure 8. STEM image of a few ED-USY grains along with EDXS spectra from three USY grains (PF1) and from a point on a dark band (S-1) and a point within a mesopore (S-2). The dark bands are found to be highly enriched in Al. Within mesopores, Al is slightly deficient. Reproduced with permission from reference 22.

the mesopore itself shows a slight Al deficiency, with a slightly higher Si/Al than that for PF1. These narrow regions of enriched Al are resistant to mild acid treatment. They appear to represent an ultimate fate of the high temperature steam-induced migration of nonframework Al, following its ejection from the crystalline lattice. An EDXS spectrum from an arbitrary location within an individual USY grain, S-3 (not shown), shows a Si/Al equivalent to that of PF1.

Steam Treated USY Catalyst. The USY grains are well mixed with the matrix in the ED-YCAT. A transmission electron microscope (TEM) image of the overall structure of this sample is shown in Figure 9. The amorphous matrix is evidenced as large lighter regions, while the crystalline (USY) grains are more dense. As in all low magnification TEM images, cracks induced by the ultramicrotoming are readily seen as fracture lines approximately parallel to each other. Mesopores are revealed as lighter domains, typically concentrated in the interior of the grains. Some grains appear to have fewer mesopores than others. In comparison with the ED-USY material, the ED-YCAT sample shows a greater departure from homogeneity with fewer mesopores overall. It is surprising that, within the ED-YCAT, some grains exhibit a significantly lower incidence of defects than others (Fig. 9), since all microspheroidal catalyst particles are subjected to very similar conditions during the fluidizing steam treatment.

Detailed analyses of HREM images of the ED-YCAT showed features similar to those seen in the ED-USY material. Low density mesopores, 15 to 50 nm in dimension, are clearly seen in the USY grains in Figure 10. In regions with high mesopore concentration, cracks have evolved from the coalesced mesopores, as seen in Figure 11. A region of coalesced mesopores comprising a precursor crack is marked by an arrow in Figure 10.

In contrast with the large cracks produced by ultramicrotoming (Fig. 9), the cracks which evolve from coalescence of mesopores (Fig. 11) show no preferred orientation. In addition there is a dark band associated with each mesopore-induced crack. These dark bands are Al enriched, similar to those observed in the ED-USY sample. The several small crystals seen in Figure 11 have broken from a single original USY grain. From inspection of the lattice patterns, it is apparent that some crystal fragments remain in the same orientation while others have rotated slightly from their original orientation, probably due to relaxation of strain created by the extraction of framework Al. In some regions, evidence of extreme fracturing is observed, resulting in crystallites of only 20 to 50 nm in size as shown in Figure 12.

Discussion

^{27}Al MASNMR Investigations. The ^{27}Al MASNMR investigations reveal the averaged local environment of local (atomic) Al species. These measurements complement the extensive TEM (HREM and AEM) studies which provide detailed spatial microanalyses of the nature and evolution of localized patterns of zeolite destruction during extended dealumination. While a definitive mapping between the different Al species found by NMR to characteristic spatial defects found by TEM is beyond the scope of this study, the majority of the nonframework aluminum species detected by ^{27}AL MASNMR is associated with the extracrystalline phases

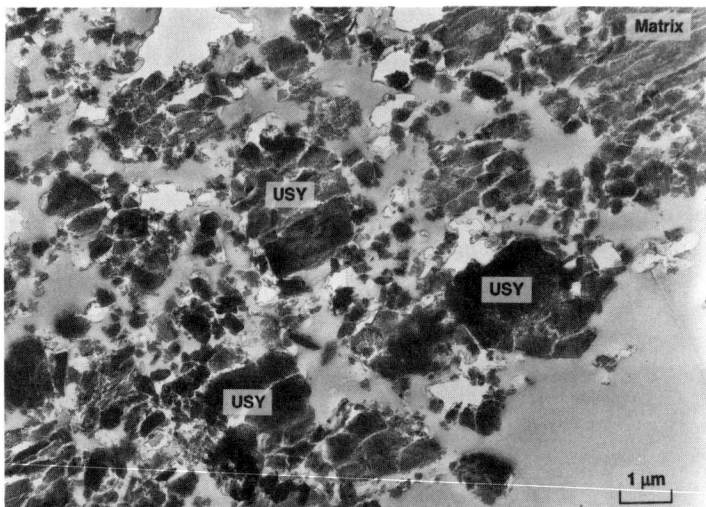

Figure 9. TEM image showing the overall structure of the ED-YCAT. The matrix appears as regions of amorphous material, the USY zeolite as denser crystalline particles as indicated. Mesopores are seen as lighter domains inside the USY grains. Large cracks due to ultramicrotoming are seen as fracture lines that are approximately parallel to each other. Reproduced with permission from reference 22.

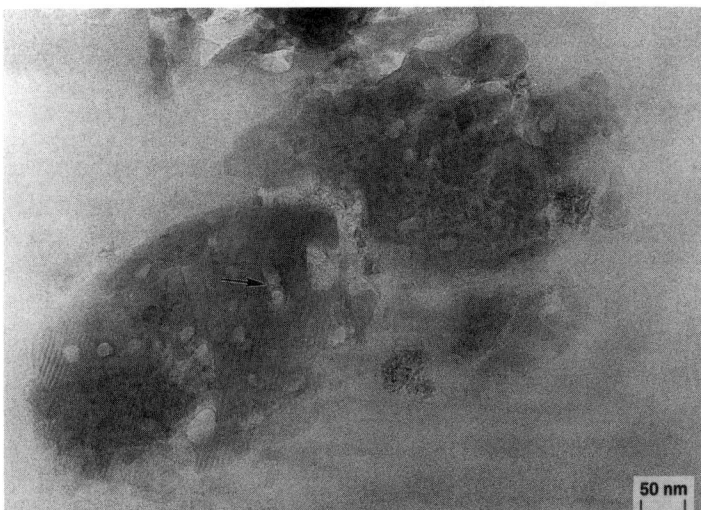

Figure 10. HREM image of several grains of the high temperature ED-YCAT showing an inhomogeneous distribution of mesopores. A coalescence of mesopores indicating an evolving fracture is indicated by an arrow. Reproduced with permission from reference 22.

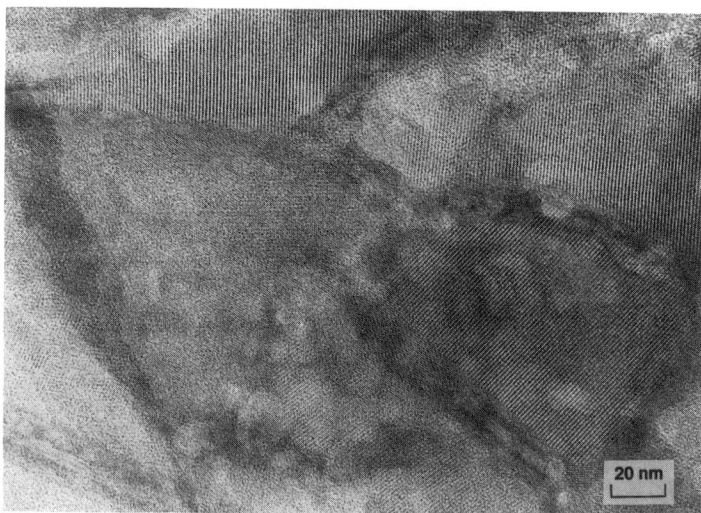

Figure 11. HREM image showing the fracture of a USY grain in the ED-YCAT into several smaller grains. Each fractured crystallite is bounded by cracks evolved from coalescence of mesopores. Similar to the case for the ED-USY material, the dark band seen along each crack is Al enriched. Reproduced with permission from reference 22.

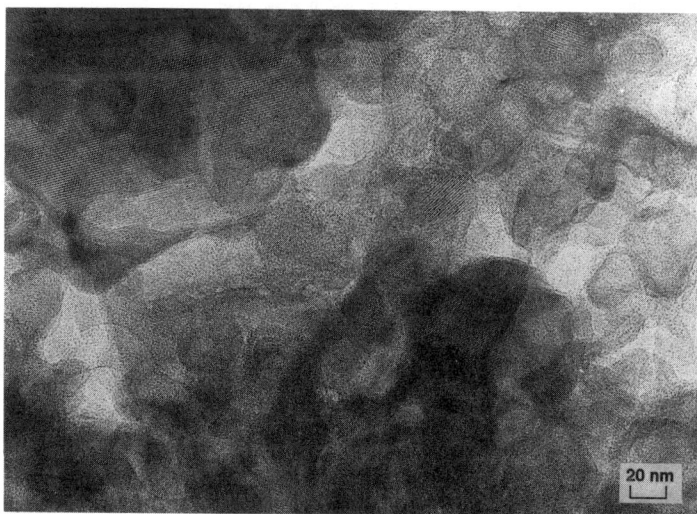

Figure 12. HREM image showing the extreme case of zeolite fracturing caused by high temperature steam treatment of the ED-YCAT. Numerous small crystallites, as small as 20 nm in size, have broken away from an originally large single grain. Reproduced with permission from reference 22.

represented by the dark bands found just outside the crystallite fracture boundaries. On the basis of previous structural studies which showed a tetrahedral species within the sodalite cage, a portion of the broadened tetrahedral resonance in ED-USY is expected to be contributed by nonframework species in the zeolite crystal (17). This assignment is consistent with the conclusions of recent kinetic studies that dealumination leads to the presence of "beneficial" nonframework Al species that exist as isolated entities within the small cages (18,19).

TEM Investigations. Consistent with previous TEM studies, the present work demonstrates that hydrothermal dealumination of zeolite Y, with or without subsequent acid treatment, leads to the formation of mesopores (1-9). Contrary to the reports by Patzelová, et al, and by Lynch, et al, the present investigations of hydrothermally aged USY materials give clear evidence for an inhomogeneous distribution of mesopores (2-4). For ED-USY materials, a marked inhomogeneity in the extent of dealumination and concomitant zeolite destruction was observed among different USY grains. Individual grains showed a markedly inhomogeneous distribution of mesopores or defects. Such inhomogeneities were even more pronounced for a higher temperature steam-deactivated commercial USY cracking catalyst.

The pronounced inhomogeneity in the character of defect formation may be a direct result of the accelerated, laboratory aging procedures. Such inhomogeneity might be expected since the dealumination by high temperature steam treatment is an accelerated, nonequilibrium process. Ejection of framework Al and subsequent zeolite stabilization require significant migration of Al and Si, each of which is a rate controlled process. Increased severity of hydrothermal treatment leads to a more rapid dealumination/fracturing, a further departure from equilibrium, and an even more inhomogeneous distribution of defects. This accounts for the more pronounced inhomogeneities in patterns of defect distribution in ED-YCAT. The extent to which the supporting matrix (for the ED-YCAT) or the acid treatment following steaming (ED-USY) are important factors is beyond the scope of this study.

The present HREM observations provide new insights into the formation and evolution of mesopores which lead to extensive crystallite fracturing. In regions with high defect concentration, mesopores coalesce to form channels and cracks, which ultimately lead to crystallite fracture. For the laboratory steam treatments employed in these studies, this defect-induced fracture mechanism appears to comprise the primary route to crystallite size reduction, which for extended steam treatment, can lead to crystallite fragments as small as 20 nm. Similarly small crystallite fragments were reported in a recent study of age separated "equilibrium" catalyst from a commercial FCU (9). In this case, the small crystallites found within an "old" fraction were not demarcated by fracture boundaries, raising the possibility that, for the slower catalyst deactivation in the FCU, mechanisms other than defect-induced fracture are important in crystallite size reduction. Both within individual grains and among different grains, the old fraction of age separated catalyst showed a more homogeneous pattern of dealumination and distribution of defects than those exhibited by the lab steamed catalysts, ED-USY and ED-YCAT (9).

The fate of the ejected framework Al has been investigated previously by Ward and Lunsford and also by Gross, et al (20, 21). Under influence of high temperature steam treatment, a significant surface enrichment of Al was observed, which was presumed to be associated with the migration of nonframework Al toward the crystallite surface. Ward and Lunsford found that this surface enrichment persisted following removal of up to half of the surface aluminum by NaOH (*20*). Gross, et al concluded that this surface enrichment was comprised of neutral species (*21*). Neither study was able to demonstrate whether this surface enrichment encompassed a concentration profile within a crystal grain or was solely comprised by a separate, presumably amorphous phase decorating the surface of the grain. More recently, Gélin et al reported that, based on STEM analysis of a steamed La-Y catalyst, no significant aluminum concentration profile was observed within an individual grain (*8*). Instead, a pronounced increase in aluminum concentration was observed at defects, described as "dark lines", within the zeolite crystals.

In the present study, regions with high mesopore concentration are found to be aluminum-deficient, indicating that framework dealumination and subsequent Al migration occurs concomitantly with mesopore formation. For extended hydrothermal treatment, the predominant fate of aluminum ejected from lattice sites appears to be closely associated with the dark bands, where Al is found to be highly enriched. In contrast to the observation by Gélin et al, our studies show that these noncrystalline aluminum-enriched regions are located not at the internal defects, but at or near newly formed fracture boundaries (*8*). These features are observed both for the ED-YCAT and for the ED-USY, consistent with previous studies which found the surface enrichment of Al to persist through aqueous treatments which removed substantial amounts of aluminum (*20*).

Conclusions

The present HREM and AEM study of a steam/acid treated USY material and of a high temperature steam treated USY cracking catalyst gives clear evidence for an inhomogeneous formation of mesopores which occurs concomitantly with a further zeolite dealumination. Such inhomogeneities are more pronounced for materials that are steam treated at a higher temperature and they are observed among different USY grains as well as within single grains. In regions with high defect concentration, mesopores coalesce to form channels and cracks, which ultimately define the boundaries of fractured crystallite fragments. Within a mesopore itself, aluminum is slightly deficient. The predominant fate of aluminum ejected from lattice sites appears to be closely associated with the dark bands, highly enriched in aluminum, that are often observed at crystallite fracture boundaries. The extracrystalline phases represented by these dark bands contribute the majority of the nonframework aluminum species, tetrahedral, penta-coordinate, and octahedral, that were detected in parallel ^{27}AL MASNMR studies. Some grains exhibit extreme fracturing into small crystallites of dimensions of 20 to 50 nm. These observations have, for the first time, provided insights into how the formation and evolution of mesopores lead to significant zeolite crystallite fracturing and concomitant Al migration in USY materials subjected to severe hydrothermal treatments.

The characteristics observed in the present study are clearly different than those observed by TEM in a previous study of age-separated equilibrium catalyst from a commercial FCU. For an "old" fraction of age-separated equilibrium catalyst, defects are less inhomogeneously distributed than for the ED-USY, and the small crystallites remaining are not demarcated by fracture boundaries, suggesting that the mechanisms of deactivation are different for the slower catalyst deactivation in the FCU. Both within individual grains and among different grains, the old fraction of age separated catalyst showed a more homogeneous pattern of dealumination and distribution of defects than those exhibited by the lab steamed catalysts, ED-USY and ED-YCAT. These observations raise serious questions for the practice of using measurements of catalyst activity on lab-steamed catalyst as representative of deactivation in the FCU.

Acknowledgements

The authors gratefully acknowledge Mr. John Newbury for performing the steam/acid treatments on the neat USY material and the high temperature steam deactivation of the commercial USY cracking catalyst, and Mrs. Mary Michaels for obtaining the NMR spectra. The authors would like to express their gratitude to Professor N. Otsuka at Purdue University, W. Lafayette, IN and to the staff at JEOL USA Application Laboratory, Peabody, MA, for the opportunity to use their high resolution electron microscopes.

Literature Cited

1. Maugé, F.; Auroux, A.; Courcelle, J. C.; Engelhard, Ph.; Gallezot, P.; and Grosmangin, J.; In *Catalysis by Acids and Bases*; Imelik, B., et al., Eds.; Elsevier: Amsterdam, 1985, pp.91.

2. Patzelová, V.; and Jaeger, N. I.; *Zeolites,* **1987**, 7, pp 240.

3. Lynch, J.; Raatz, F.; and Dufresne, P.; *Zeolites*, **1987,** 7, pp 333.

4. Lynch, J.; Raatz, F.; and Delalande, Ch.; In *Characterization of Porous Solids*, Unger, K. K., et al., Eds.; Elsevier: Amsterdam, 1988, pp 547.

5. Maugé, F.; Courcelle, J. C.; Engelhard, Ph.; Gallezot, P.; Grosmangin, J., Primet, M.; and Trasson, B.; In *Zeolites, Synthesis, Structure, Technology, and Application*, Držaj, B., et al., Eds.; Elsevier: Amsterdam, 1985, pp 401.

6. Maugé, F.; Courcelle, J. C.; Engelhard, Ph.; Gallezot, P.; and Grosmangin, J.; In *New Developments in Zeolite Science and Technology*, Murikami, Y., et al., Eds.; Elsevier: Amsterdam, 1986, pp 803.

7. Gallezot, P.; Feron, B.; Bourgogne, M.; and Engelhard, Ph.; In *Zeolites: Facts, Figures, Future*, Jacobs, P. A., et al., Eds.; Elsevier: Amsterdam, 1989, pp 128.

8. Gélin, P.; and Courières, T. Des; *Appl. Catal.*, **1991**, 72, pp 179.

9. Beyerlein, R. A.; Tamborski, G. A.; Marshall, C. L.; Meyers, B. L.; Hall, J. B.; and Huggins, B. J.; In *Fluid Catalytic Cracking II*, Occelli, M. L. Ed.; ACS Symposium Series 452; American Chemical Society, Washington, D. C., 1991, pp 109.

10. Scherzer, J.; In *Catalytic Materials--Relationship Between Structure and Reactivity*, Whyte, T. E., et al., Eds.; American Chemical Society, Washington, D. C., 1984, pp 157.

11.   Scherzer, J.; In *Octane-Enhancing Zeolite FCC Catalysts: Scientific and Technical Aspects*, Marcel Dekker; New York, 1990. See also Scherzer, J.; *Catal. Rev.-Sci. Eng.* **1989**, *31*, pp 254.

12.   Maher, P. K.; Hunter, F. D.; and Scherzer, J.; *Adv. Chem. Ser.* **1971**, *101*, pp 266..

13.   McDaniel, C. V.; and Maher, P. K.; In *Zeolite Chemistry and Catalysis*, Rabo, J. A. Ed.; ACS Monograph 171; American Chemical Society: Washington, D. C., 1976, pp 285.

14.   Lohse, U.; Stach, H.; Thamm, H.; Schirmer, W.; Isirikjan, A. A.; Regent, N. I.; and Dubinin, M. M.; *Z. Anorg. Allg. Chem.* **1980**, *460*, pp 179.

15.   Stach, H.; Lohse, U.; Thamm, H.; and Schirmer, W.; *Zeolites*, **1986**, *6*, pp 74.

16.   Ray, G. J.; and Samoson, A.; *Zeolites*, **1993**, *13*, pp 410.

17.   Parise, J. B.; Corbin, D. R.; Abrams, L.; and Cox, D. E.; *Acta. Cryst.*, **1984**, *C40*, pp 1493.

18   Beyerlein, R. A.; McVicker, G. B.; Yacullo, L. N.; and Ziemiak, J. J.; *J. Phys. Chem.*, **1988**, *92*, pp 1967.

19.   Carvajal, R.; Chu, Po-Jen; Lunsford, J. H.; *J. Catal.*, **1990**, *125*, pp 123.

20.   Ward, M. B.; and Lunsford, J. H.; *J. Catal.*, **1984**, *87*, pp 524.

21.   Gross, Th.; Lohse, U.; Engelhardt, G.; Richter, K. H.; and Patzelová, V.; *Zeolites*, **1984**, *4*, pp 25.

22.   Choi-Feng, C.; Hall, J. B.; Huggins, B. J.; and Beyerlein, R. A.; *J. Catal.*, **1993**, *140*, pp 395.

RECEIVED June 17, 1994

Chapter 8

Catalyst Fouling by Coke from Vacuum Gas Oil in Fluid Catalytic Cracking Reactors

P. Turlier, M. Forissier, P. Rivault, I. Pitault, and J. R. Bernard

Génie Catalytique des Réacteurs de Raffinage, Centre National de la Recherche Scientifique–ELF, Centre de Recherche ELF Solaize, B.P. 22, 69360 Saint Symphorien d'Ozon, France

Microactivity test experiments (MAT) show that the final carbon content of the catalyst remains constant for a given catalyst-feedstock couple when varying catalyst-to-oil ratio (C/O). This content increases with higher pressures, but it remains practically unaffected by reactor temperature. This suggests that coking reactions become strongly self-inhibited beyond a certain level of coke on catalyst.

Catalyst sampling at three elevations along an industrial riser confirms these observations since coke is made in the first meters of the riser. It is confirmed that MAT results of these samples depend only on their coke content and not on their time on stream.

Sampled catalysts are dissolved by known procedures and coke is analyzed. It is constituted of ca 50% alkylated polyaromatics (3-6 rings) and 50% less hydrogen rich, more polycondensated part (> 7 rings). The proportion of the latter increases with cracking temperature and duration of ulterior stripping.

The FCC plant behavior is entirely governed by coke make on the catalyst, as illustrated by oversimplified heat balance equations (no regenerator cooler):

$$-\Delta \dot{H} \text{(coke combustion)} = \Delta \dot{H} \text{(reactor requirements)} - \Delta \dot{H} \text{(flue gas -air)} \qquad 1)$$

$$-\Delta \dot{H} \text{(coke combustion)} = \text{Delta coke*Specific coke heat combustion*Circulation}_{\text{catalyst}} \qquad 2)$$

$$\Delta \dot{H} \text{(reactor requirements)} = Cp_{\text{catalyst}} * \Delta T \text{(regen.-reactor)} * \text{Circulation}_{\text{catalyst}} \qquad 3)$$

Equation 1 shows that for given conditions of coke combustion and feedstock cracking, the enthalpic rate of coke combustion is constant and the coke yield depends only on its specific combustion heat which is related to its hydrogen content. This

0097–6156/94/0571–0098$08.00/0

yield is very close to 5% wt in classical conditions (complete CO combustion). Equation 2 shows that coke build-up on the catalyst in the riser-reactor section (the delta coke) governs the catalyst circulation rate or the catalyst-to-oil ratio (C/O). Equation 3 sets the regenerator temperature.

As for other catalytic reactors, riser performances depend on catalyst hold-up, catalyst activity and heat and mass transfers. These parameters are not accessible in FCC risers since only C/O can be calculated but it is known that catalyst hold-up increases with catalyst circulation for a given gas velocity so that delta coke has a direct influence on this hold-up. In addition, equation 3 shows that a higher C/O results in a lower regenerator temperature so that unselective thermal cracking is minimized at the riser bottom.

Moreover, high catalyst hold-up and high instantaneous activity because of low delta coke enhance conversion. It is very probable that these conditions also favor better octane numbers because gasoline composition tends to become more aromatic and more isoparaffinic. It is also known from MAT experiments *(1)* that a low delta coke favors gasoline selectivity at a given conversion.

These general remarks show that coke make on catalyst governs plant behavior, conversion and selectivity. This phenomenon is probably the most important and the least understood in catalytic cracking. The purpose of this paper is to show how coke is produced and what is exactly FCC coke, from experiments in micro reactor and in industrial riser.

There are two other reasons to study catalyst fouling. The first is to get information to establish chemical kinetics: coking from vacuum gas oil (VGO) is a catalytic reaction *(1)* which decreases catalyst activity. A realistic kinetic model must take it into account to be independent of reactor design. The second reason is to know how this design influences conversion and yields. It is clear from equation 1, 2, 3 that the catalyst time on stream, which is mainly in relation with riser volume, may have more or less influence, depending on the kinetics of coke formation: if this latter was progressive along the riser, a volume increase would decrease C/O with negative consequences. If the whole coke was made at the bottom of the riser, this volume should not have so much influence on C/O.

Note that this study does not deal with carbon on regenerated catalyst which is always less than 0.05%wt. Our concern is the delta coke which builds up in the reactor. It is known that its molar H to C ratio is between 1.2 and 0.4. This hydrogen is difficult to analyze because of water in catalyst, so coke contents and yields are given only on basis of carbon analysis.

Because of coke fouling, the catalytic activity inside the riser is lower than equilibrium catalyst activity. Several theoretical studies on coke formation conclude that it decreases the catalytic activity through site covering and pore plugging *(2, 3, 4, 5, 6)*. But the relative importance of these phenomena is not known in FCC, and many authors find it easier to correlate catalyst activity to time on stream.

Experimental studies of deactivation by coke are rare, particularly in the field of FCC *(7, 8)*. The more detailed study of P. Magnoux *(9, 10)* shows that the zeolitic part of the catalyst is particularly concerned by the deactivation. Well-characterized molecules are formed and blocked in the zeolite cages. But these coke molecules are observed after a long time on stream with various pure reactants and nothing is known on the kinetics of formation of coke from industrial FCC feedstocks.

Experiments in fixed bed reactors always include a stripping step at reaction temperature by an inert gas flow. Coke evolution during stripping is not negligible *(12)*, so that it is not possible to get informations on real coke when catalyst is cracking. By contrast, in industrial risers the sampling method recently described *(13)* allows one to get catalyst samples at various levels and to observe coke formed without any change provoked by stripping.

Experimental

MAT technique

This fixed bed, plug flow nonstationary reactor was recently described *(14)*. The well-known microactivity test is modified to work between 1 and 5 bars *(15)*: 1 g of feedstock is injected during 50s with 40 cm^3/min nitrogen on various catalyst masses at various temperatures. The catalyst is then stripped in the same conditions for 14 mn.

Catalyst sampling technique

It is very similar to the one described by J.A. Paraskos et al.*(16)*. Samplings are done at three elevations (4m, 20m and 28m above the feedstock injection point) in the riser centerline. Sampling gas velocity approximates the estimated gas velocity in riser. Ca 1 to 2 kg of a mixture of oil, gas and catalyst are sampled. Its residence time in sampling line is less than 0.1 s. Moreover, in practice the mean temperature of this mixture at the end of the line never exceeds 350° C, so that further cracking in sampling device is assumed to be negligible.

Hydrocarbons are soxhlet extracted with methylene chloride (CH_2Cl_2) for 36 hours at 40° C and the remaining catalyst is vacuum dried at ambient temperature for 36 hours and at 270° C for 24 hours. This procedure defines the coke on catalyst in riser before any stripping.

We verified, with various artificial mixtures of coked catalyst and hydrocarbons or feedstocks, that extraction and solvent elimination are quantitative and do not increase catalyst coke content. Moreover, when soxhlet extraction is repeated, the carbon content does not change. It confirms that the so-defined coke is strongly bound to the catalyst.

The coked catalyst can be then stripped at 500° C under nitrogen flow during one hour. This further step defines ultimate delta coke after a perfect stripping.

Some indications on the experimental conditions are given in Table I. The catalyst and feedstock characteristics are given in Tables II and III, respectively.

Table I : Riser conditions

Regenerator dense phase temperature	733° C
Feedstock preheat temperature	220° C
Riser outlet temperature	540° C
Reactor absolute pressure	2.8bars

Table II : Catalyst characteristics (ADVANCE equilibrium catalyst)

Average particle size	75 μm
Surface area	110 m^2 g^{-1}
Apparent bulk density	930 kg m^{-3}
Ni content	602 ppm
V content	401 ppm
coke (carbon) content	0.04 wt%

Table III : Feedstock characteristics

Feedstock	ASTM	Manji	Aramco	Nigeria
K UOP	12.02	11.93	11.78	11.59
density at 288K	0.8894	0.9153	0.9316	0.9409
Conradson carbon %	0.18	0.25	0.84	0.60
distillation	ASTM	simulated TBP	ASTM	ASTM
10%	573K	653K	678K	664K
50%	678K	728K	726K	714K
90%	793K	812K	807K	794K

Coke analysis

The coke analysis is done by a "LECO Carbon determinator (CR 12)" which gives the carbon weight percentage deduced from carbon dioxide evolved under oxygen flux at a temperature higher than 1,400° C. It must be pointed out that coke contains also hydrogen that is not analyzed in this study. Coke contents of catalysts are given on the basis of carbon analysis.

The chemical analysis of coke is done with the Magnoux method *(9)*: the silica alumina catalyst is first dissolved by dilute hydrofluoric acid; the so-called "soluble coke" is extracted from the acid solution with methylene chloride; the "non soluble coke" is collected. The soluble coke is analyzed by various chemical techniques: gas chromatography, proton NMR spectroscopy and direct injection mass spectrometry. The boundary between soluble coke and non soluble coke is in the range of 7 to 8 polycondensated aromatic rings.

Results

Catalyst coke content from MAT experiments

Figure 1 shows the measured coke yield versus catalyst mass in the reactor. For every feedstock-catalyst couple, at 483° C or 530° C, the coke yield is proportional to the catalyst hold-up. This means that the delta coke is constant as a first approximation over the length of the reactor. As MAT is an unsteady plug flow reactor, conversion increases from the inlet to the outlet, so that delta coke is independent of conversion. These results suggest that coke builds up quickly on catalyst, from feedstock and gasoline, and its production strongly self-inhibits beyond a certain level of coke.

Table IV shows the influence of total pressure on the final catalyst coke content.   As already observed *(15)* the delta coke observed in MAT experiments increases significantly with total pressure. The variation is approximately linear and the slope depends on the nature of catalyst and feedstock. Aromatic feedstocks give more coke, also at higher pressure *(2)*.

Table IV, Effect of total pressures on MAT catalyst delta coke c at 530° C, 1g feedstock, 6g catalyst, 50s injection duration

catalyst	Feedstock	Abs. total pressure Bar	delta coke wt%
SUPER D	ASTM	1	0.72
"	"	2	0.80
"	"	3	0.92
"	"	4	1.10
NOVA D	Nigeria	1	1.32
"	"	2.5	1.97
"	"	4	2.34
"	Aramco	1	1.55
"	"	2.5	2.05
"	"	4	2.54

Figure 2 shows the effect of reaction temperature on delta coke content. In the range 430-530° C, there is no evident effect of temperature. Between 540° C and 650° C, a significant increase is observed, but these conditions lead to unusual conversions (83%-87%) where other mechanisms may play a role.

Pressure has much more influence on coke formation. These observations give information on coke content but not on its formation kinetics.

Carbon contents of riser samples

Table V shows the sampling results.

Coke content on catalyst does not vary very much between 4m and the riser top: more than 90% of the coke is made in the first 4m of the riser.

Such results were also found by Paraskos et al.*(16)* on a cracker with amorphous catalyst.

Former radioactive tracer studies on this riser *(18)* show that catalyst flow may be roughly assimilated to a dispersed plug flow. Peclet numbers between 5 and 16 have been found. This indicates that the coke axial profile is not essentially the result of catalyst axial backmixing, but that coking kinetics govern that phenomenon. As suggested by MAT results, coking is quick and then self-inhibits. Nevertheless, this behavior is probably enhanced because cracking is endothermal and temperature decreases along the riser. Thus coking kinetics are first more rapid, and self-inhibition may become then more important because of lower temperature.

The coke content evolution between dried and stripped catalyst is not negligible. This phenomenon already observed in MAT reactor *(12)* is explained by coke evolution during stripping : i) cracking of paraffins and alkyl chains on aromatic cycles, ii) condensation of polyaromatic hydrocarbons and dehydrogenation of naphthenes giving H_2, iii) migration of polyaromatic molecules out of the zeolite

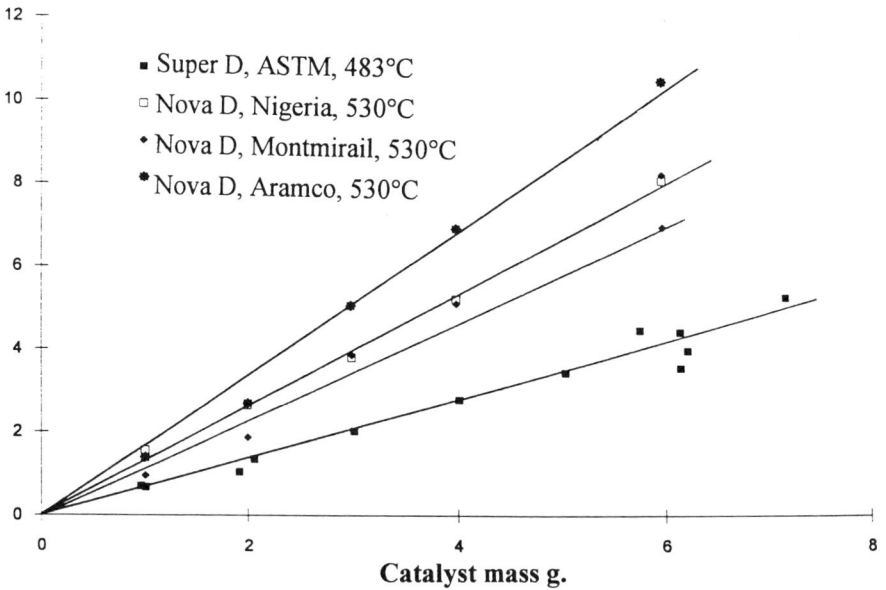

Figure 1 : MAT carbon yield (wt%) versus catalyst mass for various catalysts, feedstock and reaction temperature.

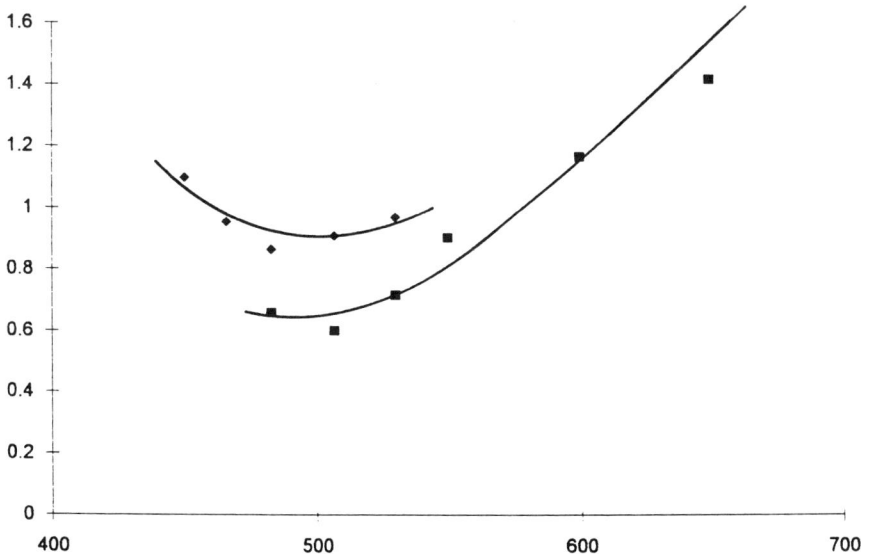

Figure 2 : Catalyst carbon content after MAT experiment (wt%) versus reaction temperature (° C) Nova D catalyst and Manji feedstock (♦), SuperD catalyst and ASTM feedstock (■).

structure *(20)*. An evolution of the same order of magnitude is observed in the MAT reactor between 1 and 15 minutes of stripping *(12)*.

Table V, Results of catalyst sampling

Sampling position,	equilibrium catalyst	4m	20m	28m
Mean catalyst time on stream s	0	0.6	1.9	3.6
Mean gas residence time s	0	0.4	1.4	2.3
carbon content dried (wt%)	0.04	1.34	1.44	1.45
carbon content stripped (w%)	0.04	1.07	1.16	1.16
ASTM MAT conversion %	80.5	68.4	66.7	67.8
Second order ASTM activity (conversion/(1-conversion))	4.13	2.16	2.00	2.10

The carbon loss during stripping is similar for all samplings (Table V). This suggests that coke evolution along the riser is not very important compared to evolution due to stripping.

The catalyst flowing in the riser is twice less active than the equilibrium catalyst.

Soluble and insoluble coke ratio

The proportion of soluble coke is given by Table VI. Very little is known about insoluble coke. Its H/C atomic ratio is close to 0.4 *(10)*. As soluble coke contains up to 7-8 polycondensed aromatic rings, insoluble coke should have at least this degree of polycondensation. Gallezot et al. *(20)* studied coke which could have been formed inside the micropores and then emerged out of the zeolite. This may also happen in FCC catalysts, but this coke may also be formed in the mesopores in the matrix.

Table VI : Soluble coke percentage

Sampling position,	4m	20m	28m	28m
state of the catalyst	dried	dried	dried	MAT stripped
soluble coke (wt%)	55	50	50	<10

This percentage does not depend on the sampling position. It dramatically decreases after 500° C stripping. This observation shows that an evolution of coke in the stripper must be taken into account for total coke quantity (Table V) and for coke chemical nature (Table VI).

The proportion of soluble coke probably depends on temperature: In a premilinary essay, an industrial catalyst (HEZ 55) was sampled downstream of the stripper. The riser end temperature was 505° C. Although this catalyst was industrially stripped, more than 90% of the coke was soluble, with the same analysis as the one described below. Higher temperatures should produce more refractory coke.

Soluble coke composition

Several analytical methods are used to characterize soluble coke evolution in the riser. Table VII gives results obtained by proton nuclear magnetic resonance.

Table VII. Soluble coke analysis by proton NMR

Sampling position (m)	4	20	28
HAR %	10	10	10
HAA %	30	40	40
HAL %	60	50	50
CH_2/CH_3	5.8	5.0	4.5

HAR : Aromatic hydrogen atom percent
HAA : Alkyl aromatic hydrogen (in α position from an aromatic cycle) atom percent.
HAL : Aliphatic hydrogen atom percent.
CH_2/CH_3 : ratio of hydrogen atom in CH_2 groups to hydrogen atom in CH_3 groups in aliphatic chains.

Table VII exhibits a small evolution of molecules blocked in zeolite cage along the riser. The alkylation degree HAA/HAR (number of aliphatic chains bound to aromatic cycle) remains important, about 4. The mean aliphatic chain length (CH_2/CH_3) is also important. They correspond to alkyl aromatics and paraffins which are also detected by gas chromatography (see below).

Figure 3 gives results of soluble coke gas chromatography. Most of the peaks appearing above the continuous response are characteristic of linear paraffins. It appears that the composition of soluble coke changes between 4m and 28m: There are more $C_{20}+$ paraffins in the 4m sample, which is closer to a feedstock sample, although this coke is trapped in the catalyst. During the 5 s time on stream between 4m and 28m, it is clear that there are some chemical transformations with conversion of heavy paraffins and production of some lighter products which remain very likely trapped by polyaromatics in the zeolite structure.

Soluble coke is also analyzed by mass spectrometry. Figure 4 shows a typical spectrum of such products. The large peaks superimposed on the continuous response are also detected when coking HUSY with n-heptane *(9, 10)*. The same products are also found in the study of stripping in MAT reactor *(12)*.

At 4m, the largest peaks correspond to the molar masses 178 and 192 and molecules such as (I):

where $(CH_2)_n$ represents various types of chains : for instance n(-CH_3), -$(CH_2)_{n-1}$-CH_3.
Molar masses 202, 216, 230, 244, 258, 272, 286 are also observed (II):

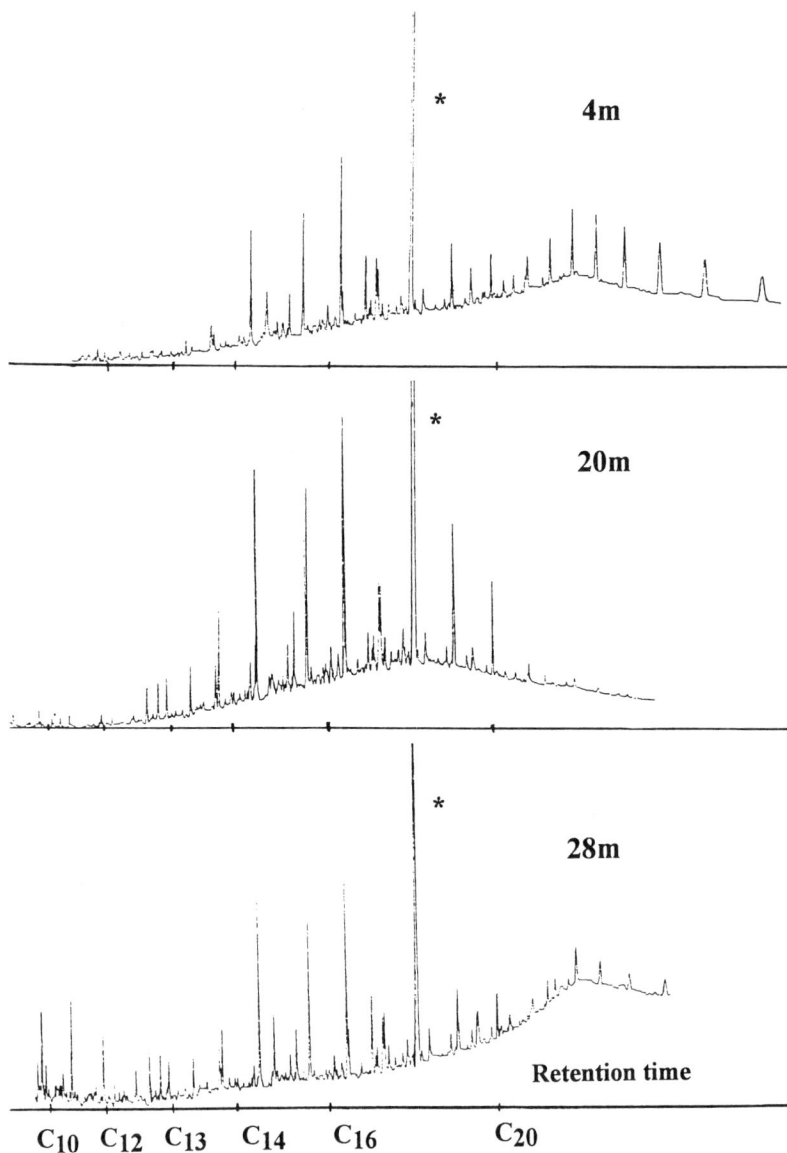

Figure 3 : Gas chromatography response versus retention time for the soluble coke from riser samples. The carbon numbers of paraffinic molecules is indicated along retention time axe. The peak with (*) is a methylene chloride impurity.

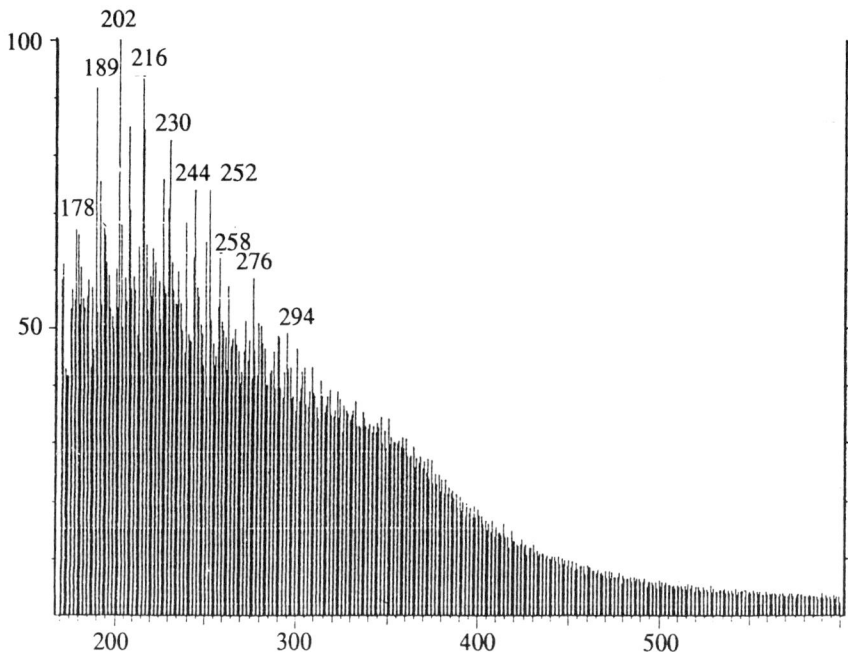

Figure 4 : Mass spectrometry analysis of soluble coke from the 20m riser sample. Relative intensity versus mass number.

or $0 < n < 6$

At 20m, the ratio of heavy compounds increases. At 28m, the group (I) disappears and (II) decreases. New groups with molar masses 268, 282 (III):

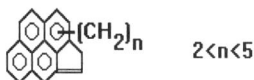

$2 < n < 5$

and 318, 332 appear (IV):

Discussion-Conclusions

Coke formation in FCC conditions is extremely rapid. Its rate cannot be measured by conventional MAT tests. We can estimate a minimum rate from riser sampling results since coke on catalyst increases of 1% wt. in 0.63 s. This rate is larger than 0.016 $kg_{coke}/kg_{catalyst}/s$.

In conventional FCC or MAT conditions, there is a maximum coke make on catalyst so that coking rate is initially very high and drops very rapidly to zero.

This delta coke depends on the feedstock-catalyst couple. It is known that cracking catalysis is limited by diffusion within zeolite crystals *(21)*. Each coked zeolite cage restricts more of the traffic of molecules, especially when they are large and are coke precursors, so that coking self-inhibits much earlier than cracking. The final delta coke is probably the result of competition between reactivity and diffusion within the crystal. But a part of the coke may also be located in the matrix.

Higher pressures increase the delta coke, probably because this competition is modified. In the range 500-550° C, cracking temperature does not influence the final carbon content on catalyst. But at higher temperatures, coke increases, perhaps because of thermal cracking in catalyst porosity.

The delta coke does not represent the maximum coke content which can be deposed on catalyst. It is still possible to increase it by operating with a very large excess of feedstock compared to classical C/O. But this operation leads to very low conversion at classical temperatures, and to consider a constant delta coke in industrial cracking conditions is a realistic approximation. This approximation could be refined if hydrogen of coke was considered.

In riser conditions, delta coke is produced after a very short time on stream. Its analysis shows that at 530° C half of this coke is constituted and aromatic rings whose polycondensation is between 3 and 7. These rings are alkylated. Paraffins are also detected, especially in the riser bottom sample.

Methylene chloride does not extract this coke, but it dissolves it when the catalyst is destroyed by acid. This proves that half of the coke remains trapped in the zeolite microporosity. The coke analysis does not change much during the 4 s. of time on stream in the riser, although there is a trend to dealkylation and aromatization.

These soluble coke molecules are characteristic of Y zeolites rather than feedstock since the same molecules are found when coking with n heptane.

The second half of the coke is not soluble in methylene chloride; it is less hydrogen rich and more polycondensed. It may be produced directly on the matrix or by polycondensation of soluble coke and extrusion from the zeolite.

About 20% of the total coke can be removed by cracking with an ulterior stripping. The remaining coke is then much more refractory. This shows that stripping is not only flushing of spent catalyst and desorption to remove hydrocarbons, but also chemical cracking of the coke.

These results allow to better understand the influence of important design parameters like riser and stripper volume. A larger riser volume will increase the catalyst hold-up without modifying delta coke and C/O, in first approximation. Thus, conversion will rise and gasoline selectivity will drop since it is an intermediate product. A stripper volume increase will favour chemical stripping (coke cracking) even if physical stripping (flushing, desorption) is already completed. Thus C/O will rise with the positive effects described hereabove, but delta coke and catalytic activity within the riser will not change.

Acknowledgements

The authors gratefully acknowledge Dr. Patrick Magnoux (Laboratoire de catalyse en chimie organique, Université de Poitiers, France) for coke chemical analysis and fruitful discussions. R. Barberet and D. Gadolet are thanked for their contribution in industrial sampling and MAT measurements.

Literature Cited

(1) Forissier, M.; Bernard, J. R., In *Catalyst deactivation 1991*, C. H. Bartholomew and J. B. Butt Eds.; Elsevier, Amsterdam, 1991. pp 359-366.

(2) Beeckman, J. W.; Froment, G. F., *Ind. Eng. Chem. Fundam.* 1979, 18, 245-256.

(3) Beeckman, J. W.; Froment, G. F., *Chem. Eng. Sci*, 1980, 33, 805-815.

(4) Haynes, H.W.; Leung, K., *Chem. Eng. Commun.*, 1982, 23, 161-179.

(5) Yortsos, Y. C.; Tsotsis, T. T., *Chem. Eng. Commun.*, 1984, 30, 331-342.

(6) Zhdanov, V. P., *Catalysis Letters*,1991, 9, 369-376.

(7) Wolf, E. E.; Alfani, F., *Catal. Rev. Sci. Eng.*, 1982, 24, 329-371.

(8) Bhatia, S.; Beltramini, J.; Do, D. D., *Catal. Rev. Sci. Eng.*, 1990, 31, 431-480.

(9) Magnoux, P.; Cartraud, P.; Mignard, S.; Guisnet, M., *J. Catal.*, 1987, 106, 235-241.

(10) Guisnet, M.; Magnoux, P., *Appl. Catal.*, 1989, 54, 1-27.

(11) Froment, G. F., In *Catalyst deactivation 1991*, C. H. Bartholomew and J. B. Butt Eds., Elsevier, Amsterdam, 1991 pp 53-83.

(12) Forissier, M.; Magnoux, P.; Rivault, P.; Bernard, J. R.; Guisnet M., To be published 1993.

(13) Turlier, P.;. Bernard, J. R, *Entropie*, 1992, 170, 24-28.

(14) Pitault, I.; Forissier, M.;. Bernard, J. R., To be published 1994.

(15) Forissier, M.; Formenti, M.; Bernard, J. R.,*Catalysis Today*, 1991, 11, 73-83.

(16) Paraskos, J.A.; Shah, Y.T.; Hulling, G.P.; Mc Kinney, J.D., *Ind. Eng. Chem.*, *Process Des. Dev.*, 1977, 16, 89-94.

(17) Schuurmans, H. J. A.; *Ind. Eng. Process Des. Dev.*, 1980, 19, 267-271.

(18) Martin, M.P.; Turlier, P.;. Bernard, J. R; Wild, G., *Powder Technology*, 1992, 70, 249-258.

(19) Martin, M.P.; Derouin, C.; Turlier, P.; Forissier, M.; Wild, G.; Bernard, J. R., *Chem. Eng. Science,*, 1992, 47, 2319-24.

(20) Gallezot, P.; Leclerq, C.; Guisnet, M.; Magnoux, P., *J. Catal.*, 1988, 114, 100-111.

(21) Rajagopalan, K.; Peters, A. W.; Edwards, G. C., *Appl. Catalysis*, 1986, 23, 69-80.

(22) Bibby, D.M.; Mc Lellan, R. F., *Appl. Catal. A. General*, 1992, 93, 1-34.

RECEIVED June 17, 1994

Chapter 9

Coke Formation in Reactor Vapor Lines of Fluid Catalytic Cracking Units
A Review with Some New Insights

E. Brevoord[1] and J. R. Wilcox[2,3]

[1]Akzo Chemicals bv, Stationsplein 4, P.O. Box 247, 3800 AE Amersfoort, Netherlands
[2]Akzo Chemicals, Houston, TX

Coke formation in the Fluid catalytic cracking unit reactor vapour line is still a problem of big concern to some refineries. The commercial experience is summarized and the coking process has been simulated in a pilot riser. Catalyst and feed quality effects have been investigated. It is concluded that coke formation is caused by thermal cracking at high temperatures at the bottom of the riser, resulting in the formation of di-olefins. Di-olefins are very reactive and form with other large convertable components, probably cyclic molecules with an extremely high molecular weight, which, if not converted, condense downstream of the riser. The influence of unit design, unit conditions, feedquality and catalyst formulation is discussed. Recommendations are given to prevent coking.

Coking in the reactor vapour line is not a new subject. In the seventies and the eighties several units suffered from coking in the vapour line, resulting in a considerable pressure drop, and causing severe fouling in downstream equipment. By changing unit design and catalyst formulation, usually the coking rates could be reduced to acceptable levels. With the introduction of new reactor/riser designs, aiming for a reduction of thermal cracking, the problems seem to arise again. So far several theories exist to explain the reaction mechanism:

1. Large molecules in the feed do not evaporate under reactor conditions and leave the riser unconverted. They form a fine mist and condense in dead zones and/or the vapour line *(1)*.

2. High hydrogen transfer results in the formation of aromatics, which can polymerize and form coke *(2)*.
3. Other types of large molecules are formed which are not converted in the riser and as a result, settle in the vapourline, where they dehydrogenate and form solid coke.

[3]Current address: Refining Process Services, 1708 Pittsburgh Street, Suite 1, Cheswick, PA 15024

0097–6156/94/0571–0110$08.00/0
© 1994 American Chemical Society

It remains an interesting question which catalyst design (matrix and zeolite activity, pore size distribution) can reduce the coking rates. A high zeolite or matrix activity may result in conversion of large polymers, a large macropore volume may allow the catalyst to absorb the coke precursors.

To determine which of the three mechanisms mentioned above is the correct one and how coking can be prevented, the commercial experience is summarized. A pilot riser program was performed, during which coking could be simulated and catalyst and feed quality effects investigated.

Commercial experience

The coking problem is clearly associated with the introduction of the modern all-riser concept dating from the mid-seventies. At comparable feedstock qualities, reactor temperatures and conversions, units started to face coking problems only after the modification from a bedreactor to a reactor riser. Units however, still operating on an older type low density, high porosity type zeolite catalyst, seemed to suffer less than units using the improved high density, zeolite containing grades *(2)*. Most units were able to cope with the problem, by taking the following actions *(1)*:

* improved insulation of the vapour line
* prevent deadzones in the reactor dome
* higher superficial velocities in the vapour line (35-45 m/s)

During the eighties some changes in operating strategy and unit design took place. To reduce fuel oil production, the feedstock became heavier; the feed end point increased and more metals were introduced. In a few cases coking rates were related with the amount of resid processed and some refineries claimed that only problems are noticed when high nickel-feeds are processed. The latter would indicate that dehydrogenation activity plays a role in the coking process. However, the influence of the feed endpoint on coking rates is not very clear. Units processing clean feeds with an endpoint below 530°C, are known to have coking problems as well. Consequently it is unlikely that coking is only caused by insufficient conversion of large molecules, which do not vaporize in the reactor riser (mechanism 1 in the introduction).

To be able to process resids, often the reactor risers were modified. Especially the feed introduction had a lot of attention. Improved feed dispersion and feed/catalyst mixing resulted in more selective cracking. It appeared that a considerable reduction in coking rates could be established as well.

In 1984 it was thought that a reduced H-transfer activity would help to reduce coking, as the formation of polyaromatics via H-transfer was a proposed route for coke formation *(2)*. The leadphase down followed by an increase in octane requirements resulted in a high demand for low RE_2O_3 octane catalysts, having a low H-transfer activity. So far however, no relation between hydrogen transfer activity and coking rates have been found, making it unlikely that coking takes place via the formation of aromatics (mechanism 2 of introduction).

Recently short contact time risers and closed cyclone systems became popular, the latter reducing the residence time in the dilute phase. A reduction of thermal cracking and delta coke allowed the refiner to operate at higher severities. Though many modified units run without any problems, some cases are known where coking started after the modification.

To investigate the coking phenomena in more detail, at one refinery several actions were taken while the coking rate was monitored:

*   all unit conditions were varied, especially the operating severity. Only a high catalyst activity and high cat/oil ratio seemed to have beneficial effects. (variations in RXT may have been too small to see a significant effect)

*   as it was thought that coking is caused by certain large molecules, a switch to a high matrix catalyst was made to convert them. Though bottoms yield reduced with 4% at constant unit conditions, coking rates remained high.

*   A major reduction in coking rates was established after increasing the dispersion steam rate. The effect of feed dispersion is illustrated in figure 1. During the first period in this graph, feed quality remained relatively constant. The coking rates (monthly averages) were calculated by measuring the weight of the vapour line. The values were checked using a radioactive probe. Initially the dispersion steam rate was doubled resulting in a considerable reduction in coking rates. After a turnaround pressure drop over the feednozzles and steamrate could be increased furthermore.

An improved feed/catalyst mixing results in a better temperature distribution at the bottom of the riser. Temperatures exceeding 560°C are prevented and as a result thermal cracking is reduced.

A clear relation was found between the coking rates and the degree of catalytic cracking (C_4/LPG-ratio), enhanced by a higher catalyst activity, cat/oil ratio and improved feed dispersion (figure 2). The C4/LPG ratio can be seen as a measure of catalytic cracking. A value of 0.5 wt/wt means that thermal cracking dominates, 0.7 wt/wt means mainly catalytic cracking *(3)*.

Finally the problem was solved by a further improvement in feed introduction, while also the feed quality changed considerably (last 3 data points in figure 1). As a result a further increase in catalytic cracking could be achieved.

We may conclude that feed introduction and the degree of catalytic cracking play an essential role.

Pilot plant simulation

A pilot plant study was performed to investigate the coking phenomena.
Three feeds were used, two being very paraffinic and easy to convert, one being aromatic.

As it was assumed that coking is enhanced if more thermal cracking takes place or if an unit operates at a low severity, steamed fresh catalyst was diluted with inert steamed equilibrium catalyst, to achieve low catalyst activities and low

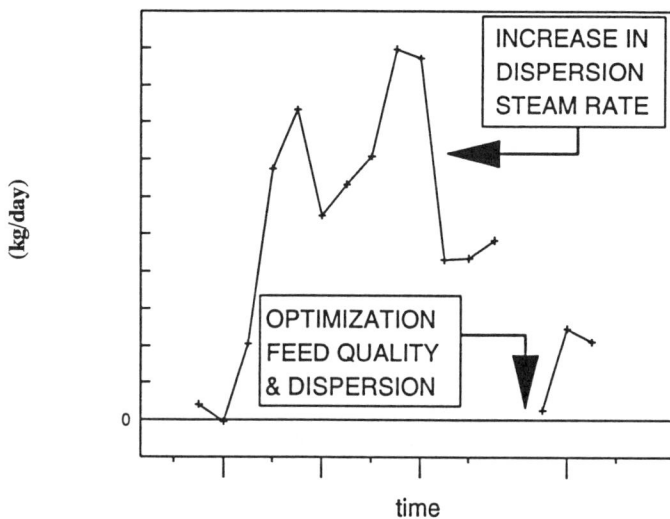

Figure 1. Coking rates dropped as a result of better feed dispersion.

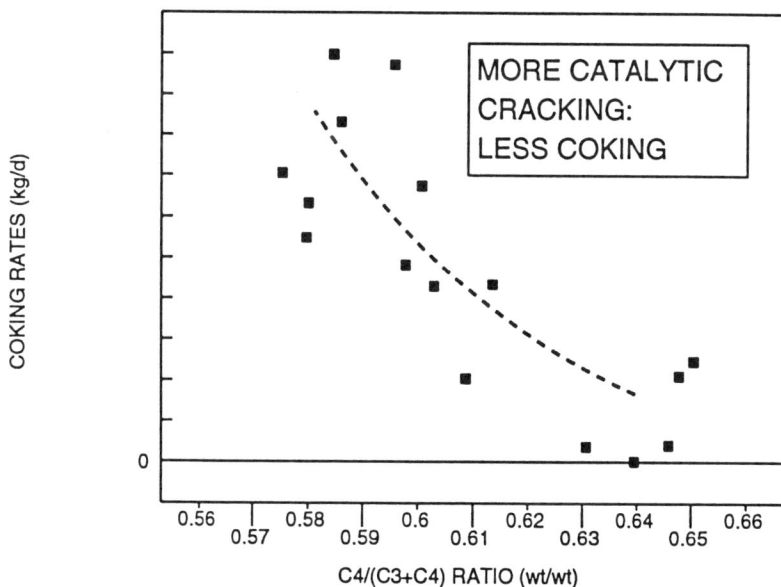

Figure 2. Coking rates are high if a large degree of thermal cracking takes place.

conversion. Several catalyst grades were used with a variety of matrix activity and pore size distributions. It appeared that also in the pilot riser (figure 3) coking took place in the transfer line. This only occurred when operating at low conversion levels and when using the waxy feed. At high conversions, the coke precursers were apparently converted.

Coking resulted in blockages in the vapour line and consequently in an unsteady catalyst circulation. As a measure of the coking tendency the stability of the operation could be used, being monitored using the temperature of the catalyst liftline at the cooler outlet.

A high matrix activity did not seem to have a substantial effect, but a catalyst with a very open structure and high accessibility appeared to eliminate the formation of coke, even at low conversion levels (<60 wt%). The pore size distribution of this catalyst versus another catalyst is shown in figure 4. Apparently the accessible catalyst is able to convert or to absorb the coke precursers.

This also explains why units featuring a short reactor residence time, are more sensitive to coking. The conversion of coke precursers is a diffusion limited process and in the riser or dilute phase sufficient time is required to bring them into contact with the catalyst.

To get more information on the mechanism of coke formation and to find tools to monitor the risk of coke formation, the feed and bottomproduct of the pilot riser runs were analysed for:

* thermal stability
* conradson carbon
* liquid chromatography (HPLC)

The thermal stability and conradson carbon did not give any valuable information. The liquid chromatography just showed that the bottoms product from the waxy feed is more paraffinic and that it becomes more aromatic with conversion.

The gasoline samples were analysed by PIANO gaschromatography. It appeared that the diolefins content of the C_5-C_9 fraction was high when using the paraffinic feed and at low conversion levels or cat/oil ratio's, exactly in line with the conditions at which coking takes place (figure 5).

Diolefins are known to be a product of thermal cracking. As commercial experience also shows that coking rates increase with an increasing degree of thermal cracking, we may conclude that thermal cracking plays an essential role. This seems contradictory with the statement that short contact time risers and closed couple cyclones may give higher coking rates, as these sophisticated systems have resulted in less thermal cracking. One should remember however, that only at the bottom of the riser temperatures up till 750°C take place and that only in these parts of the reactor a considerable amount of thermal products like di-olefins are formed.

Especially at low conversion levels, large molecules are available which can react with diolefins, resulting in components which can easily have a molecular

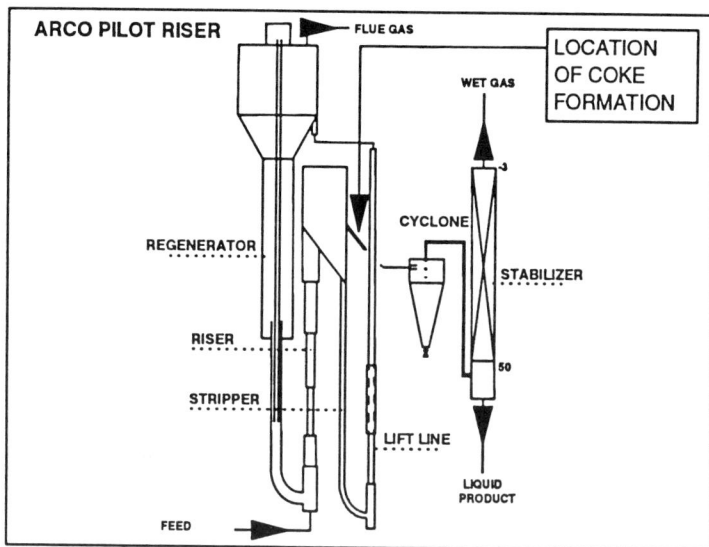

Figure 3. Arco Pilot Riser.

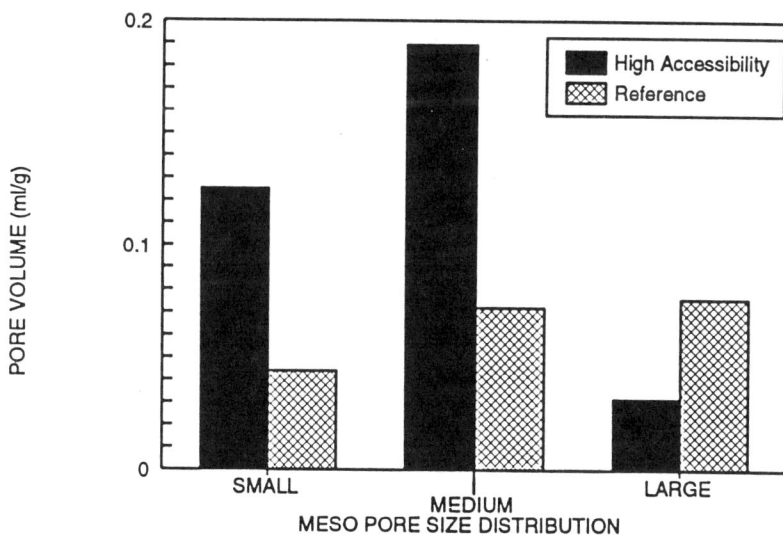

Figure 4. Accessibility of catalysts tested.

Pilot Riser Gasoline, PIANO analyses.

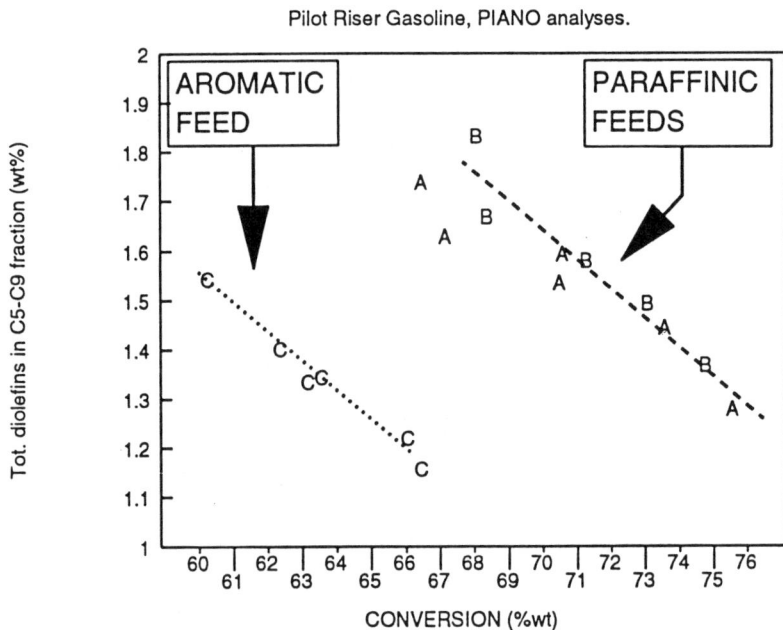

Figure 5. Diolefins production is high when processing paraffinic feed at low severity.

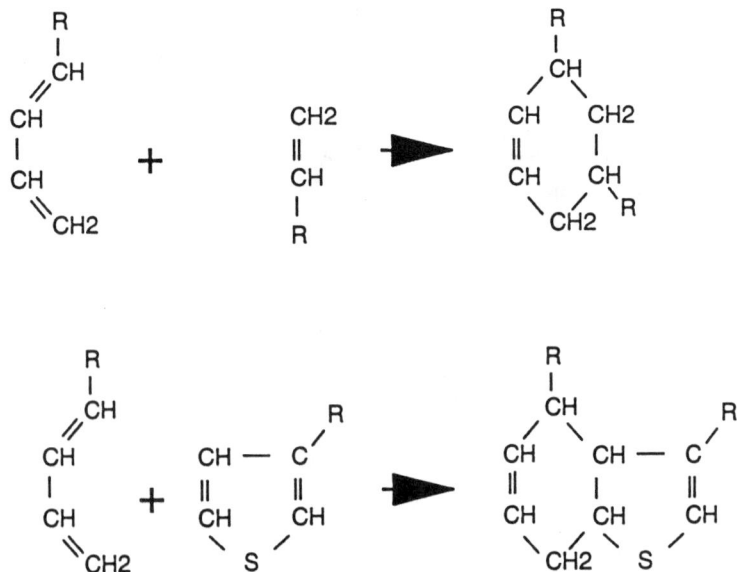

Figure 6. Examples of Diels-Alder Reactions.

weight of more than two times the average molecular weight of the feed. The reaction could take place via the Diels-Alder mechanism, being a thermal reaction as well (figure 6). As diolefins most likely react with olefins, a dehydrogenation step catalyzed by nickel may play a role.

Conclusions

Coking in the vapour line is caused by thermal cracking reactions taking place at the bottoms of the riser and resulting in the formation of di-olefins.
The reactive products of thermal cracking result in the formation of large molecules, which can easily condense in the vapour line.

The conversion of coke precursers by the catalyst is a diffusion limited process. A very accessible catalyst or sufficient residence time in the riser and/or dilute phase is required.

Apart from traditional methods as mentioned in the literature *(1,2)*, coking can be reduced by reducing the degree of thermal cracking in the bottom of the riser. This can be achieved by improving the mixing of the feed and catalyst as well as improving the feed dispersion. Very high local temperatures should be avoided. Hardware modifications may be required to achieve this goal. A higher cat/oil ratio and catalyst activity can play an important role and should be maximized.
If these measures have insufficient effects, it can be recommended to switch to a catalyst with an increased accessibility, allowing the absorption or conversion of coke precursers. A change in feed quality may be considered as well.
Options are:

* reduction in feed paraffinicity
* feed hydroprocessing
* reduce the nickel in feed, use a more nickel resistant catalyst or passivate the nickel activity

To guide the process of reducing coking rates, the di-olefins content of the gasoline or the degree of catalytic cracking can be used as a monitoring tool.

Literature Cited

1. *Akzo Catalysts Symposium 1991 Q&A Session, FCCU, page 10-15*
2. McPherson L.J., *"Reactor coking problems in Fluid Catalytic Cracking Units", Ketjen Catalyst Symposium 1984, Oil & Gas Journal, September 1984.*
3. Nieskens M.J.P.C, *"The assessment of carbenium-ion and the radical cracking in FCC", Akzo workshop, "The MON of FCC naphtha", June 1987.*

RECEIVED June 29, 1994

Chapter 10

Limitations of the Microactivity Test for Comparing New Potential Cracking Catalysts with Actual Ultrastable-Y-Based Samples

A. Corma, A. Martínez and L. J. Martínez-Triguero

Instituto de Tecnología Química, Universidad Politécnica de Valencia–Consejo Superior de Investigaciones Científicas, Camino de Vera s/n, 46071 Valencia, Spain

The gasoil cracking activity of a USY zeolite and a SAPO–37 zeotype, both having the faujasite structure, has been measured at 755 K using an automated MAT unit. MAT experiments performed at conventional (60 s) and less conventional (23 s) TOS showed a higher activity for the SAPO–37 catalyst, despite the USY sample has Brönsted sites of a higher acid strength. However, the kinetic parameters obtained for both catalysts from a complete kinetic study including experiments at very low TOS (5 s – 120 s) demonstrated that USY is more active for cracking gasoil than SAPO–37, but the former decays much faster. It is then concluded that cracking conversions obtained in conventional MAT experiments may be inadequate to compare the activity of catalysts having very different activity–decay behavior.

The new rules on gasoline composition established by the Clean Air Act (CAA) are making refiners revise and optimize their gasoline–producing units. One of the main gasoline–producing units is the FCC. This unit is facing demands for higher-octane, lower-sulfur gasoline, while higher production of C_4 and C_5 olefins is strongly desired. One way of achieving this is by using a catalyst containing zeolites other than faujasite. Introduction of ZSM–5 as an additive to faujasite based catalysts reduces gasoline production, even though the gasoline left has a higher RON (1–3). Meanwhile, the introduction of ZSM–5 produces an increase in the yield of propylene, but the increase in the highly desired i–butene and i–amylenes, which are raw materials for production of MTBE and TAME, is small (3,4). It appears that zeolites with bigger pores than those of ZSM–5 will be required to obtain the desired C_4 and C_5 olefins, and in this sense, cracking catalysts based on Beta zeolite look promising (5).

Searching for new zeolites as FCC catalysts obligates researchers to prepare and study a large number of samples whose catalytic activity and selectivity need to be measured and compared. To do that, one may use the cracking of pure compounds as a reaction test. However, it has been shown that the results obtained with pure compounds cannot always be extrapolated to what one sees when cracking a gasoil feed (*6*). For this reason, refiners and FCC catalyst manufacturers prefer to test cracking catalysts using gasoil as feed. Thus, it is obvious that when performing explorative research not all the catalysts can be tested in riser pilot plants which closely match the FCCU, and alternative, less–expensive and less time–consuming catalyst tests have to be used. It is along these lines that Micro Activity Test units (MAT) (ASTM D3907/86) were designed and are widely used to compare FCC catalysts. However, everybody is aware that it is not possible to extrapolate conversions and selectivities obtained in the MAT unit to commercial units. Indeed, MAT units are fixed–bed reactors which work at much higher contact times than riser units. Nevertheless, MAT experiments have been very useful for establishing activity and selectivity trends when different FCC catalysts are compared. For instance, they have been very successful in predicting the influence of zeolite Y unit cell size on activity and selectivity (*7*). If the trends work well when comparing similar type of catalysts, things can drastically change when materials with very different activity and decay characteristics have to be compared.

In this work, we will show the limitations that conventional, and even less conventional, MAT experiments have when the gasoil cracking behavior of a USY zeolite and a SAPO–37 zeotype, both materials having faujasite structure but differing in framework composition, are compared.

Experimental

Materials.   A SAPO–37 sample with a chemical composition given by $(Si_{0.17}Al_{0.47}P_{0.36})O_2$ and having 33 Si/u.c. was synthesized following the procedure described in ref.(*8*). Taking into account the proportion of the different mechanisms (SN2 and SN3) for Si substitution found for this sample (*9*), the expected number of Brönsted acid sites would be lower than a USY zeolite with equivalent number of Al/u.c. For this reason, a USY sample with 21 Al/u.c. has also been prepared. The USY zeolite was obtained by steam calcination of a partially ammonium–exchanged NaY zeolite (SK–40 from Union Carbide) at atmospheric pressure and 923 K for 3 hours. After steaming, the sample was again exchanged twice and finally calcined at 773 K for 3 hours. In this way, a USY sample with a unit cell size of 2.442 nm and 90% crystallinity (referred to the original NaY zeolite) was obtained. The sodium content of the final catalyst was seen to be below 0.15 %wt as Na_2O.

I.r. spectroscopic measurements were carried out on a Nicolet 710 FTIR spectrometer equipped with data station. Wafers of 10 mg cm^{-2} were introduced into a conventional greaseless i.r. cell and pretreated overnight in vacuum at 673 K and 873 K for USY and SAPO–37 samples, respectively. The acidity of the catalysts was measured by pyridine adsorption and desorption at increasing temperatures and vacuum. Figure 1 shows the i.r. spectra in the pyridine region for

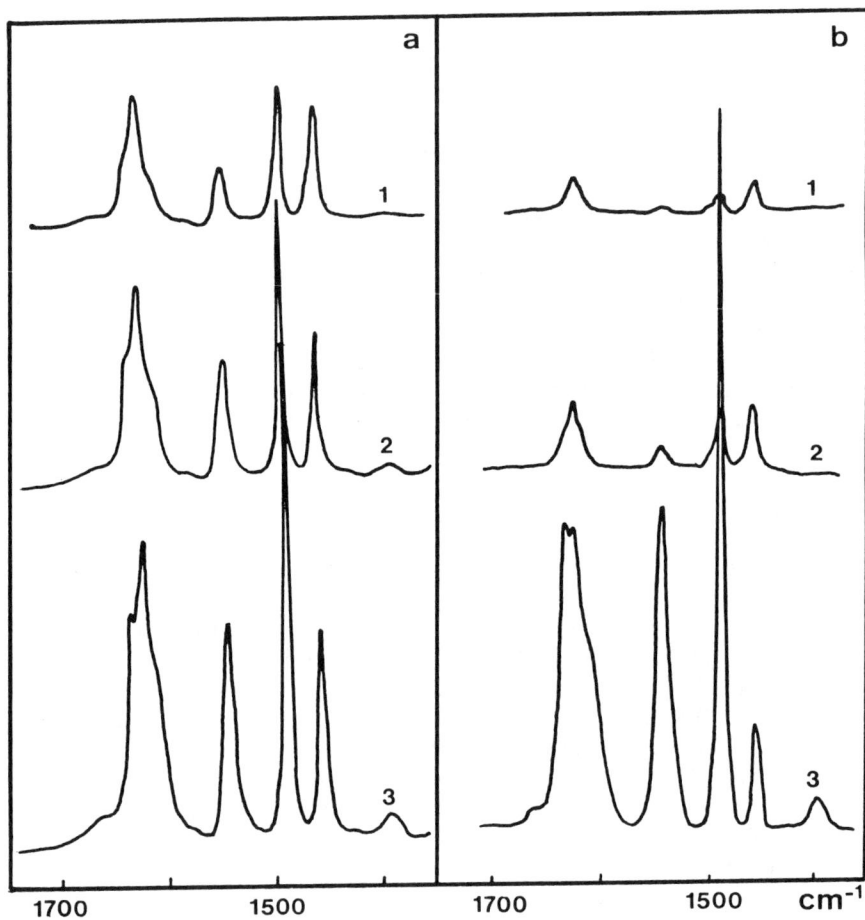

Figure 1. I.r. spectra of pyridine adsorbed at room temperature of: (a) USY zeolite and (b) SAPO–37, and desorbed at (1) 400°C, (2) 350°C and (3) 250°C.

both USY and SAPO–37 catalysts. SAPO–37 has a higher amount of total Brönsted acid sites than USY but most of them are of weaker strength.

Reaction Procedure.   The gasoil cracking reactions were performed in an automated MAT unit able to carry out several cycles of reaction–regeneration experiments. Reaction conditions (reaction and regeneration temperatures, TOS, cat/oil ratio) can be changed from one experiment to another by introducing the corresponding parameters in the computer before starting a series of runs.

The samples of catalysts were diluted in a silica matrix and activated "in situ" before the first experiment. Thus, the SAPO–37 catalyst was carefully activated following the protocol described in ref. (8), while the USY catalyst was activated at 803 K under air flow.

The feed was a vacuum gasoil whose physicochemical characteristics are given in Table I. A reaction temperature of 755 K was used in all the experiments. Different conversion levels were obtained by changing the catalyst–to–oil ratio. Moreover, to get reliable kinetic data the time on stream was also varied in the range of 5–120 sec. After each experiment, the catalysts were regenerated "in situ" by passing 100 ml min^{-1} of air through the reactor at 803 K for 5 hours.

Table I. Characteristics of Vacuum Gasoil

Density (g/cc)	0.873	Conradson Carbon (wt%)	0.03
API gravity	30.6°	MeABP (°C)	366
Nitrogen (ppm)	370	K–UOP	12.00
Sulfur (%)	1.65	Viscosity (c.s. at 50°C)	8.249

Distillation Curve (°C)

IBP	5	10	20	30	40	50	60	70	80	90	FBP
167	245	281	304	328	345	363	380	401	425	450	551

Gases were analyzed by GC and separated in two columns: hydrogen and stripping nitrogen in a molecular–sieve–packed column with argon as carrier gas and a thermal conductivity detector, and C_1–C_6 hydrocarbons in a plot alumina semicapillar column with helium as carrier gas and a flame ionization detector. Liquids were analyzed by simulated distillation. Conversion is defined here as the sum of gases, gasoline (483 K), diesel (593 K), and coke.

Results and Discussion

Table II presents the cracking results obtained on USY and SAPO–37. From these results, one would conclude that SAPO–37 is more active for gasoil cracking than the USY zeolite studied here. Indeed, cracking catalysts containing SAPO–37 were claimed (*11*) to give a higher gasoil conversion than those containing USY zeolites. It should also be noticed from the results presented in Table II that SAPO–37 gives a higher gasoil conversion at 60 sec. TOS, which is the typical TOS used in MAT experiments, and also at shorter (23 sec.) TOS. If one uses these data to discuss the catalytic properties of the two samples, one may conclude that SAPO–37 has a higher number of acid sites active for gasoil cracking, since large differences in gasoil accessibility to the cavities of the two zeolites are not expected, and, if any, they would work in favor of the USY sample in which mesopores are formed during dealumination (*10*).

Table II. Gasoil Conversions and Product Yields on USY and SAPO–37 Catalysts at a g.cat/g.oil Ratio of 0.211

T.O.S. (s)	23		60	
	USY	SAPO–37	USY	SAPO–37
CONVERSION	58.17	72.21	52.29	71.16
YIELDS				
Gasoline	31.33	46.31	35.12	44.83
Diesel	10.68	9.84	9.19	10.36
Gases	14.63	13.46	13.60	13.83
Coke	1.53	2.60	1.38	2.14

However, both i.r. spectra of adsorbed pyridine (Fig. 1) and catalytic cracking of n–heptane (*8*) do show that USY has a higher number of strong acid sites, which are believed to be active for gasoil cracking, while SAPO–37 has a higher number of weaker acid sites. At this point, and with the results generated here, one could be tempted to conclude that while short–chain alkanes need stronger acid sites to crack, gasoil cracking may also be catalyzed by weaker acid sites. This would explain why SAPO–37, while having a smaller number of strong acid sites, gives a higher gasoil cracking conversion than USY. However, it must be taken into account that in a MAT unit one obtains cumulative average conversions, and that, even when working at 23 TOS, conversions become determined not only by catalyst activity but also by catalyst decay. Then, under these circumstances it may happen that a zeolite catalyst (A) with a higher initial activity than another (B) can show a lower acummulated average conversion after 23 seconds TOS, if A decays much faster than B.

To check this possibility, a full kinetic study was carried out by using a three–lumps gasoil cracking mechanism (Fig. 2), and the methodology described elsewhere (*12–15*). Notice that the kinetic study done here includes experiments performed at very short time on stream, and also that several of them have been duplicated and triplicated to minimize and to calculate the errors associated with the experiment (Fig. 3). When the kinetic and decay parameters are compared for the two catalysts (Table III), results indicate that contrary to what could be deduced

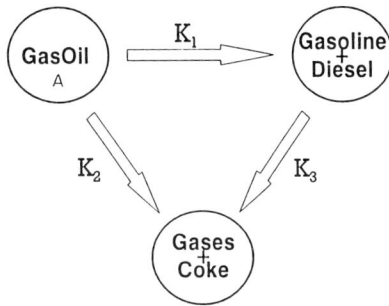

$$K_0 = (K_{10} + K_{20}) (1 + G t)^{-N} \left(\frac{C_A}{C_{A0}} \right)^W$$

Figure 2.   Three–lumps kinetic model as in ref (15). K_0 is the first order kinetic constant, G and N decay parameters, C_{A0} the initial concentration of gasoil in the gas phase and W the refractoriness index.

Table III. Kinetic Parameters at 95% Confidence Level Obtained by Fitting the Experimental Data

		USY	SAPO–37
	K_0 (s^{-1})*	869±285	336±80
	G (s^{-1})	0.46±0.25	0.09±0.1
	N	1.21±0.15	2.17±0.65
	W	0.62±0.13	0±0.13
	K_d (s^{-1})	0.55	0.19
	m	1.83	1.46
Error parameters			
	SD	0.011	0.042
	F_{FISHER}	362	155
	Φ_{EXNER}	0.16	0.24

* K_0 = kinetic rate constant; K_d = decay constant; m = order of decay where G = (m−1)K_d and N = 1/m +1; and W = refractoriness index, as in ref. (15).

AVERAGE CONVERSION (%)

AVERAGE CONVERSION (%)

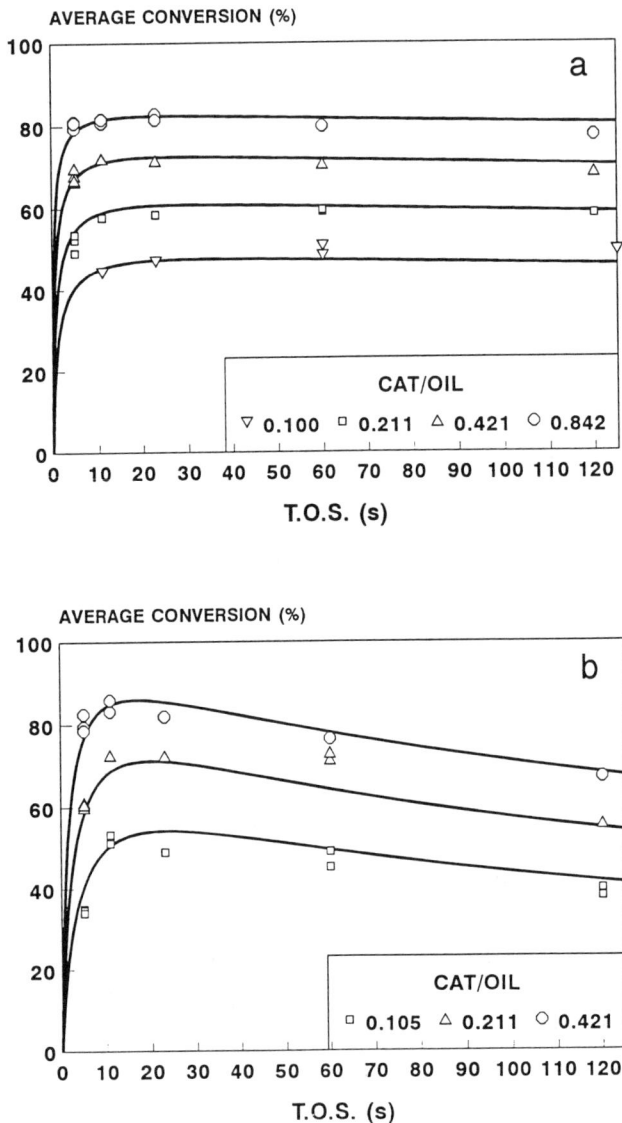

Figure 3. Experimental (symbols) and simulated (lines) average conversions obtained over a) USY zeolite and b) SAPO-37, at different time-on-stream and catalyst-to-oil ratios. Average conversions include thermal cracking.

when comparing averaged conversions, USY zeolite is more active than SAPO-37. However, the decay parameters indicate that USY decays faster than SAPO-37.

Instantaneous conversions have been generated using the kinetic parameters found for the two catalysts, and the curves are given in Figure 4 for different time on streams. It can be seen there that for units working at TOS of 2 seconds or

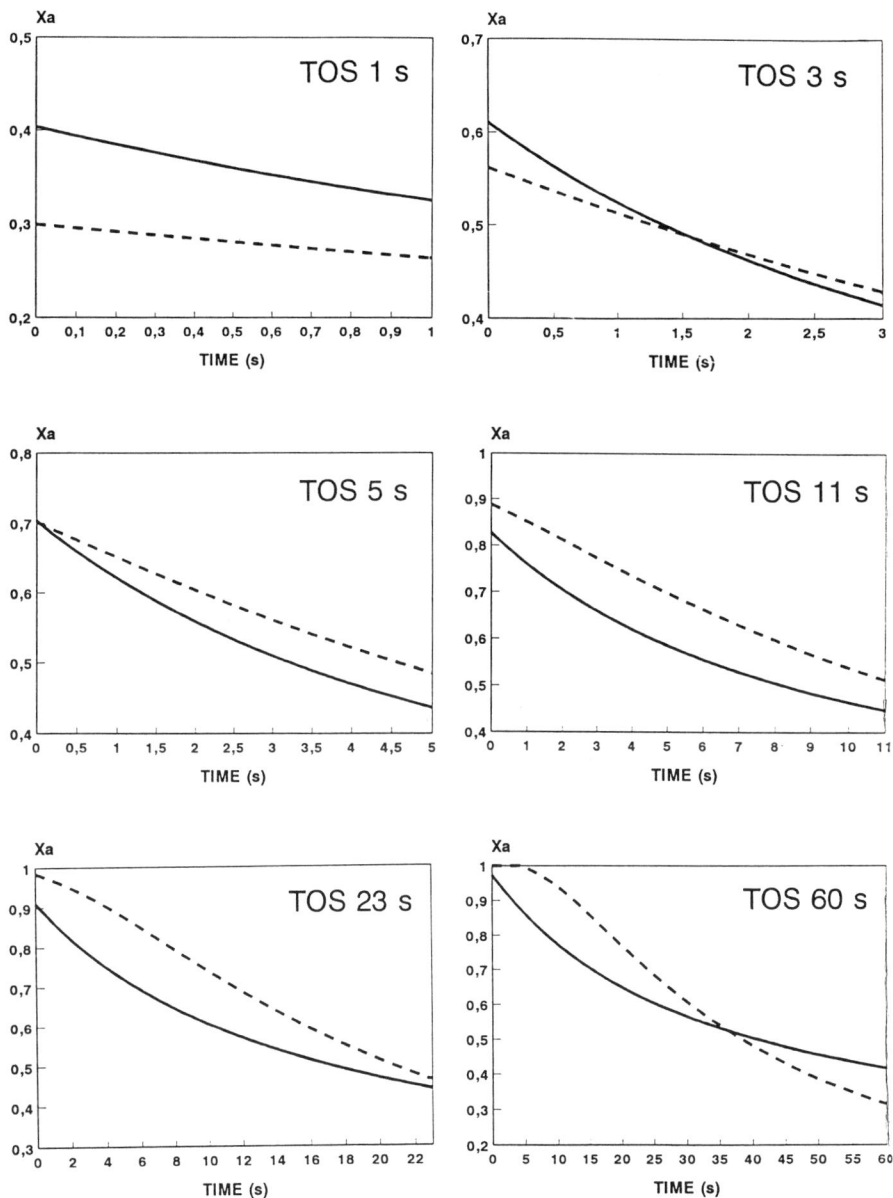

Figure 4. Instantaneous conversions plotted by computer simulation using the kinetic parameters obtained for USY (———) and SAPO–37 (– – –).

lower, which are the actual times in modern FCC, USY will give higher conversion than SAPO-37. However, for any other unit which may work at TOS higher than that, SAPO-37 may give a higher conversion.

Conclusions

In conclusion, it appears that the gasoil conversion criteria obtained in conventional MAT units may be inadequate to compare FCC cracking activity of materials with very different activity–decay behavior. This may be so even if the comparison is done on the bases of experiments performed at low TOS.

In those situations, our recommendation would be to compare the catalyst activities on the bases of kinetic parameters. The relative higher number of experiments needed in this case can be obtained inexpensively and in a short time by means of automated MAT units, such as the one used in this work.

Acknowledgements. Financial support from the Spanish CICYT, project MAT 91–1152, is gratefully acknowledged.

Literature Cited

(1)   Schipper, P.H.; Dwyer, F.G.; Sparell, P.T.; Hizrahi, S.; Herbst, J.A. *ACS. Symp. Ser.*, **1988**, 375, 64.

(2)   Elia, M.F.; Iglesias, E.; Martínez, A.; Pérez, M.A., *Appl. Catal.*, **1991**, 73, 195.

(3)   Biswas, J.; Maxwell, I.E., *Appl. Catal.*, **1990**, 58, 1.

(4)   Ragalopajan, K.; Young, G.W., *ACS Symp. Ser.*, **1988**, 375, 34.

(5)   Bonetto, L.; Camblor, M.A.; Corma, A., Pérez–Pariente, J., *Appl. Catal.*, **1992**, 82, 37–50.

(6)   Corma, A.; Fornés, V.; Martínez, A.; Melo, F.; Pallota, O., *Stud. Surf. Sci. and Catal.*, **1988**, 37, 495.

(7)   Corma, A.; Fornés, V.; Martínez, A.; Orchillés, A.V., *ACS Symp. Ser.*, **1988**, 368, 542.

(8)   Corma, A,; Fornés, V.; Franco, M.J.; Mocholí, F., *ACS. Symp. Ser.*, **1991**, 6, 79.

(9)   Corma, A.; Fornés, V.; Franco, M.J.; Melo, F.; Pérez–Pariente, J., *Proceedings of the 9th Int. Cong. of Zeol.*, **1992**, Vol. 2, 343.

(10   Lohse, V.; Stach, H.; Thamm, H.; Schirmer, W.; Isirikjan, A. A.; Regent, N. I.; Dubunin, M.M., *Z. Anorg. Chem.*, **1980**, 460, 179.

(11)   Edwards, G.C.; Gilson, J.P.; McDaniel, C.V., U.S. Patent 4 681 864, **1988**.

(12)   Campbell, D.R.; Wojciechovsky, B.W., *Can. J. Chem. Eng.*, **1969**, 47, 413.

(13)   Wojciechovsky, B.W., *Can. J. Chem. Eng.*, **1968**, 46, 48.

(14)   Viner, M.R.; Wojciechovsky, B.W., *Can. J. Chem. Eng.*, **1982**, 60, 127.

(15)   Wojciechovsky, B.W.; Corma, A., in *"Catalytic Cracking. Catalysts, Chemistry and Kinetics"*, M. Dekker, New York, **1986**; Chapter 6, 210.

RECEIVED June 17, 1994

Chapter 11

Development of a New Coefficient To Predict Feedstock Cracking

Marcelo A. Torem, Maria F. Cunha, and Vandilea Aleksitch

Petrobras Research Center, Ilha do Fundao, quadra 7, Cidade Universitaria, 21949–900 Rio de Janeiro, Brazil

Predicting the chemical composition and crackability of FCC feedstocks is difficult due to their complex structures. The development of spectrometric techniques has allowed the determination of the hydrocarbon families present in these feeds and some of their structural parameters. The major obstacle to the application of such sophisticated methods to refinery routine is their high cost, so a group of equations that uses simple analyses (as aniline point, viscosity) was developed to quantify the following groups: paraffins, cycloparaffins, monoaromatics, polyaromatics and aromatic sulfur compounds. Mass spectrometry composition data were used as reference to this development, as well as nuclear magnetic resonance to secondary applications. These equations were compared with other ones (found in the literature), showing better statistical quality. Based on this work and some other concepts, a new parameter called PFC (Petrobras Feedstock Coefficient) is defined to predict the crackability of FCC feedstocks. This coefficient has been compared with commercial conversions and shows good results.

The chemical composition of FCC feedstocks is one of the key variables to predict the industrial performance concerning product yields and quality. The standard analyses usually performed in refinery laboratories are not sufficient to give a good quantitative prediction on chemical composition or crackability of a given feed.

In this work, the development of equations and simplified models to estimate chemical composition and crackability of usual FCC feedstocks, based on simple traditional analyses, is presented. Mass spectrometry (MS) and nuclear magnetic resonance (NMR) data were used to support the generation of these equations.

Chemical Composition Estimation

Samples. Sixty-five FCC feedstocks from Petrobras refineries were selected to develop the equations for estimating their chemical composition. These samples

0097–6156/94/0571–0127$08.00/0

represent a variety of compositionally distinct vacuum gas oils. Some of them are blended with other petroleum process streams, such as coker gas oil, vacuum residue and deasphalted oil with different contents up to 30% volume.

The oils processed in Petrobras refineries are from Brazil, as well as from Middle East and in minor proportions, from Nigeria, China and Venezuela.

Table I shows the value range of some physical and chemical properties of the 65 Petrobras FCC feedstocks selected.

The analyses performed to create the data base were those shown in table I and also include ASTM D1160 distillation, asphaltene, nickel, vanadium and sodium contents, group-type analysis by MS (*1*) and combined method to quantify several structural parameters by NMR (*2*).

In respect of the ASTM D1160 distillation, it was observed that some of the feedstocks were very heavy, resulting in values of end boiling points up to 555°C with 70 % volume recovery.

The molecular weight magnitude of the analyzed samples was 450, calculated by Riazi's correlation (*3*).

Table I. Physical and Chemical Proprities of FCC Feedstocks Selected

Property	Range
Density, 20 °C / 4 °C	0.8715 - 0.9562
Kinematic viscosity (cSt)	
@ 50 °C	16.80 - 399.50
@ 100 °C	6.46 - 29.54
Refractive index @ 70 °C	1.4677 - 1.5607
Aniline point (°C)	72.0 - 109.8
Basic Nitrogen (wppm)	335 - 1500
Total Nitrogen (wppm)	400 - 4300
Sulfur (wt%)	0.12 - 3.3
Ramsbottom Carbon Residue (wt%)	0.3 - 4.2

Equations. The new equations to predict the chemical composition of FCC feedstocks were developed by multiple linear regression analysis using the data of 65 feedstocks.

The empirical equations shown in Table II as mathematical functions were obtained by using the least square method. They estimate the volumetric percent of naphtenic (NH), mono- and polyaromatic hydrocarbons (MAH and PAH) and aromatic sulfur compounds (ASC), as functions of five parameters, routinely obtained in refinery laboratories: density (d), viscosity (μ), sulfur content (S), aniline point (AP) and basic nitrogen content (BN). The content of paraffinic hydrocarbons (PH) is obtained by subtracting from 100 all the other hydrocarbon groups. The total aromatic (TA) content is calculated by adding MAH and PHA contents.

Table I shows the ranges of the independent variables for which the equations are valid.

Table II. Feedstock Chemical Composition Equations Expressed as Mathematical Functions

1 -	NH	(%vol.)	f { d, S, μ, BN }
2 -	MAH	(%vol.)	f { d, μ }
3 -	PAH	(%vol.)	f { d, AP, BN }
4 -	ASC	(%vol.)	f { S,BN }
5 -	PH	(%vol.)	= 100 - {(1)+(2)+(3)+(4)}
6 -	TA	(%vol.)	= (2) + (3)

Table III shows the minimum and maximum values of each compound group measured by MS and also their average and maximum absolute and relative deviations from their predicted values determined through the developed equations. These deviations are defined as:

$$\text{Deviation (dev.)} = x_{predicted} - x_{measured}$$

$$\textit{Average absolute deviation}(\%vol) = \sum_{n=1}^{n} | \textit{deviation} | / n$$

(avg.abs.dev.)

$$\textit{Average relative deviation}(\%) = \frac{\sum_{n=1}^{n} (100 * | deviation | / x_{measured})}{n}$$

Where n is the number of data.

Table III. Compositional Ranges and Deviations of the Various Chemical Groups

Compound Groups	MS Analysis (%vol.)		Abs. dev. (%vol.)		Rel. dev. (%)	
	Min	Max	Avg	Max	Avg	Max
NH	29.1	57.2	2.1	5.8	5.0	16.5
MAH	9.0	20.7	1.0	4.4	7.0	43.0
PAH	9.8	22.3	1.2	4.2	7.5	30.5
ASC	1.5	22.0	1.4	5.4	20.0	185
TA	18.4	42.0	1.5	5.2	5.0	18.5
PH	0.0	37.9	2.5	8.5	22.0	97.5

Evaluation of Proposed Equations. Many methods have been suggested in the literature to characterize petroleum fractions (*3,4,5,6*). Some of them require the determination of the refractive index at 20 °C. Such a measure cannot be performed in the majority of Petrobras FCC feedstocks, due to their dark color. Besides, some methods are valid within composition ranges different from those of Petrobras feedstocks.

To evaluate our proposed equations, we compared them to those developed by Riazi (3) because they are valid for composition ranges comparable to ours and also express the same type of results, i. e., the volumetric percentages of hydrocarbon groups.

Riazi has compared his results with that obtained by employing the equations proposed in the n-d-M method (4) and in the API-TDB procedure (5). He concluded that his equations were superior to those, because they showed less average and maximum absolute deviations. Table IV shows the evaluation presented in Riazi's paper.

Table IV. Riazi's Evalution of Models to Predict the Composition of Petroleum Fractions (3)

Hydrocarbon Groups	(*)	Models		
		Riazi (72 feeds)	API-TDB (72 feeds)	n-d-M (70 feeds)
PH	Range	10-81	10-81	30-81
	avg.abs.	3.1	4.3	6.4
	max.abs.	20.0	15.1	23.8
NH	Range	13-64	13-64	13-64
	avg.abs.	4.0	5.9	8.6
	max.abs.	17.9	26.6	33.5
TA	Range	0-31	0-31	0-31
	avg.abs.	2.0	4.1	5.9
	max.abs.	8.9	12.2	27.9

(*) Range   - values for which the equations are valid (%vol.)
    avg.abs.   - average absolute deviation (%vol.)
    max.abs.   - maximum absolute deviation (%vol.)

The composition ranges of Petrobras and Riazi feedstocks were not exactly the same, so to get a better statistical comparison of these results, the relative deviations were also calculated in addition to determining the average absolute deviations.

Table V shows the statistical comparison between Petrobras and Riazi's equations.

Table V.   Comparison of Petrobras' and Riazi's Models to Predict PNA Analysis of Heavy Petroleum Fractions

Hydrocarbon Groups	(*)	Models	
		Petrobras (65 feeds)	Riazi (72 feeds)
PH	Range	0-38	10-81
	avg.abs.	2.5	3.1
	max.abs	8.5	20.0
	avg.rel.	22	9
	max.rel	98	190
NH	Range	29-57	13-64
	avg.abs.	2.1	4.0
	max.abs	5.8	17.9
	avg.rel.	5	17
	max.rel	17	86
TA	Range	19-42	0-31
	avg.abs.	1.5	2.0
	max.abs	5.2	8.9
	avg.rel.	5	16
	max.rel	18	144

(*) Range   - values for which the equations are valid (%vol.)
    avg.abs. - average relative deviation (%vol.)
    max.abs. - maximum relative deviation (%vol.)

It can be observed from Table V that for all the hydrocarbon groups, Petrobras equations show lower deviation values do than those of Riazi.

The same observation can be seen in Table VI, which shows the comparison of mono- and polyaromatic hydrocarbon estimations.

It can be concluded that for FCC feedstocks processed by Petrobras units, the new group of equations gives better chemical composition estimation.

Crackability Prediction

After developing the equations to estimate FCC feedstock compositions, a new parameter, generated by a model, was created to predict the crackability of these petroleum fractions. The way to build the structure of this model is to consider some of the classical cracking mechanism steps as empirical mathematical function blocks and join them in just one parameter called Petrobras Feedstock Coefficient or PFC.

The idea used to formulate PFC was to develop a group of equations based on simple feed chemical and physical analyses to generate a parameter comparable in magnitude to commercial FCC conversion, i.e., that could discriminate feedstocks related to their trends to yield converted products. Coke is not considered here because it is strongly related to FCC unit heat balance.

The PFC model is constituted of four blocks, as are described below:

FUNCTION 1 - POLYAROMATICS (PAH)

According to the literature, polyaromatic molecules are not crackable to noble products (7) and would be the main precursors of unconverted material (light cycle and decanted oils) and coke. The polyaromatics's function, previously demonstrated, is :

$$PAH \ f \ (d \ , \ AP \ , \ BN) \tag{1}$$

FUNCTION 2 - SIDE CHAINS CONTRIBUTION (SCC)

The mass spectrometry data, used to estimate the polyaromatic content in FCC feed, quantify the aromatic rings and their side chains. Under cracking conditions, these side chains are dealkylated to gasoline or lighter products. The average number of carbon atoms of the side chain linked to aromatic rings is estimated based on NMR data (2). This function expresses the side chain average size (SCAS) and is given by:

$$SCAS \ (as \ CARBON \ ATOMS) \ \ f \ (AP \ , \ d) \tag{2a}$$

The target of this part of the PFC model is to account for dealkylated side chain contribution to converted products. One of the possible approaches to the problem is to consider the dealkylation of the long original side chain linked to the polyaromatics, resulting in an average hypothetical polyaromatic molecule that would have just three small remaining alkyl groups (two methyl and one ethyl) (8). Based on this hypothesis, it was estimated that the size of this polyaromatic structure would represent a calculated mean of di-, tri- and tetraromatics, called estimated polyaromatic size (EPS):

$$EPS \ (as \ CARBON \ ATOMS) \ f \ (SCAS) \tag{2b}$$

Joining all these concepts, we obtain a general function using expression 2a and 2b, being represented as follows:

$$SCC \ f \ (SCAS \ , \ EPS \ , \ PAH) \tag{2c}$$

FUNCTION 3 - UNCONVERTED SULFUR COMPOUNDS (USC)

The purpose of this block is to quantify the sulfur compounds that would be in the unconverted products (cycle oils) and were originally present in the feedstock.

Wollaston and co-workers (9) reported that the amount of feed sulfur compounds remaining in the cycle oil range (LCO and decanted oil) varies according to the conversion level. Based on this study and in according to the average conversion level of Petrobras FCC units, we estimate the total of sulfur feedstock compounds that are not converted to coke, gasoline and lighter fractions. Therefore, the function that represents the unconverted sulfur compounds is:

USC f (S , BN , FCC UNIT CONVERSION LEVEL)   (3)

FUNCTION 4 - DEACTIVATION BY BASIC NITROGEN COMPOUNDS
The effect of basic nitrogen present in FCC feedstocks has been extensively studied by many authors in the literature (*10,11*). The main consequences of basic nitrogen contamination, on the catalytic cracking process, are reduced catalytic activity (resulting in lower conversions due to catalyst acid sites neutralization) and poor gas and coke selectivities.

Petrobras also studied (*12*) the influence of a feedstock doped with increasing amounts of quinoline, a strong nitrogenated poison, on diferent commercial catalysts, in MAT scale. From this work, we developed the equation that describes the deactivation process in terms of conversion retention related to basic nitrogen content. It was taken into account the effect of the scale (MAT x commercial) to correct possible distortions due to longer contact time of the catalyst in MAT reactor as compared to commercial riser. Then, it was decided to introduce a parameter called Basic Nitrogen Deactivation Factor (BNDF) that could correct our crackability index (PFC) taking into consideration this issue. The type of equation that represents this poisoning process is:

BNDF = 1 / (A + B . BASIC NITROGEN CONTENT)   (4)

where A and B are constants depending on the catalytic system being used in FCC unit. For the commercial example that will be shown, it was applied the constants corresponding to a representative catalyst grade, used in most Petrobras FCC units, designed for high octane-barrel potential and moderate bottoms cracking activity.

After describing the four functions that represent some of the main cracking process phenomena, it was necessary to join them in a single expression that could link all parts in a logical and applied sense. It was decided that the best approach would be to work in weight basis, considering 100 as reference (due to mass balance reasons), and then subtracting the blocks that could not be accounted as converted products, excluding coke yield due to the reasons already mentioned above. Therefore, the general expression for PFC is:

PFC = {100 - (PAH - SCC) - USC} . BNDF

Discussion. To validate the PFC model, one of the Petrobras FCC units was chosen for a test with commercial data. The commercial conversion was selected as a comparison parameter and its classical definition is:

COMMERCIAL CONVERSION = 100 - LCO - DECANTED OIL (wt%)

The selected unit is an UOP STACKED model and the days used, in the validation phase, were spread in time over a period of two years. During this period, the operating conditions and catalysts were not the same and the feedstock quality changed many times with   different types of crudes and resid contents. Figure 1 shows the commercial conversion fluctuations and the previous coefficient used by Petrobras. The latter is given by:

Table VI. Evaluation of Petrobras' and Riazi's Models to Predict Monoaromatic and Polyaromatic Hydrocarbons

Hydrocarbon Groups	(*)	Models	
		Petrobras (65 feeds)	Riazi (72 feeds)
MAH	Range	9-21	5-84
	avg.abs.	1.1	5.5
	max.abs	4.4	18.2
	avg.rel.	7	24
	max.rel	43	111
PAH	Range	10-22	0-66
	avg.abs.	1.2	6.5
	max.abs	4.2	24.2
	avg.rel.	7	29
	max.rel	30	127

(*)   Range -   values for which the equations are valid (%vol.)
       avg.abs -   average relative deviation (%vol.)
       max.abs -   maximum relative deviation (%vol.)

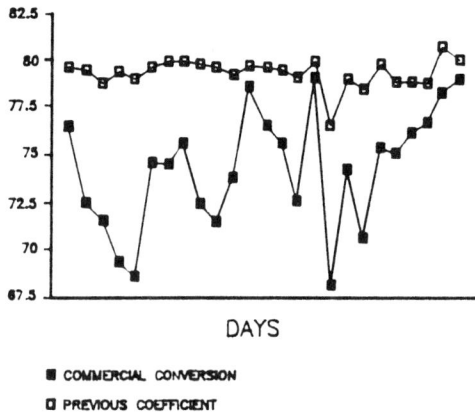

DAYS

■ COMMERCIAL CONVERSION
□ PREVIOUS COEFFICIENT

Figure 1 - Previous Coefficient Profile.

PREVIOUS COEFFICIENT = 100 - %C aromatic,

where %C aromatic is calculated based on a correlation between NMR data and simple feed analysis.

The main observation is that the previous coefficient does not follow the changes in feedstock quality.

In this case, one can verify that the new coefficient follows the commercial conversion profile. This fact confirms the great influence of feedstock quality on FCC unit performance, in spite of the many interferences due to operational variations. The results from Figure 2 validates PFC as a useful parameter to be applied in different activities in the FCC area, such as:
  - operational optimization of existing units;
  - commercial evaluation of catalysts or design modifications, by considering feedstock influence;
  - design of new units;
  - development of global mathematical models.

Conclusions

A group of equations having better statistical quality compared to others described in the literature, was developed to estimate the detailed chemical composition of usual FCC feedstocks.

The Petrobras Feedstock Coefficient gives a practical and quantitative estimation of the feedstock crackability, presenting a good agreement with commercial FCC conversion profile.

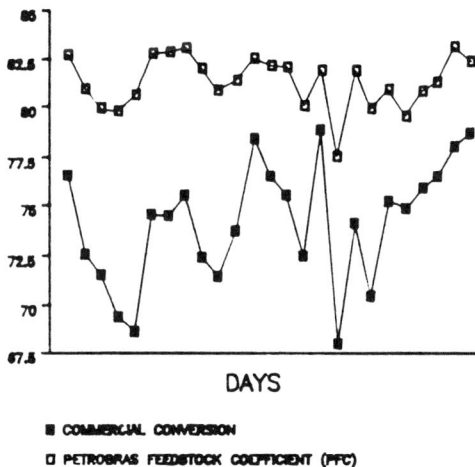

DAYS

■ COMMERCIAL CONVERSION
□ PETROBRAS FEEDSTOCK COEFFICIENT (PFC)

Figure 2 - PFC Profile.

Literature Cited

(1) TEETER, R. M., *Mass.Spectr. Rev.*, 4, 1223-1433 (1985).

(2) HASAN, M.V., LI, M.F., BUKHARI, A. *Fuel, 62,* 518-523 (1983).

(3) RIAZI, M. R., DAUBERT, T. E. *Ind. Chem. Process Des. Dev., 25,* 1009-1015 (1986).

(4) AMERICAN SOCIETY FOR TESTING OF MATERIALS. *Standard D3238-82, calculation of carbon distribution and structural group analysis of petroleum oil by n.d.m method,* [s.l.], 1982.

(5) DAUBERT, T.E., DANNER, R.P. (Eds) *Technical data book:* petroleum refining. 4. ed. Washington, D.C.: American Petroleum Institute, 1982.

(6) DHULESIA, H. *Oil & Gas J., 84,*2 (1986).

(7) KRISHNA, R., *Erdol und kohle, Erdgas Petroch,* 194-199 (may 1989).

(8) FISHER, J.P. Appl. Catal., *65,* 189-210 (1990).

(9) WOLLASTON, E. G., FORSYTHE, W.L., VASALOS, I.A. *Oil & Gas J., 69,* 1, 64-69 (1971).

(10) FU, C.M., SCHAFFER, A.M., *Ind. Eng. Chem. Prod. Res. Dev, 24,* 68-75 (1985).

(11) HO, T.C., KATRITZKY, A.R., CATO, S. J. *Ind. Eng. Chem. Res., 31,* 1589-1597 (1992).

(12) TOREM, M.A., RAWET, R. Influence of nitrogenated feedstock on FCC commercial catalyst performance. In: BRAZILIAN CATALYSIS SEMINAR, 6, [s.l.], 1991.

RECEIVED June 17, 1994

Chapter 12

Breakthrough in Microactivity Test Effluent Analysis

Detailed Analytical Characterization of Gasolines, Light-Cycle Oil, and Bottom Cuts

N. Boisdron, M. Bouquet, and C. Largeteau

Total Raffinage Distribution, Research Center, B.P. 27, 76700 Harfleur, France

MicroActivity Tests (MAT) are commonly used to evaluate FCC feedstocks and FCC catalysts properties. Evaluation is rapid and requires only small amounts of catalyst and feedstock. However, until now, the small quantity of recovered liquid effluent didn't allow to characterize completely the whole effluent. Only two chromatographic analyses could be made on a routine basis :
1) TBP chromatographic analysis,
2) PIANO determination and octane estimation on light part of the effluent.
Improved techniques of distillation and separation have been built up to characterize the whole effluent and more specifically the heaviest cuts.
Using appropriate device requiring small amount of whole effluent, the method elaborates fractions such as gasoline, LCO and bottom cuts by distillation and enables to perform molecular and structural analysis on each cut. The set of analyses is devoted to such characterization and requires very small quantity of effluent fraction sample. One of the separation techniques (liquid chromatography) has been developed at purpose.
Examples of applying such new concept of microanalysis are given and illustrated in this paper.

The first technique, called "microdistillation" allows to split MAT liquid effluents into four cuts. All desirable analyses can be performed on each of these cuts, provided the analysis doesn't require more product than available by microdistillation.

Special emphasis has been put on the characterization of bottoms fractions. On one hand, Fisher mass spectrometry (1), which identifies aromatics and sulfur

0097–6156/94/0571–0137$08.00/0

compounds, can be directly applied to microdistilled bottoms fractions ; on the other hand Hood/O'Neal-MS (2) provides data on saturated compounds but requires a further step of separation, achieved by liquid chromatography. So, to perform O'Neal analysis on microdistilled bottoms, a microtechnique of liquid chromatography has been developed.

This paper presents these two new techniques still under development. The first analytical results obtained an MAT effluents are also shown. An overview of the experimental procedures is given, and the technical feasibility of microdistillation and liquid microchromatography is demonstrated. Finally, an application of microdistillation will be illustrated by Fisher-MS results obtained on bottoms fractions from MAT effluents.

Experimental

Microdistillation. Experiments were run on a modified GKR-51 Buchi apparatus. A home-made ice-cooler improvement was added to the cooling section to reach complete condensation.

Typical sequences applied to MAT total effluent were based on the following steps :
- CS_2 + light ends including light gasoline head were obtained at 95°C and atmospheric pressure during 40 minutes,
- light and heavy gasolines were obtained at 170°C and reduced pressure (typical value : 100 mm Hg) during 1 hour,
- light cycle oil (LCO) was obtained at 200°C and low pressure (typical value : 1 mm Hg) during 1 hour,
- bottoms as residue in the ball tube.

The feed is typically 2.4 g, including 1.8 g of carbon disulphide (CS_2) and 0.6 g of effluent from the MAT pilot (CS_2 is added as solvent to fully recover the effluent).

Liquid microchromatography. The microchromatography is not very different from classical chromatography, albeit the constraints of miniaturization dealing with :

Column	30 cm length / 10 mm e.d. / 6 mm i.d.
Microvalve	Hamilton 86777
Silica	Davison CG2
Detector	Waters RI
Pump..	Milton Roy (1 ml/min)
Syringe.............................	Exmire Microsyringe MS GIL 050

The bottoms sample is typically 100 mg and diluted with 200 µl of cyclohexane. The sample is injected in the column and after separation on the silica fed to a refractometer to detect the cut point of the run.

Much care must be taken while evaporating the solvents to avoid the bubbles formation during the gentle heating of small vials (0.5 ml). The yield of the separation has to be close to 100 %, never less than 95 %.

Rationale of the microanalysis

Microdistillation. The microdistillation is based on the Buchi glass tube oven GKR 51. The oven allows a separation of the total effluent from MAT experiment, into four fractions.

First of all, microdistillation feasibility was checked on a pilot plant effluent with no addition of CS_2. The pilot provided enough product to perform classical ASTM D-1160 distillation. The following table compares the results given with optimized settings for the GKR 51 Buchi with ASTM distillation yields. There is a good agreement between the two methods but we still have some difficulties to split 0.6 g sample into four fractions.

Microdistillation yields

	Pilot plant effluent / ASTM D-1160	Pilot plant effluent Buchi	
		2 g feed	0.6 g feed
Light gasoline (IBP-160°C)	40.4	41.1	⎫
Heavy gasoline (160-220°C)	15.7	16.4	⎬ 80.3
LCO (220-350°C)	25.6	22.8	⎭
Bottoms (350+°C)	18.3	19.7	19.7
350-°C fraction in bottoms (weight %)	3	13	7

Then, using CS_2 on a common basis to fully recover the effluent at the end of MAT test, new trials were performed with small amounts of the pilot plant effluent and additional CS_2. Under these conditions, the effluent is split into a CS_2 rich fraction including light gasoline head and three other fractions. The settings of the Buchi are optimized such as these fractions overlap standard LCO and bottoms cuts.

The settings of the Buchi are presented in the experimental section. They have been chosen with yields derived from standardized cuts, i.e.

$$\left\{ \begin{array}{l} \text{IBP - 95°C heads} \\ \text{95 - 220°C gasolines} \\ \text{220 - 350°C LCO} \\ \text{350}^+\text{°C bottoms} \end{array} \right.$$

Feasibility was proved on bottoms fractions with a separation step lasting about 3 hours. A weight of 0.6 g of total effluent was required. LCO fraction separation is not so accurate. Repeatability was examined, particularly on bottoms yields.

A set of five runs (0.6 g effluent + 1.8 g CS_2) were performed with the microdistillation apparatus under similar conditions. Results with bottoms yields are illustrated on next table.

Set of five runs

Run n°	1	2	3	4	5
Microdistillation bottoms yield (weight %)	18.95	18.84	18.56	19.88	19.28
350-°C in bottoms (weight %)	15	15	9	16	16

Mass spectrometry (1) was used to determine the molecular structure of the compounds. Table I displays the full results of this analysis. The compounds identification from the five runs is consistent. The agreement with the reference sample is fairly good.

We determined that CS_2 has no incidence on the results.

Table I. Fisher results on bottoms fractions

Yields (weight %)	Microdistillation run					Statistics Average	σ	Reference Distill. ASTM D-1160
Bottoms	18.95	18.81	18.56	19.80	19.28	19.1	0.49	18.34
Effluent R2R test n°	1	2	3	4	5			
Saturates	**29.7**	**29.6**	**29.3**	**30.0**	**29.5**	**29.6**	**0.22**	**28.7**
C_nH_{2n+2}	12.19	12.27	10.69	11.78	12.25	11.8	0.60	11.39
$C_nH_{2n, -2, -4, -6}$	17.49	17.29	18.60	18.19	17.24	17.8	0.54	17.35
Aromatics	**60.9**	**61.0**	**61.1**	**60.5**	**61.0**	**60.9**	**0.20**	**62.2**
Monoaromatics								
C_nH_{2n-6}								
C_nH_{2n-8}								
C_nH_{2n-10}								
Diaromatics	**2.4**	**3.3**	**2.5**	**4.3**	**4.5**	**3.4**	**0.89**	**1.3**
C_nH_{2n-12}	1.02	1.02	1.01	1.23	1.19	1.1	0.10	0.73
C_nH_{2n-14}	0.88	1.22	0.94	1.44	1.62	1.2	0.28	0.46
C_nH_{2n-16}	0.49	1.01	0.50	1.61	1.72	1.1	0.53	0.09
Triaromatics	**16.4**	**17.2**	**17.3**	**17.7**	**18.0**	**17.3**	**0.55**	**16.8**
C_nH_{2n-18}	7.46	7.96	8.19	8.87	8.86	8.2	0.50	8.23
C_nH_{2n-20}	8.89	9.20	9.09	8.80	9.30	9.1	0.19	8.35
Tetraaromatics	**22.7**	**22.1**	**23.3**	**21.7**	**21.3**	**22.2**	**0.71**	**22.1**
C_nH_{2n-22}	13.39	13.00	14.12	13.17	12.99	13.3	0.42	13.46
C_nH_{2n-24}	4.90	4.78	4.90	4.46	4.24	4.7	0.26	3.54
C_nH_{2n-26}	4.37	4.29	4.28	4.08	4.05	4.2	0.13	5.12
Pentaaromatics	**9.2**	**8.9**	**8.9**	**8.3**	**8.3**	**8.7**	**0.37**	**9.5**
C_nH_{2n-28}	6.24	6.00	6.08	5.68	5.52	5.9	0.27	6.09
C_nH_{2n-30}	2.99	2.93	2.80	2.65	2.75	2.8	0.12	3.41
Hexaaromatics	**6.8**	**6.4**	**6.3**	**5.8**	**6.0**	**6.3**	**0.35**	**7.7**
C_nH_{2n-32}	5.05	4.71	4.82	4.3	4.40	4.7	0.27	5.62
C_nH_{2n-34}	1.79	1.66	1.48	1.49	1.59	1.6	0.11	2.06
Heptaaromatics	**3.4**	**3.2**	**2.9**	**2.7**	**3.0**	**3.0**	**0.25**	**4.8**
C_nH_{2n-36}	1.46	1.43	1.26	1.20	1.37	1.3	0.10	1.67
C_nH_{2n-38}	1.97	1.77	1.65	1.51	1.63	1.7	0.16	3.13
Sulphur compounds	**9.4**	**9.5**	**9.5**	**9.5**	**9.5**	**9.5**	**0.03**	**9.1**
Dibenzothiophenes								
$C_nH_{2n-16}S$	2.31	2.37	2.46	2.53	2.43	2.4	0.08	2.01
$C_nH_{2n-18}S$	0.65	0.67	0.69	0.69	0.66	0.7	0.02	0.54
$C_nH_{2n-20}S$	0.99	1.00	1.04	1.01	0.99	1.0	0.02	0.90
Naphtalenothiophenes								
$C_nH_{2n-22}S$	2.69	2.67	2.78	2.68	2.64	2.7	0.05	2.54
$C_nH_{2n-24}S$	0.95	0.94	0.85	0.93	0.95	0.9	0.04	0.93
$C_nH_{2n-26}S$	0.78	0.81	0.72	0.75	0.82	0.8	0.04	0.76
$C_nH_{2n-28}S$	0.47	0.46	0.45	0.46	0.47	0.5	0.01	0.51
Disulphurised compounds	0.58	0.54	0.50	0.48	0.51	0.5	0.03	0.91

Liquid microchromatography. A liquid microchromatography method was optimized to separate quantitatively the "saturated species" from the "aromatics + sulphur compounds" in the bottom fraction from the microdistillation device.

Settings were selected according to the described experimental details and applied on four samples from bottom fraction. The following tables present the yields obtained and exhibit the composition of these "saturated species" according to O'Neal method.

Saturated compounds yields

Sample (mg)	Saturates (weight %)	
79.1	32.8	
80.3	32.4	Average = 33.6 %
77.5	34.1	σ = 1 %
83.9	35.2	

Saturated compounds composition

(weight %)	1	2	3	4
Paraffins	32.3	28.5	31.8	32.6
Mononaphtenes	18.8	16.1	16.6	18.3
Dinaphtenes	16.2	15.8	16.5	15.4
Trinaphtenes	18.4	21.0	18.5	16.5
Tetranaphtenes	11.1	13.5	11.4	11.0
Pentanaphtenes	3.2	5.1	5.2	5.2
Hexanaphtenes	0.0	0.0	0.0	1.0

A good agreement between the samples is reached, either for "saturated" yields or for "saturated" compositions in this selected set of results.

Analytical limitations

Both microanalyses described in this paper are relatively delicate and require skilled operators. The fractionation quality obtained with microdistillation is not as good as the one obtained with a standardized 30-trays distillation. However, it is sufficient to have repeatable and significant mass spectroscopy results on bottoms fractions.

Our first interest in this work is to be able to characterize heavy fractions, boiling up to 750°C. Analyses are limitated by Fisher and Hood mass spectrometry techniques. Fisher characterization is devoted to petroleum cuts ranging from 250 to 700°C. Repeatability problems occur on fractions analysis when the feedstock contains a large amount of resid. Applied to the heaviest part of bottoms fractions, Hood-MS is used on the outer upper limit of its application, as this technique is normally used for cuts boiling from 250 to 500°C.

Example of application to MAT effluents characterization

MAT tests were performed on two different feedstocks to evaluate bottoms upgrading capability of two USY catalysts. The pilot used for these tests is a R2R MAT pilot, developed by TOTAL and I.F.P. (3). Both feeds contain some atmospheric resid. Feed 2 has a higher basic nitrogen level than feed 1.

Catalysts comparison was made at constant coke yield as the potential industrial unit is considered to be airblower limited.

1) FEEDSTOCK 1 : Yields at 5.7 weight % coke yield

		Catalyst A	Catalyst B
Cat to oil		4.7	3.8
Fuel gas	(weight %)	2.6	3.0
LPG	(weight %)	23.0	23.8
Gasoline (C_5-220°C)	(weight %)	49.0	41.3
LCO (220-350°C)	(weight %)	11.4	11.5
Bottoms (350+°C)	(weight %)	8.3	14.7
Coke	(weight %)	5.7	5.7
Delta coke		1.21	1.50

2) FEEDSTOCK 2 : Yields at 6.9 weight % coke yield

		Catalyst A	Catalyst B
Cat to oil		6.4	6.4
Fuel gas	(weight %)	3.1	3.4
LPG	(weight %)	19.6	20.9
Gasoline (C_5-220°C)	(weight %)	40.6	36.7
LCO (220-350°C)	(weight %)	12.4	9.7
Bottoms (350+°C)	(weight %)	17.4	22.4
Coke	(weight %)	6.9	6.9
Delta coke		1.08	1.08

These results show clearly that Catalyst A is better than Catalyst B for bottoms upgrading at constant coke yield, on both feedstocks. However, we were not able to determine molecular species cracked by Catalyst A and not cracked by Catalyst B. Therefore we have performed microdistillation on these four MAT effluents to characterize bottoms by Fisher-MS analysis. Bottoms yields given by microdistillation and Fisher results are given hereafter.

	Feedstock 1		Feedstock 2	
	Catalyst A	Catalyst B	Catalyst A	Catalyst B
Effluent weight (g)	0.525	0.741	0.451	0.438
Bottoms weight recovered (g)	0.084	0.189	0.118	0.165
Bottoms weight expected based on TBP effluent analysis (g)	0.088	0.193	0.136	0.175
350 °C in bottoms (TBP analysis) (weight %)	8	7	6	8

Fisher results, given on Tables II and III, are expressed in terms of weight % related to bottoms and in terms of weight % related to feedstock.

On both feedstocks, Catalyst B does not crack all monoaromatic species and leaves more saturated and diaromatic species in the heavy fraction than Catalyst A. These results appear consistent with the difference observed in yield structures.

Table II. Feedstock 1

Detailed HT-MS Fisher analysis	Catalyst A		Catalyst B		Feedstock analysis
	Weight %	Weight % × bottoms yield	Weight %	Weight % × bottoms yield	
Saturates	**22.5**	**1.87**	**24.6**	**3.63**	**43.0**
C_nH_{2n+2}	4.85	0.40	5.42	0.80	13.25
$C_nH_{2n, -2, -4, -6}$	17.60	1.46	19.14	2.82	29.75
Aromatics	**68.3**	**5.68**	**68.4**	**10.09**	**52.9**
Monoaromatics			**4.39**	**0.65**	**21.77**
C_nH_{2n-6}			0.82	0.12	8.07
C_nH_{2n-8}			1.34	0.20	8.31
C_nH_{2n-10}			2.23	0.33	5.39
Diaromatics	**3.9**	**0.32**	**11.7**	**1.73**	**10.6**
C_nH_{2n-12}	1.07	0.089	3.62	0.53	3.73
C_nH_{2n-14}	1.24	0.103	3.92	0.58	3.81
C_nH_{2n-16}	1.55	0.129	4.18	0.62	3.02
Triaromatics	**18.9**	**1.57**	**17.0**	**2.51**	**5.7**
C_nH_{2n-18}	10.72	0.89	10.33	1.52	3.30
C_nH_{2n-20}	8.18	0.68	6.63	0.98	2.36
Tetraaromatics	**21.6**	**1.80**	**16.5**	**2.43**	**6.2**
C_nH_{2n-22}	13.27	1.10	9.31	1.37	2.90
C_nH_{2n-24}	3.94	0.33	3.33	0.49	1.51
C_nH_{2n-26}	4.42	0.37	3.85	0.57	1.82
Pentaaromatics	**10.6**	**0.88**	**7.8**	**1.15**	**3.2**
C_nH_{2n-28}	7.18	0.60	4.86	0.72	1.73
C_nH_{2n-30}	3.45	0.29	2.98	0.44	1.49
Hexaaromatics	**8.3**	**0.69**	**6.3**	**0.93**	**2.5**
C_nH_{2n-32}	5.99	0.50	4.41	0.65	1.67
C_nH_{2n-34}	2.30	0.19	1.87	0.276	0.80
Heptaaromatics	**5.0**	**0.42**	**4.7**	**0.693**	**3.0**
C_nH_{2n-36}	1.86	0.15	1.53	0.226	0.71
C_nH_{2n-38}	3.18	0.26	3.16	0.466	2.30
Sulphur compounds	**9.2**	**0.77**	**7.1**	**1.05**	**4.1**
Benzothiophenes					0.56
Dibenzothiophenes					
$C_nH_{2n-16}S$	1.81	0.15	1.26	0.19	0.35
$C_nH_{2n-18}S$	0.49	0.041	0.32	0.05	0.11
$C_nH_{2n-20}S$	0.75	0.062	0.49	0.072	0.16
Naphtalenothiophenes					
$C_nH_{2n-22}S$	2.75	0.23	2.01	0.30	0.89
$C_nH_{2n-24}S$	1.06	0.088	0.86	0.13	0.43
$C_nH_{2n-26}S$	0.86	0.072	0.70	0.10	0.38
$C_nH_{2n-28}S$	0.65	0.054	0.53	0.078	0.29
Disulphurized compounds	0.82	0.068	0.91	0.13	0.92

Table III. Feedstock 2

Detailed HT-MS Fisher analysis	Catalyst A		Catalyst B		Feedstock analysis
	Weight %	Weight % × bottoms yield	Weight %	Weight % × bottoms yield	
Saturates	21.9	3.82	25.4	5.69	41.4
C_nH_{2n+2}	3.23	0.56	3.30	0.74	10.0
$C_nH_{2n, -2, -4, -6}$	18.70	3.26	22.10	4.95	31.40
Aromatics	70.1	12.22	67.2	15.05	50.4
Monoaromatics			3.33	0.75	10.89
C_nH_{2n-6}			0.70	0.16	3.06
C_nH_{2n-8}			1.05	0.24	4.15
C_nH_{2n-10}			1.58	0.35	3.68
Diaromatics	8.7	1.52	11.7	2.62	12.0
C_nH_{2n-12}	2.00	0.35	3.63	0.81	3.58
C_nH_{2n-14}	3.57	0.62	4.11	0.92	4.24
C_nH_{2n-16}	3.10	0.54	3.94	0.88	4.13
Triaromatics	17.7	3.09	15.3	3.43	8.5
C_nH_{2n-18}	8.97	1.56	7.92	1.77	4.79
C_nH_{2n-20}	8.73	1.52	7.40	1.66	3.73
Tetraaromatics	20.6	3.59	17.0	3.81	10.4
C_nH_{2n-22}	11.58	2.02	8.45	1.89	5.14
C_nH_{2n-24}	4.31	0.75	4.19	0.94	2.76
C_nH_{2n-26}	4.74	0.83	4.31	0.97	2.53
Pentaaromatics	9.9	1.73	8.6	1.92	4.8
C_nH_{2n-28}	5.99	1.04	4.85	1.09	2.69
C_nH_{2n-30}	3.90	0.68	3.73	0.84	2.13
Hexaaromatics	7.7	1.34	6.5	1.46	2.2
C_nH_{2n-32}	5.35	0.93	4.45	1.00	1.50
C_nH_{2n-34}	2.32	0.40	2.00	0.45	0.70
Heptaaromatics	5.6	0.98	4.9	1.10	1.6
C_nH_{2n-36}	1.83	0.32	1.57	0.35	0.56
C_nH_{2n-38}	3.75	0.65	3.33	0.75	1.06
Sulphur compounds	7.9	1.38	7.4	1.66	8.1
Benzothiophenes					1.76
Dibenzothiophenes					
$C_nH_{2n-16}S$	1.29	0.22	0.97	0.22	0.72
$C_nH_{2n-18}S$	0.34	0.059	0.25	0.056	0.19
$C_nH_{2n-20}S$	0.54	0.094	0.42	0.094	0.32
Naphtalenothiophenes					
$C_nH_{2n-22}S$	2.33	0.41	2.26	0.51	2.03
$C_nH_{2n-24}S$	1.01	0.18	1.01	0.23	0.95
$C_nH_{2n-26}S$	0.80	0.14	0.80	0.18	0.75
$C_nH_{2n-28}S$	0.58	0.10	0.57	0.13	0.45
Disulphurized compounds	1.05	0.18	1.11	0.25	0.97

Conclusion and perspectives

An original analytical tool has been developed. It consists on :
• A microdistillation, using a Buchi apparatus, to separate and provide fractions from a MAT total effluent, i.e. gasolines, LCO and bottoms from 0.6 g sample.
• A liquid microchromatography, which requires as little as 100 mg sample. It is particularly dedicated to the bottoms fractions to separate "saturated species" from "aromatics + sulphur compounds".
• The microfractions collected are then examined for molecular analysis, using mass spectrometry.
• The results on bottoms fractions from these microtechniques are in good agreement with those derived from more classical techniques.
  Information from both Fisher and O'Neal molecular analyses can help us to meet the challenge of ever higher bottoms upgrading, by a better understanding of the cracking reactions involved.
• We are still working to improve the repeatability on lighter fractions (gasolines and LCO). The table hereafter gives all the analyses we hope to be able to perform on MAT effluents in the near future.

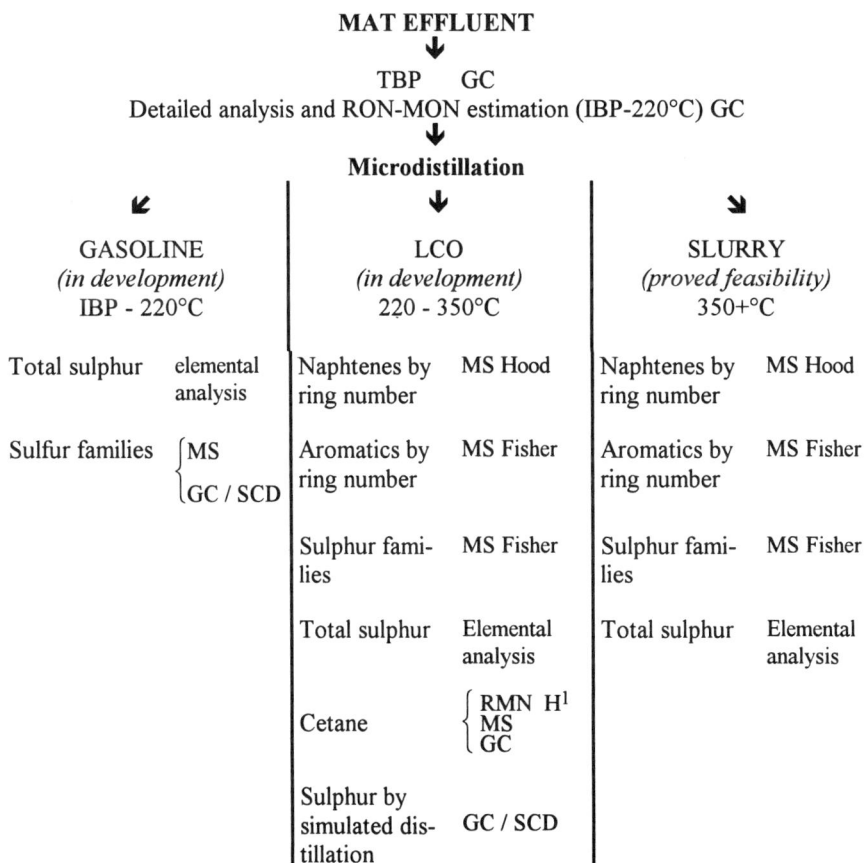

MAT EFFLUENT
↓
TBP   GC
Detailed analysis and RON-MON estimation (IBP-220°C) GC
↓
Microdistillation

GASOLINE *(in development)* IBP - 220°C		LCO *(in development)* 220 - 350°C		SLURRY *(proved feasibility)* 350+°C	
Total sulphur	elemental analysis	Naphtenes by ring number	MS Hood	Naphtenes by ring number	MS Hood
Sulfur families	$\begin{cases} MS \\ GC / SCD \end{cases}$	Aromatics by ring number	MS Fisher	Aromatics by ring number	MS Fisher
		Sulphur families	MS Fisher	Sulphur families	MS Fisher
		Total sulphur	Elemental analysis	Total sulphur	Elemental analysis
		Cetane	$\begin{cases} RMN\ H^1 \\ MS \\ GC \end{cases}$		
		Sulphur by simulated distillation	GC / SCD		

Acknowledgments

The authors aknowledge Gilbert Becquet and Michel Haÿs for performing the micro-analyses.

Literature cited

(1)   Bouquet, M. and Brument, J., Fuel Science and Technology Int'l, $\underline{8}$ (9), 961-986 (1990)
(2)   O'Neal, A., Anal. Chem., $\underline{31}$, 164 (1959).
(3)   Mauleon, J.L., and Courcelle, J.C., Oil & Gas Journal, Oct. 21, 1985.

RECEIVED June 17, 1994

Chapter 13

Solid-State Interaction Between NaY Zeolite and Various Oxides

J. Thoret, P. P. Man, and J. Fraissard

Laboratoire de Chimie des Surfaces, Unité de Recherche Associé
au Centre National de la Recherche Scientifique 1428, Université Pierre
et Marie Curie, 4 place Jussieu, tour 55, 75252 Paris Cedex 05, France

In this work we have shown the behavior of two types of oxide in their solid-state interaction with NaY zeolite. These differences in behavior are related to their distinct physical properties (melting point, solubility in hot water and possibly the vapour pressure at the interaction temperatures considered). Thus, the oxides with low melting points and soluble in hot water (V_2O_5 or MoO_3) migrate into the pores of the zeolite lattice when they are calcined in air at atmospheric pressure, within well-defined temperature and composition ranges. On the other hand, oxides with high melting points and insoluble in hot water (WO_3, ThO_2, UO_2, Nb_2O_5 or Ta_2O_5) do not enter the lattice pores under the same conditions.

In the last 20 years, many researchers have extensively modified zeolites. Initially, it was a question of replacing the Si or the Al of the lattice by another element (*1-8*). The reason for such studies is that for a low degree of substitution the modified zeolites can acquire interesting specific catalytic properties even if the position of the heteroatoms is uncertain, whether they are in the lattice or the cavities.

In his monograph (*9*), Rabo reported the first studies of solid-state reactions between zeolites, mainly Y zeolite, and some salts. These studies revealed either ion exchange or reversible occlusion of the salt. In cases of occlusion, the salt anion (halide, nitrate, or oxygenated chlorine anion) was usually located in sodalite cavities.

Recently, Karge et al. reported the solid-state ion exchange between alkaline chlorides and HZSM-5 or NH_4ZSM-5 (*10*) and between alkaline earth chlorides and mordenite (*11*). Evidence for exchange was obtained by i.r. spectroscopy, thermogravimetric analysis, titration of the evolved gases and mass spectroscopy associated with temperature-programmed gas evolution. The mass spectroscopic technique enabled them to distinguish between low- and high-temperature exchange reactions. Subsequently, they studied the solid-state interaction of HZSM-5 with several oxides (MnO, Cu_2O, CuO and NiO) and chlorides of the same metals (*12*). The degree of exchange of zeolite protons was considerably lower for the oxides than for the corresponding chlorides.

Kucherov and Slinkin reported solid-state reactions of H-mordenite and HZSM-5 zeolites with metallic oxides such as CuO (*13*), Cr_2O_3, MoO_3 and V_2O_5 (*14-18*). The

0097–6156/94/0571–0147$08.00/0

resulting samples were studied by e.p.r. spectroscopy. The authors showed that the metal cations migrate to cationic sites, where they are coordinately unsaturated.

Other solid-solid reactions have been observed, but in the presence of a stream of gas. Thus, Petras and Wichterlowa (*19*) studied the interaction of V_2O_5 and HZSM-5 at high temperature in the presence of nitrogen. Kanazirev and Price (*20*) doped HZSM-5 with gallium in the form of Ga_2O_5 in the solid state in the presence of hydrogen. Rouchaud and Fripiat (*21*) prepared zeolites containing Mo by impregnation. Dai and Lunsford (*22*) introduced Mo ions into Y zeolite by heating mixtures of NaY and $MoCl_4$ at 673 K, but with loss of NaY crystallinity. Johns and Howe (*23*) prepared mordenites loaded with Mo by adsorption of a $MoCl_5$ vapor phase into the mordenite, without loss of crystallinity. Recently, Huang, et al. (*24*) introduced vanadium in the form of $(VO)^{3+}$ into Y zeolite by heating a mixture of HY and V_2O_5 at 700 K in a stream of air containing water vapour. Jirka et al. (*25*) showed that the extent of Cu_2O insertion into NH_4Y in the solid state increased with the degree of hydration. Fyfe et al. (*26*) observed cationic exchange at ambient temperature in a mixture of LiA and NaA by simple physical contact between the zeolite crystals. Borbely, et al. (*27*) exchanged hydrated NaY with different chlorides by mechanical mixing at ambient temperature.

These zeolites, modified either by ion exchange or by occlusion of salts or oxides, often have interesting physicochemical and catalytic properties. We shall cite some of those: cracking of n-decane on Y catalysts with lanthanum (*28*); toluene alkylation by ethanol is considerably facilitated by ZSM-5 containing boron (*29*); the conversion of light paraffin gases to aromatic is increased by ZSM-11 loaded with Zn (*30*); gas phase octanol amination on Y impregnated with $(UO_2)^{2+}$ cations (*31*); Y zeolites loaded with ytterbium or europium have high catalytic activity for but-1-ene isomerization and for ethene hydrogenation depending on the pretreatment conditions (*32*) and Cu-exchanged Y zeolite (*33*) induces 90% selectivity in the conversion of acrylonitrile to acrylamide.

In this article we continue the study begun with the $NaY-V_2O_5$ system (*34*) by extending it to $NaY-MoO_3$, $NaY-WO_3$, $NaY-ThO_2$, $NaY-UO_2$, $NaY-Nb_2O_5$ and $NaY-Ta_2O_5$.

Experimental Procedures

The starting materials used are V_2O_5, MoO_3, WO_3, ThO_2, UO_2, Nb_2O_5 and Ta_2O_5 (Prolabo) and NaY zeolite (LZY 52 UOP). Before mixing with each of the above oxides, the NaY zeolite is put in a dessicator containing solution of NH_4NO_3 to saturate it with water. All the mixtures are defined by the ratio R = number of M atoms/ number of (Al + Si) atoms in the NaY lattice, where M = V, Mo, W, Th, U, Nb or Ta.

To simplify the text, we shall denote the different ratios as $R_{V_2O_5}$, R_{MoO_3}, R_{WO_3} We prepared about 2 g of each NaY-oxide mixture with various values of R. After grinding in an agate mortar for 15 min, the mixtures were deposited in flat rectangular refractory nacelles as a bed about 1 mm thick and 15 cm^2 in area. They were then raised to a temperature T between 550 and 950 K (heating rate 120 K/h) and held for 24 h. After the heat treatment the samples were left in the heating element until it cools to room temperature. They were then put back in the dessicator containing NH_4NO_3 until completely rehydrated, being weighed every 24 hr until the mass remains constant. At this stage, we can then evaluate the reduction of the mass of the mixture by thermal treatment; this can be explained by the obstruction of the zeolite pores by the oxide or by an amorphous phase which partially inhibits rehydration.

After these treatments, the samples were studied by X-ray analysis and by ^{29}Si Magic-Angle-Spinning (MAS) n.m.r. spectroscopy. To check the evolution of the void volume of the zeolite, xenon gas adsorption isotherm and ^{129}Xe n.m.r. were performed.

X-ray diffractograms were obtained with a Philips PW 1025/30 apparatus using the Kα radiation of cobalt. The value of the unit cell parameter (a_0) was determined to $\pm$ 5 10^{-3} Å by means of an internal silicon standard and an average of 25 reflections with known Miller indexes. ^{29}Si MAS n.m.r. spectra were obtained with a Bruker MSL 400 apparatus at 79.5 MHz. The framework Si/Al ratios were calculated by the usual method (*35*) after simulation of the experimental spectra.

Prior to xenon adsorption and ^{129}Xe n.m.r. experiments, the samples were pretreated in deep bed conditions at 393 K and evacuated to 10^{-4} torr for 15 h to ensure the elimination of most of the water of hydration. Moreover, Gedeon et al. (*36*) showed that when NaY is less than 15% hydrated the remaining water molecules are in the sodalite cages. Such a small amount of water cannot affect xenon adsorption, which occurs only in the supercages.

Measurements of xenon gas adsorption isotherms, N = f(P), were performed at temperatures between 273 and 299.5 K (this latter temperature being that of the ^{129}Xe n.m.r. probe). N is the number of Xe atoms adsorbed per gram of zeolite and P the equilibrium xenon pressure.

^{129}Xe n.m.r. experiments were performed according to previously described methods (*36*) on a Bruker CXP 100 apparatus at 24.3 MHz. The chemical shift (δ) of adsorbed xenon is given relative to the value for gaseous ^{129}Xe extrapolated to zero pressure.

Results

X-ray Diffraction at Ambient Temperature

Y-V$_2$O$_5$. In the NaY-V$_2$O$_5$ system, when the temperature T is below 600 K the initial phases coexist regardless of the composition and the heating time. After heat treatment and return to ambient temperature in the presence of water vapor we observe no variation of the mass of the system compared to its original state. This proves that rehydration has not been hindered by any obstruction of the cages, thus confirming the diffractogram results.

In the range 680 K < T $\le$ 750 K, the results depend on the ratio $R_{V_2O_5}$:

(i) For $R_{V_2O_5} \le 0.05$, the diagram contains only the lines characteristic of NaY diffraction, the lattice parameter going from 24.649 $\pm$ 5 10^{-3} Å for NaY ($R_{V_2O_5}$ = 0) to 24.673 $\pm$ 5 10^{-3} Å for $R_{V_2O_5}$ = 0.05 (Table I). This slight increase in a_0 (0.10%) may be explained by the introduction of the vanadium oxide into the pores, confirmed by the decrease of the mass of the system after thermal treatment under the same hydration conditions.

(ii) In the range 0.05 < $R_{V_2O_5} \le 0.10$, we detect a few V$_2$O$_5$ lines as well as those of NaY, less stronger than before because of a slight amorphization of the zeolite lattice (Table I).

(iii) Finally, for 0.1 < $R_{V_2O_5} \le 0.6$, at T = 750 K several phases are detected: small amounts of NaY, vanadium oxide, sodium vanadate NaV$_5^V$V^{IV}O$_{15}$, resulting from solid-solid reaction between NaY and V$_2$O$_5$ (Table II). After treating the $R_{V_2O_5}$ = 0.6 mixture at 870 K for 9 h, we detect a new monoclinic phase and vanadium-rich: $Na_5V^V_{11}V^{IV}O_{32}$ (Table II) and an amorphous phase which contains Si and Al.

Table I. X-ray Data of NaY-V_2O_5 System at 720 K for $R_{V_2O_5}$ = 0.05 and 0.10

$R_{V_2O_5}$ = 0.05			$R_{V_2O_5}$ = 0.10			
$d_{h,k,l}$ (Å)	I^a	NaY h,k,l	$d_{h,k,l}$ (Å)	I^a	NaY h,k,l	V_2O_5 h,k,l
14.247	vs	111	14.251	vs	111	
8.721	m	220	8.724	m	220	
7.437	m	311	7.438	m	311	
5.661	s	331	5.661	s	331	
4.747	m	511	4.748	m	511	
4.361	s	440	4.361	s	440	001
3.901	m	620	4.080	m		101
3.762	s	533	3.901	m	620	
3.561	w	444	3.764	s	533	
3.454	m	551	3.562	w	444	
3.296	s	642	3.455	w	551	
3.212	m	731	3.407	w		110
3.014	m	733	3.297	s	642	
2.908	m	660	3.211	w	731	
2.849	s	555	3.014	m	733	
2.758	m	840	2.908	m	660	440
2.707	w	911	2.849	s	555	

^{a} I for intensity and vs, s, m, and w mean very strong, strong, medium and weak intensity, respectively

Y-MoO3. As for the previous system, in the NaY-MoO3 mixture when T is below 600 K the starting phases coexist, whatever the composition. After heating and leaving in air at ambient temperature until saturated with water, there is no change in the mass of the system compared to the initial mass, which shows that the cages are not obstructed.

As for NaY-V_2O_5, when T is between 680 and 750 K the results depend on R_{MoO3} (Table III):

(i) For $R_{MoO3} \leq 0.075$, the X-ray diffractogram reveals a monophase region consisting of pure NaY whose unit cell parameter goes from $24.649 \pm 5 \; 10^{-3}$ Å for NaY to $24.691 \pm 5 \; 10^{-3}$ Å for R_{MoO_3} = 0.075. After heating then return to ambient temperature under the same conditions as above, the loss of mass is approximately proportional to R_{MoO3}, indicating that rehydration is limited, probably because of partial obstruction of lattice pores by MoO3.

(ii) In the range $0.075 < R_{MoO3} \leq 0.125$, only the structure of NaY remains, but with weakened X-ray intensities (slight amorphization).

(iii) For $0.125 < R_{MoO_3} \leq 0.2$, the zeolite structure is partially destroyed (strong amorphization). Moreover, we observe the appearance of one or several new crystalline phases which were not identified.

At T = 790 K, the amorphization of the zeolite begins when $R_{MoO_3} > 0{,}075$ and is completely destroyed for $R_{MoO_3} = 0{,}125$.

Table II. X-ray Data of NaY-V_2O_5 System for $R_{V_2O_5}$ =
0.2 at 790 K and $R_{V_2O_5}$ = 0.6 at 870 K

$R_{V_2O_5} = 0.2$ at 750 K				$R_{V_2O_5} = 0.6$ at 870 K		
$d_{h,k,l}$ (Å)	I^a	NaY h,k,l	V_2O_5 h,k,l	NaV_6O_{15} h,k,l	$d_{h,k,l}$ (Å)	I^a	$Na_5V_{12}O_{32}$ h,k,l
7.272	s			0 0 2	11.62	m	1 0 0
7.013	w			1 0 $\bar{2}$	6.98	vs	0 0 1
5.661	m	331			5.82	m	2 0 0
4.752	m	511		2 0 0	5.34	w	1 0 1
4.366	m	440	001		3.883	m	3 0 0
3.857	w			1 0 $\bar{4}$	3.653	w	1 0 $\bar{2}$
3.761	m	533			3.503	m	0 0 2
3.633	w			0 0 4	3.460	s	1 1 0
3.482	m			2 0 2	3.213	vs	0 1 1
3.391	s			1 1 $\bar{1}$	3.11	s	1 0 2
3.301	m	642			3.078	s	2 1 0
3.20	m			1 1 1	3.01	s	4 0 $\bar{1}$
3.07	s			1 0 4	2.988	s	2 1 $\bar{1}$
2.928	s			3 0 $\bar{4}$	2.653	m	3 1 $\bar{1}$
2.852	m	555			2.639	m	4 0 $\bar{2}$
2.76	w	840			2.577	m	1 1 $\bar{2}$
2.728	m			2 1 $\bar{3}$	2.442	w	4 0 1

^{a} I for intensity and vs, s, m, and w mean very strong, strong, medium and weak intensity, respectively

For T = 850 K the lattice remains but only for R_{MoO_3} values much smaller than before. Thus, we obtain:

(i) For $R_{MoO_3} \leq 0.075$, the system is monophasic, consisting of pure NaY whose X-ray intensities become weaker and weaker as R_{MoO_3} increases and appearance of an amorphous phase, there is no MoO_3 (Table III).

(ii) For $R_{MoO_3} > 0.075$, the zeolite structure is destroyed, with the appearance of new diffraction lines corresponding to one or several unidentified phases.

Table III. X-ray Data of NaY-MoO$_3$ System for R_{MoO_3} = 0.075 and 0.125 at 750 K and R_{MoO_3} = 0.05 at 850 K

$R_{MoO_3} = 0.075$ 750 K		$R_{MoO_3} = 0.100$ 750 K		$R_{MoO_3} = 0.05$ 850 K		
$d_{h,k,l}$ (Å)	I	$d_{h,k,l}$ (Å)	I	$d_{h,k,l}$ (Å)	I	NaY h,k,l
14.260	vs	14.263	m	14.264	m	111
8.731	m	8.736	w	8.729	m	220
7.445	m	7.449	w	7.446	m	311
5.662	s	5.663	m	5.661	m	331
4.751	m	4.754	vw	4.748	m	511
4.365	s	4.367	m	4.364	m	440
3.904	m	3.905	w	3.904	w	620
3.766	s	3.767	w	3.763	m	533
3.563	w			3.563	w	444
3.458	m	3.460	w	3.458	w	551
3.299	s	3.301	m	3.297	m	642
3.215	m	3.215	w	3.214	w	731
3.016	m	3.017	w	3.015	w	733
2.910	m	2.911	m	2.910	w	660
2.851	s	2.851	m	2.850	m	555
2.760	m	2.761	w	2.759	w	840

Y-WO$_3$, ThO$_2$, UO$_2$, Nb$_2$O$_5$, Ta$_2$O$_5$. For each oxide with NaY and up to T = 850 K the two initial phases coexist for R ≤ 0.05. Beyond this composition, happens a slight progressive amorphization of the lattice. Beyond this temperature and for $R_{WO_3} > 0.050$ the zeolite lattice is completely destroyed with appeareance of new unidentified phases.

Adsorption and n.m.r. The ^{29}Si MAS n.m.r. spectrum indicates that the Si/Al ratio of NaY zeolite is 2.46 ± 0.15. This ratio was not significantly modified for the NaY-V$_2$O$_5$ ($R_{V_2O_5}$ = 0.05) and NaY-MoO$_3$ (R_{MoO_3} ≤ 0.125) samples. As for thetreated samples containing an amorphous phase, the ^{29}Si MAS n.m.r. signal could not be used since the resolution is too low

The xenon adsorption isotherms were determined at 273 and 299.5 K for NaY and the following systems treated at 700 K: NaY-V$_2$O$_5$, NaY-MoO$_3$ and NaY-WO$_3$ with $R_{V_2O_5}$ = 0.05, R_{MoO_3} = 0.05 and 0.10 and R_{WO_3} = 0.05, respectively. For all these samples the log N = f(log P) plots, are almost parallel straight lines, those for the above mentioned samples lying below that of Y zeolite. For NaY-MoO$_3$ with two R_{MoO_3} values, the two adsorption isotherms are practically parallel, and the intercept decreases with R_{MoO_3} (Figure 1). These observations show that part of the zeolite lattice has been affected by the presence of the oxide: some cavities are blocked or their diameter is reduced.

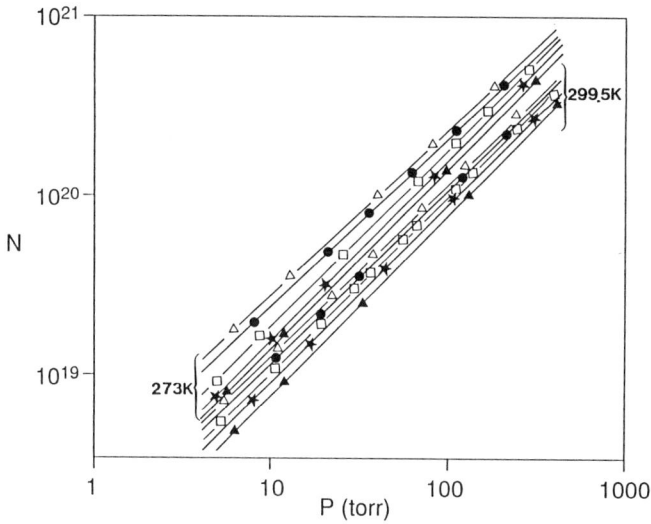

Figure 1. The xenon gas adsorption isotherm curves, or log-log plot of N (number of adsorbed xenon per gram of anhydrous zeolite) versus P (xenon pressure), for NaY, NaY-V$_2$O$_5$, NaY-MoO$_3$ and NaY-WO$_3$ at 273 and 299.5 K.

At 273 K: $\triangle$ NaY, $\bullet$ NaY-WO$_3$ (R$_{WO_3}$ = 0.05), $\square$ NaY-V$_2$O$_5$ (R$_{V_2O_5}$ = 0.05), * NaY-MoO$_3$ (R$_{MoO_3}$ = 0.05), and $\blacktriangle$ NaY-MoO$_3$ (R$_{MoO_3}$ = 0.10).

At 299.5 K: $\triangle$ NaY, $\bullet$ NaY-WO$_3$ (R$_{WO_3}$ = 0.05), $\square$ NaY-V$_2$O$_5$ (R$_{V_2O_5}$ = 0.05), * NaY- MoO$_3$ (R$_{MoO_3}$ = 0.05), and $\blacktriangle$ NaY-MoO$_3$ (R$_{MoO_3}$ = 0.10).

The specific heats of adsorption found for the systems treated at 700 K, NaY-MoO$_3$ and NaY-WO3 are close to those of NaY. Table IV summarizes the different values of the specific heats of adsorption for the systems studied between 273 and 299.5 K.

Table IV. Specific Heats of Adsorption of NaY, NaY-V$_2$O$_5$, NaY-MoO$_3$ and NaY-WO$_3$ Systems

Systems	ΔH (kJ/ mole)
NaY	18.81 ± 0.80
NaY-V$_2$O$_5$	14.71 ± 0.80
$R_{V_2O_5} = 0.05$	
NaY-MoO$_3$	17.14 ± 0.80
$R_{MoO_3} = 0.05$	
NaY-MoO$_3$	17.71 ± 0.80
$R_{MoO_3} = 0.10$	
NaY-WO$_3$	17.97 ± 0.80
$R_{WO_3} = 0.05$	

For NaY-V$_2$O$_5$, the value of the specific heat of adsorption indicates that the affinity of the xenon for the zeolite surface is weakened, particularly with $R_{V_2O_5} = 0.05$. This phenomenon could be due to a modification of the internal surface of the zeolite after thermal treatment of the zeolite-V$_2$O$_5$ mixtures. The [129]Xe n.m.r. spectra of xenon adsorbed at 299.5 K on the NaY-V$_2$O$_5$ ($R_{V_2O_5} = 0.05$), NaY-MoO$_3$ ($R_{MoO_3} = 0.05$ and 0.10) and NaY-WO$_3$ ($R_{WO_3} = 0.05$) systems consist of a single signal whose chemical shift, δ, is linearly dependent on N (xenon adsorption by the oxides is negligible, possibly zero), Table V. The extrapolation of the chemical shift to zero concentration (δ_S) is correlated with the mean free path of xenon in the supercages. The similar values of δ_S which are obtained for the various systems (59 ± 2 ppm), prove that the cavities accessible to xenon have been little modified (Figure 2).

Table V. Chemical Shift to Zero Concentration (δ_s) of NaY, NaY-V$_2$O$_5$, NaY-MoO$_3$ and NaY-WO$_3$ Systems

System	Intercept δ_s (ppm)	Slope of δ (ppm/10^{18}at.)
NaY	59.6 ± 2	0.0391
NaY-V$_2$O$_5$ $R_{V_2O_5} = 0.05$	58.4 ± 2	0.0451
NaY-MoO$_3$ $R_{MoO_3} = 0.05$	60.3 ± 2	0.0481
NaY-MoO3 $R_{MoO_3} = 0.10$	60.9 ± 2	0.0479
NaY-WO$_3$ $R_{WO_3} = 0.05$	58.7 ± 2	0.0486

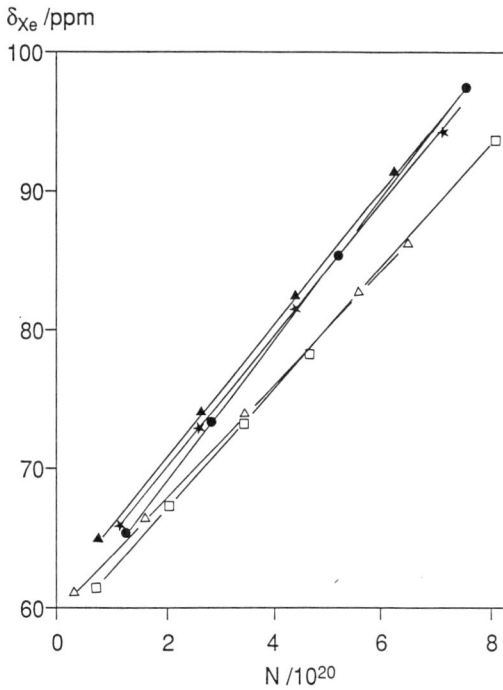

Figure 2. ^{129}Xe chemical shift δ versus N, the number of adsorbed xenon per gram of anhydrous zeolite: Δ NaY, $\bullet$ NaY-WO$_3$ (R$_{WO_3}$ = 0.05), $\square$ NaY-V$_2$O$_5$ (R$_{V_2O_5}$ = 0.05), * NaY- MoO$_3$ (R$_{MoO_3}$ = 0.05), $\blacktriangle$ NaY-MoO$_3$ (R$_{MoO_3}$ = 0.10).

The slopes of the $\delta = f(N)$ lines, referenced to 1 g of zeolite, are greater for the treated mixed systems than for pure NaY, which shows that for the same quantity of Xe per g of zeolite, the local xenon density is greater than that of NaY. Given that the value of δ_S is the same, this means that some supercages have been blocked by the oxide. The results obtained above are listed in Table V.

Discussion and Conclusions

The insertion of V_2O_5 or MoO_3 into NaY is related to their particular physical properties: relatively low melting points and solubilities in hot water (Table VI).

Up to 550 K each of these oxides coexists with NaY, whatever the composition. At about 730 K, the zeolite releases its water molecules, partially dissolving the V_2O_5 and MoO_3 oxides. We detect only pure NaY, whose diffraction line intensities are conserved for $R_{V_2O_5} = 0.05$ and for $R_{MoO_3} = 0.075$; there is an intensity decrease for $0.05 < R_{V_2O_5} \leq 0.10$ and for $0.075 < R_{MoO_3} \leq 0.125$ while weak new lines corresponding to the initial oxides appear. At the same time, the unit cell parameter increases; for example, 0.10% for $R_{V_2O_5} = 0.05$ and 0.18% for $R_{MoO_3} = 0.075$.

Table VI. Physical Properties of the Oxides

Oxides	Melting point (K)	Solubility in water (g/100 cc)
V_2O_5	963	0.85[a]
MoO_3	1063	2.05[b]
Nb_2O_5	1733	0
WO_3	1746	0
Ta_2O_5	2073	0
UO_2	2773	0
ThO_2	3323	0

[a] in cold water
[b] in hot water

Above 750 K but with higher concentrations, V_2O_5 or MoO_3 partially destroy the zeolite lattice with the formation of new phases. Thus, V_2O_5 which is a weakly acidic anhydride, reacts with Na^+ ions of NaY leading to a known crystalline phase: NaV_6O_{15} (37-39) for $R_{V_2O_5} = 0.2$ and T = 790 K (about 3% of the Na^+ ions per cell transform V_2O_5 to NaV_6O_{15}). On the other hand, with $R_{MoO_3} > 0.125$ and T = 790 K, it also appears one or several unidentified crystalline phases, which do not correspond to the sodium molybdates in the JCPDS files. Finally, at 870 K the NaY lattice is completely destroyed for much higher oxide concentrations. For example, a new phase $Na_5V_{12}O_{32}$ (*37-39*) appears ($R_{V_2O_5} = 0.6$). At this stage there are really solid-state reactions, with destruction of the initial phases and the appearance of new phases.

Tungsten oxide, thorium oxide, uranium oxide, niobium oxide and tantalum oxide (high melting point and zero solubility in hot water) behave differently with NaY. Like the other two oxides, these latter can coexist with NaY, whatever the composition, but up to higher temperatures (880 K) for $R \leq 0.05$. There is no variation of mass for each of these systems after their thermal treatment. Moreover, we detect no variation of the unit cell parameter. These facts indicate that all oxides do not penetrate into the pores of the zeolite. Above 880 K, they destroy the NaY lattice and create one or more as yet unidentified crystalline phases. Thus, these oxides are not incorporated into Y zeolite, whatever the temperature of the solid-solid reaction and the composition of the mixture.

Thus, for well defined temperature and composition ranges we have demonstrated the introduction into Y zeolite of compounds based on V or Mo in the form of species which are difficult to define: probably V_2O_5 or colloidal V_2O_5 or V_2O_4 or ions such as $(VO)^{3+}$ or $VO(OH)^+$ for V; MoO_3 or even trimers Mo_3O_9 or ions such as $(MoO_4)^{2-}$ for Mo. These different species enter probably at the beginning of thermal treatment, at about 573-673 K. At this temperature range, Y zeolite releases most of its water molecules, dissolving a significant fraction of the MoO_3 or V_2O_5 which assists their diffusion into the zeolite pores. Beyond 24 hr, the heat treatment time has no effect on the extent of insertion.

According to Karge and Beyer (*40*), the detailed mechanism of the migration of ions in NaY and of ion exchange remains a mystery. Although most insertions, even ion exchange in zeolites, are carried out in the presence of ambient water vapour, it is nonetheless true that insertions and exchanges can occur in an anhydrous atmosphere (*41*) or with compounds which are water insoluble, such as AgCl and Hg_2Cl_2 (*40*).

At solid-solid reaction temperatures (about 790 K) and with high oxide concentrations, the mechanism of the destruction of the NaY zeolite can be explained quite easily for V_2O_5 but not for MoO_3 or WO_3. When $R_{V_2O_5} = 0.2$, the X-ray patterns show the coexistence with three phases: NaV_6O_{15}, V_2O_5 and slightly amorphized NaY. At higher $R_{V_2O_5}$ and T ($R_{V_2O_5} = 0.6$ and T = 870 K), the lattice is completely destroyed with the formation of a sodium-rich vanadate: $Na_5V_{12}O_{32}$. With MoO_3 or WO_3 the destruction of the lattice does not appear to produce sodium molybdates or tungstates, indicating that those oxides appear to be weaker acids than vanadium anhydride.

It seems therefore that the melting point of the oxides and their solubility in hot water determine their interaction with Y zeolite.

In this work, we have studied zeolite-oxide interaction in the solid state and have found that V and Mo are inserted in forms which remain to be determined. Thus, in the range 680-750 K, the zeolite appears to accept levels of V or Mo corresponding to ratios $R_{V_2O_5} = 0.05$ and $R_{MoO_3} = 0.075$, without any modification of the lattice. This insertion can be explained by the physical properties of these oxides (low melting points and solubilities in hot water). These compounds based on V or Mo diffuse into the channels or supercages which they coat in the form of species which are difficult to define. It is, moreover, very likely that this insertion occurs at the very beginning of the anhydride-zeolite interaction (573-673 K). Above 750 K and depending on the increasing oxide concentration, we observe progressive lattice amorphization. Then above 790 K, we see its destruction with the formation of new phases which have been identified for V but which are so far unknown for Mo. These are true solid-state reactions in which the initial crystalline structures are destroyed and rearrangement leads to the appearance of new phases.

Tungsten oxide, thorium oxide, uranium oxide, niobium oxide and tantalum oxide, in contrast to the V_2O_5 and MoO_3 oxides, are not inserted into the zeolite, whatever the temperature of interaction or reaction and whatever the quantity of oxide involved. It would appear to be the same for all oxides with high melting points and zero solubility in hot water.

To date, no precise mechanism has been proposed for the breakdown of the lattice due to solid-state interaction of V_2O_5 or MoO_3 with NaY zeolite at ca. 750 K for compositions with R above 0.125.

Literature Cited

1   Flaningen, E. M.; Lock, B. M.; Patton, R. L. and Wilson, S. T. In *New Developments in Zeolite Science and Technology*; Murakami, Y., Iijima, A. and Ward, J. W., Eds.; Studies in Surface Science and Catalysis, Vol 28; Elsevier: Amsterdam, 1986, p. 103.

2   Guth, J. L.; Kessler, H. and Wey, R. In *New Developments in Zeolite Science and Technology*; Murakami, Y., Iijima, A. and Ward, J. W., Eds.; Studies in Surface Science and Catalysis, Vol 28; Elsevier: Amsterdam, 1986, p. 121.

3   Coudurier, G. and Védrine, J. C. In *New Developments in Zeolite Science and Technology* ; Murakami, Y., Iijima, A. and Ward, J. W., Eds.; Studies in Surface Science and Catalysis, Vol 28; Elsevier: Amsterdam, 1986, p. 643.

4   Newsam, J. M. and Vaughan, D. E. W. In *New Developments in Zeolite Science and Technology*; Murakami, Y., Iijima, A. and Ward, J. W., Eds.; Studies in Science and Catalysis, Vol 28; Elsevier: Amsterdam, 1986, p. 457.

5   Inui, T.; Miyamoto, A.; Matsuda, H.; Nagata, H.; Makino, Y.; Fukuda, K. and Okazumi, F. In *New Developments in Zeolite Science and Technology*; Murakami, Y., Iijima, A. and Ward, J. W., Eds.; Studies in Surface Science and Catalysis, Vol 28; Elsevier: Amsterdam, 1986, p. 859.

6   Mostowicz, R.; Dabrowski, A. J. and Jablonski, J. M. In *Zeolites: Facts, Figures, Future*; Jacobs, P. A. and Van Santen, R. A., Eds.; Studies in Surface Science and Catalysis, Vol 49; Elsevier: Amsterdam, 1989, p. 249.

7   Skeels, G. W. and Flaningen, E. M. In *Zeolites: Facts, Figures, Future*; Jacobs, P. A. and Van Santen, R. A., Eds.; Studies in Surface Science and Catalysis, Vol 49; Elsevier: Amsterdam, 1989, p. 331.

8   Davies, M. E.; Montes, C.; Hathaway, P. E. and Garces, J. M. In *Zeolites: Facts, Figures, Future*; Jacobs, P. A. and Van Santen, R. A., Eds.; Studies in Surface Science and Catalysis, Vol 49; Elsevier: Amsterdam, 1989, p. 199.

9   Rabo, J. A. In *Zeolite Chemistry and Catalysis*; Rabo, J. A. Ed.; ACS Monograph, Vol 171, Washington, DC, 1976, p. 332.

10   Beyer, H. K.; Karge, H. G. and Borbely, G. *Zeolites* **1988**, *8*, 79.

11   Karge, H. G.; Beyer, H. K. and Borbely, G. *Catalysis Today* **1988**, *3*, 41.

12   Wichterlowa, B., Beran, S. and Karge, H. G. In *Zeolites for the Nineties: Recent Research Reports*; Jansen, J. C., Moscou, L. and Post, M. F. M. Eds. Elsevier: Amsterdam, 1989, p. 87.

13   Kucherov, A. V. and Slinkin, A. A. *Zeolites* **1986**, *6*, 175.

14   Kucherov, A. V. and Slinkin, A. A. *Zeolites* **1987**, *7*, 38.

15   Kucherov, A. V. and Slinkin, A. A. *Zeolites* **1987**, *7*, 43.

16   Kucherov, A. V. and Slinkin, A. A. *Zeolites* **1987**, *7*, 583.

17   Kucherov, A. V., Slinkin, A. A., Beyer, H.K. and Borbely, G. In *Zeolites for the Nineties: Recent Research Reports*; Jansen, J. C., Moscou, L. and Post, M. F. M. Eds.; Elsevier: Amsterdam, 1989, p. 89.

18   Kucherov, A. V.; Slinkin, A. A.; Beyer, H. K. and Borbely, G. *J. Chem. Soc., Faraday Trans. I* **1989**, *85*, 2737.
19   Petras, M. and Wichterlowa, B. *J. Phys. Chem.* **1992**, *96*, 1905.
20   Kanazirev, V.; Price, G. L. and Dooley, K. M. In *Zeolite Chemistry and Catalysis* Jacobs, P. A.; Jaeger, N. I.; Kubelkowa, L. and Wichterlowa, B., Eds.; Studies in Surface Science and Catalysis, Vol 69; Elsevier: Amsterdam, 1991, p.277.
21   Rouchaud, J. and Fripiat, J. *Bull. Soc. Chim. Fr.* **1969**, *1*, 78.
22   Dai, P. S. E. and Lunsford, J. H. *J. Catal.* **1980**, *64*, 173.
23   Johns, J. R. and Howe, R. F. *Zeolites* **1985**, *5*, 251.
24   Huang, S.; Shan, C.; Yuan, C.; Li, Y. and Wang, Q. *Zeolites* **1990**, *10*, 772.
25   Jirka, I.; Wichterlowa, B. and Maryska, M. In *Zeolite Chemistry and Catalysis*; Jacobs, P.A.; Jaeger, N. I., Kubelkowa, L. and Wichterlowa, B., Eds.; Studies and Surface Science and Catalysis, Vol 69; Elsevier: Amsterdam, 1991, p. 269.
26   Fyfe, C. A.; Kokotailo, G. T.; Graham, J. D.; Browning, C.; Gobbi, G. C.; Hyland, M.; Kennedy, G. J. and de Schutter, C. T. *J. Am. Chem. Soc.* **1986**, *108*, 522.
27   Borbely, G.; Beyer, H. K.; Radics, L. and Sandor, P. *Zeolites* **1989**, *9*, 428.
28   Karge, H. G.; Mavrodinova, V.; Zheng, Z. K. and Beyer, H. K. *Applied Catalysis* **1991**, *75*, 343.
29   Chen, L. Z. and Feng, Y. Q. *Zeolites* **1992**, *12*, 347.
30   Annunziata, O. A. and Pierella, L. B. *Catalysis letters.* **1992**, *16*, 437.
31   Azzouz, A.; Nibou, D., Abbad, B. and Achache, M. *J. Mol. Catal.* **1991**, *68*, 18.
32   Baba, T.; Kim, T. J. and Ono, Y. *J. Chem. Soc. Faraday Trans.* **1992**, *88*(6), 89.
33   Kaushik, V. K. and Ravindranathan, M. *Zeolites* **1992**, *12*, 415.
34   Marchal, C.; Thoret, J.; Gruia, M.; Doremieux-Morin, C. and Fraissard, J. In *Fluid Catalytic Cracking II, Concepts in Catalyst Design*; Occelli, M. L. Ed. ACS Symp. ser. **452**, ACS, Washington, DC, 1991, p. 212.
35   Engelhardt, G.; Lohse, U.; Lippmaa, E.; Tarmak, M. and Magi, M. Z. *Anorg. Allg. Chem.* **1981**, *482*, 49.
36   Gédéon, A.; Ito, T. and Fraissard, J. *Zeolites* **1988**, *8*, 376.
37   Hardy, A.; Galy, J.; Cassalot, A. and Pouchard, M. *Bull. Soc. Chim. Fr.* **1965**, *4*, 1056.
38   Perlstein, J. H. and Sienko, M. J. *J. Chem. Phys.* **1968**, *48*, 174.
39   Chakrabarty, D. K.; Dipak, G. and Bismas, A. B. *J. Mat. Sci.* **1976**, *11*, 1347.
40   Karge, H. G. and Beyer, H. K. In *Zeolite Chemistry and Catalysis*; Jacobs, P. A., Jaeger, N. I.; Kubelkowa, L. and Wichterlowa, B., Eds.; Studies in Surface Science and Catalysis, Vol 69; Elsevier: Amsterdam, 1991, P. 43.
41   Chen, Q. and Fraissard, J. *Chem. Phys. Letters* **1990**, *169*, 595.

RECEIVED June 24, 1994

Chapter 14

Interactions in Alumina-Based Iron–Vanadium Catalysts under High-Temperature Oxidation Conditions

M. C. Springman[1], F. T. Clark[1], D. Willcox[2], and I. E. Wachs[3]

[1]Amoco Research Center, Naperville, IL 60566
[2]Department of Chemical Engineering, University of Illinois at Chicago, Chicago, IL 60607
[3]Zettlemoyer Center for Surface Studies, Department of Chemical Engineering, Lehigh University, Bethlehem, PA 18015

Alumina supported iron and vanadium oxides have been studied under a variety of conditions in an effort to understand the metal oxide-support interactions which occur during oxidative regeneration of spent hydrotreating catalysts. Iron and vanadium - two common contaminant metals found in petroleum feedstocks - were deposited on gamma alumina at levels of 5 to 10wt%, as oxides. The catalysts were subjected to calcination temperatures between 500°C and 900°C, and to atmospheres of dry air, wet air, NH_3, and SO_2/O_2. Characterization by XRD, N_2 desorption porosimetry, XPS, ICP, and Raman Spectroscopy was performed before and after treatment.

Under conditions of dry air calcination, vanadium oxides form hydrated surface metavanadates at low temperature, but at higher temperatures, vanadium migrates to the particle surface where crystalline vanadium pentoxide forms. Vanadium oxides catalyze the loss of surface area and a transition of the alumina phase from gamma to theta at elevated temperature. Iron oxides alone do not affect the surface area or pore structure of alumina under the conditions studied. In the presence of iron oxides, however, vanadium oxides form hydrated surface pyrovanadates at low calcination temperature, and at higher temperatures, vanadium migration and alumina pore sintering are enhanced. A transition from gamma to kappa alumina occurs when the catalyst containing both iron and vanadium oxides is calcined at elevated temperature.

Treatment at 900°C in atmospheres of wet air, NH_3, and SO_2/O_2 did not induce significant changes in the blank alumina or in catalysts containing only iron oxides. These catalysts retained the high surface area, gamma alumina phase. For catalysts containing only vanadium oxides, vanadium migrated to the surface of the catalyst particle at about the same rate regardless of atmosphere

0097–6156/94/0571–0160$08.00/0

after 900°C treatment. However, for catalysts containing both iron and vanadium oxides, the rate of vanadium migration to the particle surface after 900°C treatment increased in the order: dry air < wet air < NH_3 Surface area and pore volume are retained to a greater degree for the catalyst containing only vanadium oxides after NH_3 treatment compared with dry and wet air treatment. Analysis of the catalyst residues after high temperature treatment in the SO_2/O_2 environment indicates that exposure to SO_2 retards the destructive interactions which occur between vanadium and alumina in the absence of SO_2. These results are discussed in terms of existing models for alumina desurfacing and implications for laboratory aging of commercial catalysts.

Petroleum refining operations represent one of the most important industrial applications of catalysis. Of the three major U.S. catalyst markets in 1990, petroleum refining was the largest, representing 37% of the 1.8 billion dollar industry *(1)*. Hydrotreating catalysts are used at several stages of the refining process. The primary cause of deactivation in hydrotreating catalysts is coke deposition; accumulation of V, Fe, and Ni contribute as well. Once a catalyst has become deactivated, it must be replaced or regenerated in order to maintain activity. Regeneration is more desirable economically since it represents savings not only in catalyst replacement costs, but also in the disposal of spent catalyst, which often contains hazardous materials.

Regeneration of spent hydrotreating catalysts has been approached in a number of ways. Coke deposits can be removed by a controlled burn off, but metals such as nickel, iron, and vanadium often remain in substantial concentrations. These metals not only interfere with catalyst performance, but can be destructive to the catalyst itself, especially during the high temperature removal of coke. In regeneration studies which have focused on the effects of temperature, oxygen, and steam on the removal of coke and sulfur from spent hydrotreating catalysts *(2-7)*, it is generally concluded that full activity can be restored if coke deposits are removed under carefully controlled oxidation conditions, including moderate temperatures and the absence of steam. Vanadium, nickel and iron are deposited along with coke and sulfur and contribute to deactivation in hydrotreating catalysts by reducing hydrogenation and heteroatom removal activity, however, little is known about how these metals interact with the support and with each other under regeneration conditions.

Recent work performed at Amoco suggests that iron and vanadium are particularly harmful to hydrotreating catalysts during high temperature regeneration *(8)*. In this work, the activity of a spent resid hydrotreating catalyst for Ramscarbon removal, desulfurization, devanadation, and denitrogenation returned to near fresh catalyst levels after decoking, but the mechanical strength or attrition resistance of the catalyst deteriorated considerably. Samples of the regenerated (decoked) catalyst were subjected to an attrition test, after which the fines generated during the test and the extrudates remaining afterwards were analyzed by a number of techniques. The fines were found to be enriched in iron and vanadium compared to the extrudates, and mesopore sintering was observed in the fines while the pore size distribution of the extrudates remained similar to that of fresh catalyst.

While the removal of coke deposits has been shown effective in re-activating spent hydrotreating catalysts, the physical integrity of the catalyst must be maintained in order for the regeneration to be of practical use. In this work, we consider the interactions which occur between iron and/or vanadium and alumina under regeneration conditions as an extension of previous work which suggests that iron and vanadium play an important role in altering the physical characteristics of decoked catalysts. The metals are studied as dispersed metal oxides since this is the form they take during regeneration, and they are considered both alone (VO_x on alumina and FeO_x on alumina) and together (VO_x and FeO_x on alumina) in order to determine what effect, if any, they have on each other.

Experimental

A highly purified alumina, PHF gamma alumina (American Cyanamid,196.4m^2/g), was used as the starting material for all catalysts. Extrudates were crushed to a fine powder and dried at 538°C for 90 minutes prior to impregnation and calcination. Catalysts containing iron and/or vanadium oxides were prepared by incipient wetness impregnation. Metal loadings of 5-10 wt% as oxides were chosen because these are typical concentrations seen in spent resid hydrotreating catalysts. Spent resid hydrotreating catalysts were used in previous studies focusing on the feasibiity of regeneration *(8)* which formed the basis for the work currently being reviewed. Multiple 15 gram batches of each of the starting materials were prepared as follows:

5wt% and 10wt% Fe_2O_3: impregnation with a solution containing $Fe(NO_3)_2 9H_2O$ (EM Science) in methanol, and aging for one hour in room air, followed by drying at 121°C for 16 hours under nitrogen.

5wt% and 10wt% V_2O_5: impregnation with a solution containing vanadium triisopropoxide oxide (Alpha, 95-99%) in isopropanol and aging for one hour within a nitrogen purged glovebox, because of the moisture sensitive nature of the vanadium salt, followed by drying at 121°C for 16 hours under nitrogen.

5wt% Fe_2O_3 + 5wt% V_2O_5: impregnation with vanadium triisopropoxide oxide in isopropanol as described above. After drying, the sample was ground to its original consistency, and impregnated with $Fe(NO_3)_2 9H_2O$, also described above. The iron impregnation took place in a nitrogen purged glovebox as well.

Separate impregnations were required because the iron and vanadium salts were not soluble in the same alcohol. The vanadium impregnation was done first in order to insure dispersion of the vanadium.

Table I summarizes the preparation and nomenclature of these starting materials.

TABLE I

CATALYST PREPARATIONS AND NOMENCLATURE

CATALYST	DESCRIPTION
Blank	PHF gamma alumina
5% V_2O_5	5wt% V_2O_5 on PHF gamma alumina
5% Fe_2O_3	5wt% Fe_2O_3 on PHF gamma alumina
5% V_2O_5 + 5% Fe_2O_3	5wt% V_2O_5 + 5wt% Fe_2O_3 co-impregnated on PHF gamma alumina
10% V_2O_5	10wt% V_2O_5 on PHF gamma alumina
10% Fe_2O_3	10wt% Fe_2O_3 on PHF gamma alumina

Sample Preparation

Three gram portions of each of the starting materials, as well as blank alumina powder, were calcined at 500°C, 700°C, 800°C, and 900°C in a 60cm x 1.75cm inner diameter quartz reactor with upward air flow at 100cc/minute. The temperature was ramped to its final setting at 300°C/hour, and held there for 2 hours after which the reactor was cooled under nitrogen. This procedure was repeated for some catalysts in wet air and NH_3 environments as well. The wet air environment was created by bubbling dry air through distilled water at 100cc/minute, 25°C. The NH_3 environment was established by bubbling dry air through a 1wt% solution of NH_4OH at 100cc/minute, 25°C. Selected samples were treated at 900°C in an SO_2/O_2 environment in the course of determining SO_2 oxidation activity. Activity test results are reported elsewhere *(9)*. These catalysts were exposed to a 5% SO_2/5% O_2 (balance argon) gas mixture while the temperature was ramped to 900°C at 10°C/minute. Table II describes specifically all of the catalysts prepared.

Standards

V_2O_5 (Fisher Scientific) and Fe_2O_3 (Matheson Coleman and Bell) were used as reference materials.

Characterization

X-ray Photoelectron Spectroscopy: XPS spectra were recorded on a Surface Science SSX-100 XPS spectrometer using monochromatic Al K-α radiation.
Typical operating parameters were 600μ spot size at an analyzer pass energy of 100eV. A flood gun operating at about 2eV was used for charge

TABLE II

EXPERIMENTAL CONDITIONS

CATALYST	500°C			700°C			800°C			900°C			
	dry	wet	NH$_3$	dry	wet	NH$_3$	dry	wet	NH$_3$	dry	wet	NH$_3$	SO$_2$/O$_2$
Blank	x			x	x	x	x			x	x	x	x
5% Fe$_2$O$_3$	x			x	x	x	x			x	x	x	
5% V$_2$O$_5$	x			x			x			x	x	x	x
5% V$_2$O$_5$ + 5% Fe$_2$O$_3$	x			x			x			x	x	x	x
10% V$_2$O$_5$	x												x
10% Fe$_2$O$_3$	x												x

'x' indicates conditions at which each catalyst was treated.

neutralization during analysis. Iron and vanadium binding energies were referenced to Al 2p at 74.5eV since changes in the oxidation state of the aluminum would be negligible as a result of the oxidation experiments conducted in this study.

Inductively Coupled Plasma/X-ray Fluorescence: ICP analysis were performed on an Applied Research Laboratory Model 3560 ICP Atomic Emission Analyzer. Samples were prepared by acid digestion and microwave irradiation. Catalysts treated at 900°C were analyzed by X-ray fluorescence rather than ICP because of difficulty in dissolving them by the acid digestion technique. XRF analysis were performed on a Phillips model 1404 XRF Analyzer. Sample preparation involved fusing the catalysts with lithium borate mixtures and casting them into sample discs.

Raman Spectroscopy: Raman spectra were obtained with a Spex model 1877 spectrometer coupled to a Princeton Applied Research, model 1463 multichannel analyzer equipped with an intensified photodiode array detector (1024 pixels, cooled to -35°C). The 514.5 nm line of an argon laser (Spectra Physics) was used as the excitation source and the laser power at the sample was 10-50 mW. Samples were pressed onto KBr and spun at 2000 rpm during analysis.

X-ray Diffraction: A Scintag PAD V Powder Diffractometer equipped with an Ortec intrinsic germanium detector and 4 position sample chamber was used to obtain the XRD patterns. The instrument was configured with 2mm and 4mm incident beam slits, and 0.5mm and 0.3mm diffracted beam slits. Incident radiation was unfiltered Cu K-α radiation. Monochromatization was done in the detector. The Powder Diffraction File was used for phase identification.

Porosity: Nitrogen desorption porosimetry was performed with a Micrometrics Corporation model ASAP 2400 instrument, using the ASTM D32 procedure. Sample pre-treatment consisted of drying at 250°C in situ under vacuum for 16 hours.

Results and Discussion

Raman Spectra. Figure 1 shows the Raman spectra of the 5% V_2O_5 and 5% V_2O_5 + 5% Fe_2O_3 catalysts after calcination at 500°C. The major band at $\sim937cm^{-1}$ and weak bands at $\sim885cm^{-1}$, $\sim550cm^{-1}$, $\sim350cm^{-1}$, and $\sim220cm^{-1}$ seen in the 5% V_2O_5 spectrum are assigned to a hydrated surface metavanadate, $(VO_3)_n$, species (10). This agrees with solid state ^{51}V-NMR studies with similar catalysts (11). Broad bands at $\sim885cm^{-1}$ and $\sim800cm^{-1}$ are observed in the spectrum of the 5% V_2O_5 + 5% Fe_2O_3 catalyst after 500°C calcination. This shift in band intensity from $937cm^{-1}$ to $885cm^{-1}$ indicates a conversion of hydrated metavanadates to hydrated pyrovanadates (V_2O_7) (10) in the presence of iron oxides. Under the ambient conditions in which these spectra were obtained, the surface of the oxide support is hydrated and the surface metal oxide overlayer can be thought of as existing in an aqueous medium. As a result, the structure of the supported metal oxides follow the metal oxide aqueous chemistry. The appearance of the hydrated pyrovanadate may result from a change in the aqueous environment in the presence of iron oxides - aqueous iron oxides are basic in nature and cause an increase in the pH at which the surface possesses zero surface charge (pzc). The pyrovanadate appears to be more stable in the basic environment of the iron oxides.

Figure 2a-d shows Raman spectra of the 5% V_2O_5 catalyst after calcination at 500°C, 700°C, 800°C, and 900°C, respectively. The 500°C, 700°C, and 800°C spectra (Figures 2a,b,c) have broad bands at ~937cm^{-1}, ~825cm^{-1}, ~550cm^{-1}, and ~350cm^{-1} which are assigned to the tetrahedrally coordinated surface species (hydrated metavanadate). The 900°C spectrum (Figure 2d) has sharp bands at 994cm^{-1}, 700cm^{-1}, 525cm^{-1}, 480cm^{-1}, 400cm^{-1}, 310cm^{-1}, 290cm^{-1}, and 200cm^{-1}, which are attributed to crystalline vanadium pentoxide. A band at 250cm^{-1} also appears in the 900°C spectrum, and is assigned to a theta-alumina phase. The appearance of crystalline V_2O_5 after 900°C calcination can be attributed to two events. Migration of vanadium to the particle surface and a loss of surface area, which are discussed next, both serve to increase the surface vanadium concentration. Crystalline V_2O_5 forms when the surface vanadium concentration exceeds a monolayer (20wt% on gamma alumina) as a result of one or both of these events *(12,13,14)*.

XPS and ICP. XPS and ICP results for all of the vanadium containing catalysts are given in Table III. Table IV lists similar results for the iron containing catalysts. The V 2p3/2 XPS binding energies average 517.4 eV and are constant within the precision of the method. This binding energy indicates a +5 oxidation state, which is expected. The Fe 2p3/2 XPS binding energies for the dry air treated catalysts average 711.2 eV and are constant within the precision of the method, indicating a +3 oxidation state. The Fe 2p3/2 binding energies for the wet air and NH$_3$ treated 5% Fe$_2$O$_3$ catalysts are slightly higher at 712.4 +/- 0.1eV, which also indicate a +3 oxidation state, but suggests a possible change in speciation due to the wet air and NH$_3$ environments. The wet air and NH$_3$ treated 5% V_2O_5 + 5% Fe$_2$O$_3$ catalysts have Fe 2p3/2 binding energies of 711.7 +/- 0.1eV which lie between the values of the dry air catalysts and the wet air/NH$_3$ 5% Fe$_2$O$_3$ catalysts. This can be interpreted as another species or a combination of the 711.2 and 712.4 eV species. Mossbauer data, which are reported elsewhere *(9)*, confirm that all iron is in the +3 oxidation state, but confirmation of any change in speciation would require further characterization.

Bulk vanadium, iron, and aluminum concentrations as measured by ICP (and XRF in the case of the 900°C catalysts) remain constant within the error of the method under all conditions tested.

Figure 3 describes the V/Al and Fe/Al atomic ratios at the particle surface, as measured by XPS, relative to that at 500°C for the dry air treated catalysts.

The data in Figure 3 are based on the data in Tables III and IV. Surface iron concentration remains relatively constant at the temperatures considered, but surface vanadium concentration increases with increasing calcination temperature indicating migration of vanadium to the particle surface. The increase in surface vanadium concentration is more pronounced in the presence of iron oxides and the migration of vanadium begins at a lower temperature in

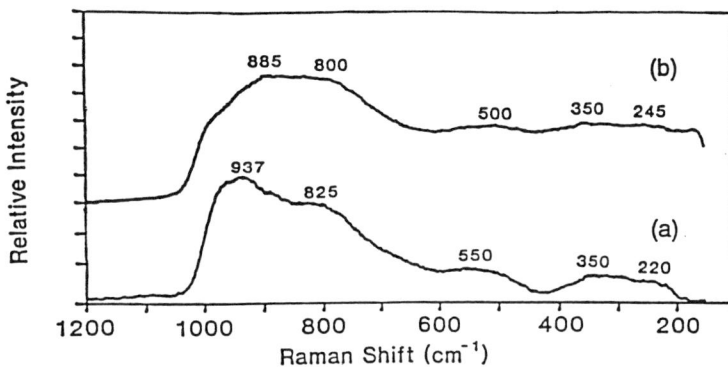

Figure 1: Raman spectra of catalysts treated at 500°C.
a) 5% V$_2$O$_5$ b) 5% V$_2$O$_5$ + 5% Fe$_2$O$_3$

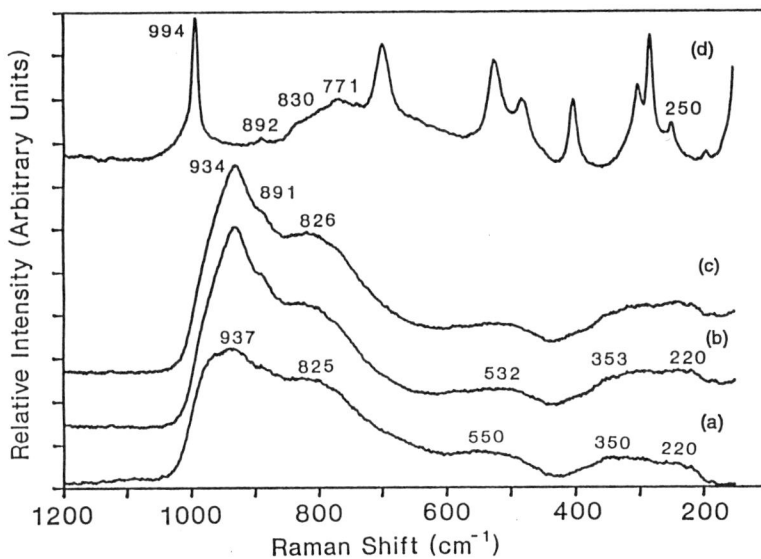

Figure 2: Raman spectra of the 5% V$_2$O$_5$ catalyst as a function of temperature.
a) 500$_o$C b) 700°C c) 800°C d) 900°C

TABLE III

XPS AND ICP RESULTS: VANADIUM

CATALYST	XPS BINDING ENERGY (V 2p3/2, eV)	V/Al	ICP V/Al
5% V_2O_5			
500°C, dry	517.1	.03	.051
700°C, dry	517.1	.03	
800°C, dry	517.3	.03	
900°C, dry	517.3	.11	.053
900°C, wet	517.5	.13	.049
900°C, NH_3	517.7	.11	.056
5% V_2O_5 + 5% Fe_2O_3			
500°C, dry	517.1	.02	.052
700°C, dry	517.2	.03	
800°C, dry	517.7	.04	
900°C, dry	517.2	.12	.054
900°C, wet	517.5	.15	.049
900°C, NH_3	517.5	.17	.056

dry= ambient air flowing at 100cc/minute
wet= ambient air bubbled through distilled water at
     100cc/minute, 25°C.
NH_3= ambient air bubbled through a 1 wt% solution of NH_4OH
     at 100cc/minute, 25°C.

TABLE IV

XPS AND ICP RESULTS: IRON

CATALYST	XPS BINDING ENERGY (Fe 2p3/2, eV)	Fe/Al	ICP Fe/Al
5% Fe$_2$O$_3$			
500°C, dry	711.2	.01	.069
700°C, dry	711.4	.01	
800°C, dry	711.3	.02	
900°C, dry	711.5	.02	.070
900°C, wet	711.5	.02	.071
900°C, NH$_3$	712.4	.02	.071
5% V$_2$O$_5$ + 5% Fe$_2$O$_3$			
500°C, dry	710.8	.01	.066
700°C, dry	711.5	.01	
800°C, dry	711.1	.01	
900°C, dry	711.1	.02	.067
900°C, wet	711.8	.02	.065
900°C, NH$_3$	711.6	.02	.060

dry= ambient air flowing at 100cc/minute
wet= ambient air bubbled through distilled water at
     100cc/minute, 25°C.
NH$_3$= ambient air bubbled through a 1 wt% solution of NH$_4$OH
     at 100cc/minute, 25°C.

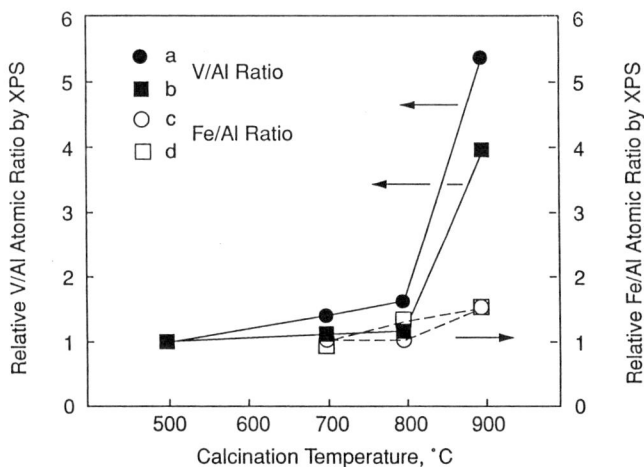

Figure 3: Particle surface concentrations as a function of temperature.
a) 5% Fe$_2$O$_3$ b) 5% V$_2$O$_5$ + 5% Fe$_2$O$_3$ c) 5% V$_2$O$_5$

the presence of iron oxides as well. The mobility of vanadium on catalyst surfaces due to the low melting temperature of vanadium oxides is well documented *(15,16,17)*. The reason for the enhanced migration of vanadium in the presence of iron is not ompletely clear, however, Raman data have shown that the surface vanadates formed in the presence and absence of iron oxides differ, and it is possible that these different vanadium oxide species vary in volatility. This would explain the onset of vanadium migration at different temperatures in the presence and absence of iron oxides.

Figure 4 shows surface V/Al ratios at 900°C in various atmospheres relative to that at 500°C in dry air. The dry air case has been discussed. The V/Al ratio at the surface of the 5% V_2O_5 catalyst particle remains relatively constant in dry air, wet air, and NH_3 environments as indicated by the relative V/Al ratios of 3.96, 4.33, and 3.67 respectively. Although some previous work suggests that vanadium migration will not occur in the absence of steam *(18,19)*, our results for this catalyst agree with studies indicating that vanadium will migrate during dry calcination at sufficiently high temperatures *(17)*. In the case of the 5% V_2O_5 + 5% Fe_2O_3 catalyst, the surface V/Al ratio increases to 6.0 times that observed at 500°C in dry air after treatment at 900°C in dry air, to 7.5 times after wet air treatment at 900°C, and to 8.5 times after NH_3 treatment at 900°C. These results again suggest that the surface vanadate formed in the presence of iron (pyrovanadate, V_2O_7) might behave differently than that formed in its absence (metavanadate, $(VO_3)_n$), since vanadium migration occurs at a lower temperature in the 5% V_2O_5 + 5% Fe_2O_3 catalyst, (Figure 3) and the migration appears more sensitive to steam and NH_3 atmospheres as well (Figure 4). Fe/Al ratios after treatment at 900°C in wet air and NH_3 were similar to those seen after 900°C treatment in dry air.

Porosity. Porosimetry results are listed in Table V. The blank alumina and 5% Fe_2O_3 catalysts are quite similar; both retain at least 80% surface area and at least 90% pore volume after 900°C calcination. The 5% V_2O_5 catalyst, however, retains only 38% of its pore volume and only 23% of its surface area after 900°C treatment, and the most dramatic changes occur in the 5% V_2O_5 + 5% Fe_2O_3 catalyst where, after 900°C calcination, only 3% and 5% of surface area and pore volume, respectively, are retained.

A similar trend was seen in the pore size distributions, which are reported in detail elsewhere *(20)*. The pore size distribution of the 5% Fe_2O_3 catalyst and blank alumina remain fairly constant over the entire temperature range, so the addition of iron oxides alone to the alumina support does not appear to have an effect on the alumina pore structure after calcination at temperatures as high as 900°C. The presence of vanadium does appear to affect the pore structure of alumina after high temperature calcination. Mesopore sintering is observed after 900°C calcination of the 5% V_2O_5 catalyst, and as early as 700°C calcination in the 5% V_2O_5 + 5% Fe_2O_3 catalyst.

TABLE V

POROSIMETRY RESULTS

CATALYST	SURFACE AREA (m^2/g)	PORE VOLUME (cc/g)	4V/A (A)
Blank			
500°C, dry	191	0.51	107
700°C, dry	183	0.50	109
900°C, dry	161	0.49	121
900°C, wet	152	0.50	132
900°C, NH_3	152	0.50	132
5% Fe_2O_3			
500°C, dry	191	0.49	103
700°C, dry	175	0.46	105
900°C, dry	157	0.44	112
900°C, wet	141	0.46	130
900°C, NH_3	139	0.46	132
5% V_2O_5			
500°C, dry	191	0.47	98
700°C, dry	189	0.47	99
900°C, dry	44	0.18	164
900°C, wet	40	0.22	220
900°C, NH_3	50	0.38	304
5% V_2O_5 +			
5% Fe_2O_3			
500°C, dry	190	0.44	93
700°C, dry	173	0.45	104
900°C, dry	6	0.02	*
900°C, wet	6	0.02	*
900°C, NH_3	4	0.08	*

* high analytical uncertainty using N_2 desorption with low surface area samples.

note: 4V/A = (4 X pore volume)/(surface area)

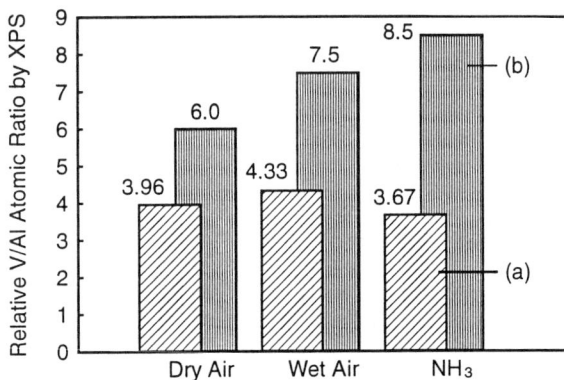

Figure 4: Surface vanadium concentrations in various atmospheres at 900°C (relative to analogous catalysts treated at 500°C in dry air).
a) 5% V_2O_5 b) 5% V_2O_5 + 5% Fe_2O_3

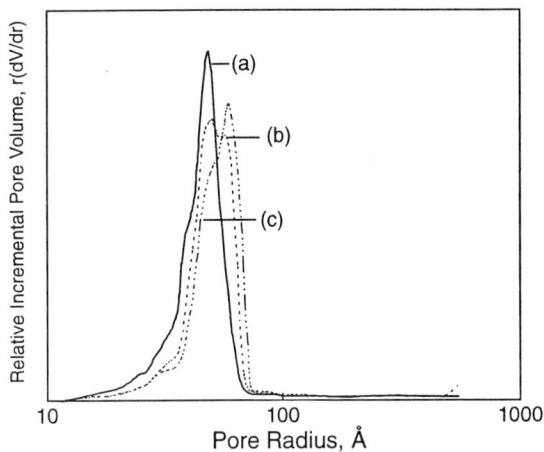

Figure 5: Effect of atmosphere on pore size distribution of the 5% Fe_2O_3 catalyst at 900°C.
a) dry air b) wet air c) NH_3

Note the correlation between the onset of vanadium migration and pore sintering. Migration of vanadium was not observed in the 5% V_2O_5 catalyst until 900°C calcination. This is the temperature at which pore sintering is first observed in the 5% V_2O_5 catalyst. In the case of the 5% V_2O_5 + 5% Fe_2O_3 catalyst, vanadium migration occurs at 700°C, the temperature at which pore sintering is first observed in this catalyst. The change in pore structure may be a consequence of either the vanadium migration itself, or the increase in local vanadium concentration which results from the migration, since the earlier the migration occurs, the earlier pore sintering occurs, and the more pronounced the migration, the more extensive the pore sintering. A recently proposed model for the mechanism of alumina pore sintering suggests that the loss of surface area occurs by successive elimination of water from hydroxyl groups residing on adjacent particles resulting in Al-O-Al linkages *(21)*. Vanadium is known to interact with alumina by reaction with surface hydroxyl groups *(22,23,24)*. In the context of the pore sintering model, the migration of vanadium might be visualized as a means of displacing hydroxyl groups and therefore accelerating the sintering process.

The pore size distributions of the 5% Fe_2O_3 catalyst at 900°C in dry air, wet air, and NH_3 atmospheres are shown in Figure 5. They are very similar to those of the blank alumina catalysts treated under the same conditions. A slight shift toward larger pores is seen in the wet air and NH_3 catalysts. This is also indicated by the 4V/A pore diameters (4 times pore volume per gram of catalyst/surface area per gram of catalyst) of 121A for the dry air catalyst and 130A and 132A for the wet air and NH_3 catalysts, respectively, as listed in Table V. Enhanced pore sintering in the presence of steam has been documented previously *(25)*. The small magnitude of the increase in pore diameters may be due to the low H_2O partial pressure used in these experiments (P_{H_2O} at 25°C = 25mmHg or 0.03 atm.). The NH_3 atmosphere appears to have very little, if any, effect on the pore structure of the blank alumina and 5% Fe_2O_3 catalyst. Figure 6 shows the pore size distributions of the 5% V_2O_5 catalyst treated at 900°C in various atmospheres. Again, a shift toward larger pores is seen in the wet air and NH_3 catalysts. In this case, the average pore diameter in the NH_3 treated catalyst (304A from Table V) is significantly larger than that of the dry air treated catalyst (164A). This could be interpreted to indicate a greater degree of pore sintering in the NH_3 treated catalyst. For the 5% Fe_2O_3, 5% V_2O_5, and 5% V_2O_5 + 5% Fe_2O_3 catalysts treated in dry air, increases in the average pore diameter were accompanied by decreasing pore volumes (Table V). However, the NH_3 treated 5% V_2O_5 catalyst retains about twice the pore volume of the dry air treated 5% V_2O_5 catalyst (0.38cc/g vs. 0.18cc/g, respectively), despite the increase in average pore diameter. Comparison of these results suggests that the type of changes in pore structure which occur with high temperature treatment vary in nature depending on the atmosphere in which the catalysts are treated. This result is consistent with a proposed mechanism for the destruction of cracking catalysts in the presence of vanadium and steam in which it is

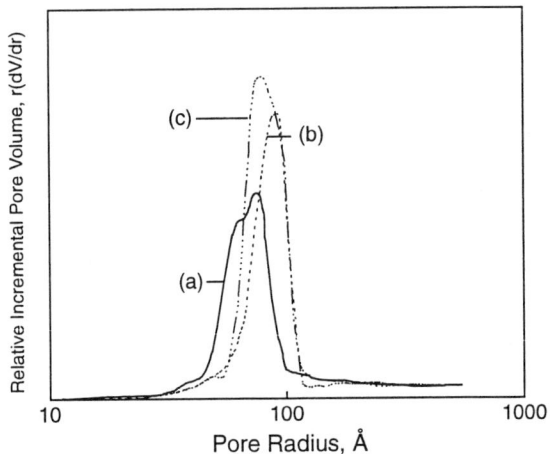

Figure 6: Effect of atmosphere on pore size distribution of the 5% V_2O_5 catalyst at 900°C.
a) dry air b) wet air c) NH_3

TABLE VI

ALUMINA PHASES AT 900°C

CATALYST	DRY AIR	WET AIR	NH_3	SO_2/O_3
Blank	gamma	gamma	gamma	gamma
5% Fe_2O_3	gamma	gamma	gamma	gamma
10% Fe_2O_3				gamma
5% V_2O_5	theta	theta	theta	gamma
10% V_2O_5				theta
5% V_2O_5 + 5% Fe_2O_3	kappa, delta	kappa, delta	kappa, delta	delta

suggested that the destruction of zeolite crystallinity results from the formation of vanadic acid and subsequent hydrolysis of the zeolite *(17)*. If this mechanism applies to the case of pure alumina, it is possible that the basic NH_3 may in some way neutralize the deleterious effect of the vanadic acid, resulting in a greater retention of pore volume.

XRD

The pore sintering and loss of surface area and pore volume shown in the previous section were accompanied by changes of alumina phase.   At temperatures as high as 800°C, all catalysts retained the gamma alumina phase. Table VI describes the alumina phases, as determined by XRD, after 900°C treatment in various atmospheres.

Comparison of the dry air phases suggests that vanadium catalyzes the conversion from gamma to higher temperature alumina phases.   This accelerated conversion of alumina phases in the presence of vanadium has been demonstrated by previous workers *(13)*. A similar effect has been observed in the conversion of titania (anatase) to titania (rutile), as well (26-29). Again, these results are consistent with the Johnson model for alumina desurfacing *(21)*. If we consider vanadium migration to be a mechanism by which surface hydroxyls are displaced, then the dehydration of the alumina and resulting phase transformation might be considered, in part, to be a function of vanadium migration.

Wet air and NH_3 environments do not affect the alumina phase transitions, however, the phase changes are less dramatic in the SO_2/O_2 environment than in the other atmospheres. For example, the 5% V_2O_5 catalyst subjected to calcination at 900°C underwent a conversion from gamma to theta alumina, but the gamma phase was retained when this same catalyst was treated at 900°C in an SO_2 environment. Conversion from gamma to theta alumina was observed in the SO_2 atmosphere only in the catalyst containing 10 wt% V_2O_5. For the 5% V_2O_5 + 5% Fe_2O_3 catalyst, calcination at 900°C resulted in conversion to a minor phase of delta alumina and a major phase of kappa alumina, but when this same catalyst was heated to 900°C in SO_2, the alumina phase changed only to delta, a less severe transition.

Mesopore sintering and the consequent change of alumina phase have been discussed in terms of vanadium migration and the Johnson model *(21)* for alumina pore sintering. The Johnson model proposes that pore sintering occurs by successive elimination of water from hydroxyl groups residing on adjacent particles. Vanadium migration may serve to displace hydroxyls and therefore accelerate the process.   It is unlikely that the SO_2 environment prevented migration since crystalline V_2O_5, which forms as a result of vanadium migration, was detected in the SO^2 treated catalysts.   A possible explanation for the apparent "buffering" effect of the SO_2 lies in the formation of aluminum sulfate, which forms under these conditions at the catalyst surface *(9)*. This aluminum sulfate surface layer may serve to reduce the number of interactions between vanadium and surface hydroxyls during vanadium migration, i.e. vanadium migration still takes place, but the interaction with surface hydroxyl groups is reduced, so that the elimination of water and resulting condensation of alumina are diminished on the presence of SO_2. This observation may find application in commercial catalyst testing. Under typical laboratory aging of FCC catalysts, the catalysts are steamed at high temperature in order to simulate regeneration conditions *(17,18,30)*. Heat treatment in an atmosphere of steam and SO_2 would be more realistic since SO_2 forms when sulfur which has deposited onto cracking catalysts is burned off in the regenerator. The results of our SO_2 oxidation experiments suggest that the SO_2 environment has a significant effect on the catalyst, and so should be included in the laboratory aging process.

Conclusions

The interactions between iron and vanadium oxides and alumina support vary with temperature and atmosphere, and in the presence or absence of additional metal oxides. A hydrated surface metavanadate forms when vanadium oxides alone are deposited on gamma alumina, and a hydrated surface pyrovanadate forms in the presence of iron oxides. At elevated temperatures, vanadium migrates to the particle surface and catalyzes pore sintering and alumina phase changes. Vanadium migration, pore sintering, and changes of alumina phase are enhanced in the presence of iron oxides, possibly due to the differing nature of the surface vanadates formed when iron oxides are present, and the deleterious effects of vanadium are diminished in the presence of SO_2. These results are consistent with an existing model for alumina desurfacing.

Literature Cited

(1) Barry, D. M., Spectrum: Chemical Industry, Focus on Catalysts: Petroleum Refining, Decision Resources (1991), 34.
(2) Arteaga, A., Fierro, J. L. G., Delanney, F., Delman, B. Applied Catalysis, **1986**, 26, 227.
(3) Furimsky, E. Applied Catalysis, **1988**, 44, 189.
(4) Arteaga, A., Fierro, J. L. G., Grange, P., Delman, B. Symposium on Advances in Hydrotreating , American Chemical Society Denver Meeting (1987).
(5) Arteaga, A., Fierro, J. L. G., Grange, P., Delman, B. Applied Catalysis, **1987**, 34, 89.
(6) Yoshimura, Y., Matsubayashi, N., Yokakawa, H., Sato, T., Shimada, H., Nishijima, A. Industrial and Engineering Chemistry Research, 30, 1091 (1991).
(7) Yoshimura, Y., Furimsky, E. Fuel, **1986**, 65, 1388.

(8) Clark, F. T., Hensley, A. L., Shyu, J. Z., Kaduk, J. A., Ray, G. J. in Bartholomew and Butt (Editors), Catalyst Deactivation 1991, Studies in Surface Science and Catalysis, Elsevier, Amsterdam, **1992**, Vol. 68, p. 29.

(9) Clark, F. T., Springman, M. C., Willcox, D., Wachs, I. E. Journal of Catalysis, **1993**, 139, 1.

(10) Deo, G., Hardcastle, F. D., Richards, M., Wachs, I.E., Hirt, A.M. in R. T. Baker and L. L. Murell (Editors), New Catalytic Materials, ACS Symposium Series 1990, Vol. 437, p. 317.

(11) Eckert, H., Wachs, I. E. Journal of Physical Chemistry, **1989**, 93, 6796.

(12) Bond, G. C., Tahir, S. F. Applied Catalysis, **1991**, 71, 1.

(13) Wachs, I. E., Jehng, J. M., Hardcastle, F. D. Solid State Ionics, **1989**, 32/33, 904.

(14) Nag, N. K., Massoth, F. E. Journal of Catalysis, **1990**, 124, 127.

(15) Pompe, R., Jaras, S., Vannerberg, N.G. Applied Catalysis, **1984**, 13, 171.

(16) Anderson, M. W., Suib, S. L., Occelli, M. L. Journal of Molecular Catalysis, **1990**, 61, 295.

(17) Wormsbecher, R. F., Peters, A. W., Masselli, J. M. Journal of Catalysis, **1986**, 100, 130.

(18) Woolery, G. L., Chin, A. A., Kirker, G. W., Huss, A. Jr. ACS Preprints, August/September 1987.

(19) Pine, L. A., Journal of Catalysis, **1990**, 125, 514.

(20) Springman, M. C. M. S. Thesis, University of Illinois at Chicago, 1993.

(21) Johnson, M. F. L. Journal of Catalysis, **1990**, 123, 245.

(22) delArgo, M., Holgado, M. J., Martin, C., Rives, V. Langmuir, **1990**, 6, 801.

(23) Sobalik, Z., Kozlowski, R., Haber, J. Journal of Catalysis, **1991** 127, 665.

(24) Bond, G. C., Konig, P. Journal of Catalysis, **1982**, 77, 309.

(25) Hopkins, P. D., Meyers, B. L. I&EC Product Research and Development, **1983**, 22, 421.

(26) Vejux, A., Courtine, P. Journal of Solid State Chemistry, **1978**, 23, 92.

(27) Saleh, R. Y., Wachs, I. E., Chan, S. S., Chersich, C. C. Journal of Catalysis, **1992**, 98, 102.

(28) Kang, Z.C., Bao, Q.X. Applied Catalysis, 26, 251 (1986).

(29) DelArgo, M., Holgado, M. J., Martin, C., Rives, V. Journal of Catalysis, 99, 19 (1986).

(30) Feron, B., Gallezot, P., Bourgogne, M. Journal of Catalysis, 134, 469 (1992).

RECEIVED June 17, 1994

Chapter 15

Activity and Deactivation in Catalytic Cracking Studied by Measurement of Adsorption During Reaction

Frank Hershkowitz, Haroon S. Kheshgi, and Paul D. Madiara

Exxon Research and Engineering Company, Route 22 East, Annandale, NJ 08801

A new research tool has been developed that allows in-situ measurement of the transient adsorption and coke deposition that occurs on zeolitic catalysts during short contact-time interactions with reactants. The tool is a microbalance pulse reactor that uses an *inertial* microbalance to accurately weigh a catalyst bed without regard to high velocities of vapor passing through. Zeolitic catalysts have been exposed to short pulses of vapor phase adsorbate carried through the catalyst bed on helium. Adsorption responses to pulses of n-decane, isopropylbenzene and triisopropylbenzene on $La^{3+}Y$ and other zeolites are reported. A regression / simulation method is used characterize the adsorption responses in terms of adsorption and reaction parameters. The deactivation of the zeolite by coke is found to follow a linear trend, as measured using a sequence of pulses.

The conversion and selectivity that are obtained in cat cracking result directly from a combination and interaction of very fast and nonuniform catalyst/oil contacting, a complex catalyst surface, a strong role of competitive adsorption and shape-selective diffusion, a broad range of feed components, and a catalyst whose activity is rapidly changing. Thus, understanding about fundamental mechanisms of cat cracking must fit within this contacting framework in order to be relevant to cat cracking. Clearly, performance of fundamental research requires some compromises in terms of experimental complexity. This paper presents a new approach that includes fewer compromises than has previously been possible.

The adsorption and diffusion of reactants in zeolitic catalysts are traditionally studied at lower temperature and with lower reactivity zeolites and then extrapolated to reaction conditions to understand their mechanistic

0097–6156/94/0571–0178$08.00/0

roles. For example, Jänchen and Stach (*1*) measure *n*-decane adsorption on sodium–Y zeolite, which is less active than the rare-earth or hydrogen forms used in cracking catalysts, and at temperatures from 290 to 570°K, which are well below the ~773°K temperatures used for cracking. Diffusion parameters may be measured by many means, including sorption, NMR, chromatographic methods (*2*), and frequency response methods (*3*), but the avoidance of reaction drives practitioners of each to use low reactivity (smaller) hydrocarbons, low reactivity zeolites, and temperatures usually below 500 °K. Some information about the role of size selectivity can also be obtained in reaction studies, a classic example being Nace's comparison of C_{16} species hexadecane and perhydropyrene (*4*).

In order to study deactivation, one must break up the reaction into deactivation increments. This may be done with any reaction variable that contributes to severity. Commonly used approaches include a combination of lower temperature, less reactive hydrocarbon, lower pressure, and very short contact time. For example, Magnoux et al. (*5*) studied deactivation using a low reactivity hydrocarbon (heptane) at moderate temperature (450°C) and pressure (30 kPa). Under these conditions deactivation could be stretched out to take between 2 minutes and 6 hours, time duration's very easily accessible in the lab. However, such conditions exclude the observation of any role of competitive adsorption or of mass-transfer limitations.

Our approach is to use a microbalance to weigh the bed of catalyst while it is being exposed to pulses of hydrocarbon vapor (*6*). A crucial, new aspect of this work is the use of an *inertial* microbalance that can determine bed mass without regard to high velocities of vapor passing through. Our reactor system is designed to capture more of the contacting complexity of cat cracking by using very short pulses of larger, more reactive hydrocarbons at near-commercial temperatures and near-commercial pressures. The microbalance aspect of the reactor permits on-line measurement of extent of hydrocarbon adsorption on the catalyst, and of coke, which is expected to be closely associated with deactivation. The rapid and transient measurement of adsorption during reaction creates the opportunity, for the first time, to deconvolve cat cracking mechanisms into their reaction, adsorption, and mass-transfer phenomena *during* reaction under commercially relevant conditions.

Reactor systems with the capability to weigh the catalyst bed have been built based on thermogravimetric analysis (TGA) instrumentation, for example by Lin et al. (*7*) and by Dean and Dadyburjor (*8*). The principal difference between our work and prior practice is that our balance is based on an inertial design. Thus, it can enclose a packed bed of catalyst through which vapor may be forced at considerable rate. This supports short contact cracking as well as sharp hydrocarbon concentration gradients, which are useful in quantifying mass transfer.

Experimental

The experimental system and methods used for this study have been described
in greater detail in Hershkowitz and Madiara (6), and will only be summarized
here. The microbalance that we use is of inertial design based on the principle
of an oscillating tapered element. This microbalance was first reported by
Patashnick et al. (9) as a device that could be used to quantify amounts of dust
suspended in a gas. Rupprecht and Patashnick Company ("R&P") custom
manufactured our TEOM microbalance (TEOM is a registered trademark of
R&P Co.) for us in 1988. A diagram of the TEOM, as it is adapted to
measuring the mass of a catalyst bed, is shown in Figure 1. At the time it
represented the first such high-temperature, high-flow, fast time resolution
microbalance in existence. The current design is comparable with R&P's
"Model 1500 Pulse Mass Analyzer." It uses mechanical energy, which is
controlled via a feedback circuit, to drive the tapered glass element to oscillate
at its natural frequency. As the bed's mass changes, so does the natural
frequency. Software and hardware provided by R&P performs the required
frequency counting and mass calculation as often as once per 0.11 seconds.

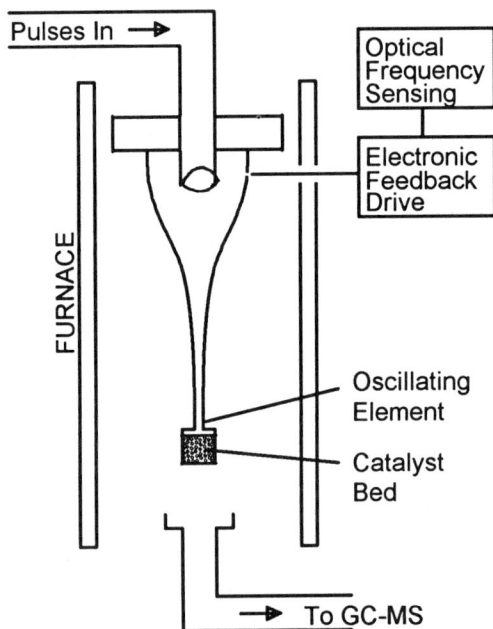

Figure 1. Schematic of the TEOM microbalance system as adapted by
R&P Co. for the measurement of the mass of a catalyst bed during
exposure to hydrocarbon pulses.

The block flow diagram for the microbalance pulse reactor (MPR) unit is shown in Figure 2. The gas manifold uses mass flow controllers to deliver metered amounts of air and helium to the unit. The carrier gas for reactions is helium that is introduced into the top of the preheat furnace. Air can also be directed to that location to regenerate the catalyst. Helium is also used as the reference gas in the thermal conductivity detector (TCD), and as the purge gas to the microbalance. Hydrocarbon is introduced into the system by syringe, through a septum located at the top of the preheat furnace. After volatilization, the pulse shape is measured by the TCD, and the pulse travels through the microbalance to the product valve. This four-way valve either directs product to vent or directly into the gas chromatograph's (GC) inlet splitter.

Zeolites used in this study include a lanthanum exchanged Y zeolite ($La^{3+}Y$) having Si/Al atomic ratio of 2.4, and LZY-82, a moderately dealuminated zeolite from Union Carbide. The LZY-82 has a Si/Al bulk atomic ratio of 2.7, with a framework Si/Al ratio closer to 5. The $La^{3+}Y$ has a sorption capacity of 14.6 g/100g octane at 100°C. Each zeolite sample has been made into larger particles by pilling at 17 MPa followed by crushing and sieving to a -170/+250 mesh size, resulting in a roughly 75 micron particle size. Each zeolite is calcined in the microbalance for several hours at 500°C prior to carrying out reactions.

Measurement of Adsorption During Reaction

With helium as carrier gas, the *gas-phase* difference in density between the helium and the hydrocarbon produces a mass response in the microbalance due to the feed pulse alone. When a bed of inert fused-quartz particles is used, the

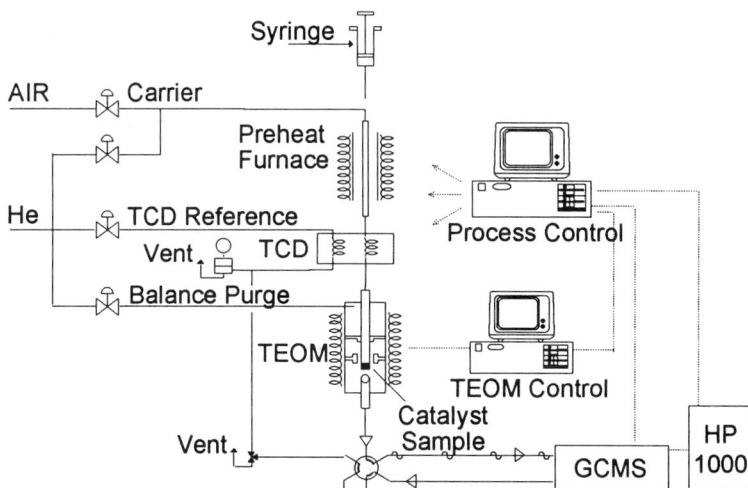

Figure 2. Flow Diagram for the Microbalance Pulse Reactor unit.

mass response quantifies the shape and size of hydrocarbon vapor pulses passing through. Although the TCD shown in Figure 2 can be used to measure pulse shape, we prefer, when possible, to measure that shape by performing a series of "blank" injections of hydrocarbon over fused quartz at the exact same conditions that are used for the adsorptive/reactive runs. This vapor peak shape, as observed in the TCD and TEOM are quite similar, as shown in Figure 3. Pulse width, as measured at half-height, is virtually identical. However, it is seen that there is an additional amount of tailing to the peak as observed in the TEOM. This tailing is minimized by appropriate design of the reactor system, including elimination of dead volumes and cold spots.

The performance of the microbalance reactor system is demonstrated by mass responses obtained with a catalyst bed containing 8 mg (dry) of $La^{3+}Y$ zeolite. The zeolite sample is diluted in 40 mg of similarly sized fused quartz prior to loading into the balance. Hydrocarbons are injected in amounts of 5μL, each over freshly regenerated (2 h with air at 500°C) catalyst. The carrier gas for the pulses is helium at a flow rate of 100 cm^3(STP)/min.

The extraordinary capability of the zeolite to rapidly sorb hydrocarbon (and of the balance to follow this sorption) is demonstrated by the response to a decane injection at 300°C, which is shown in Figure 4. Also shown is the

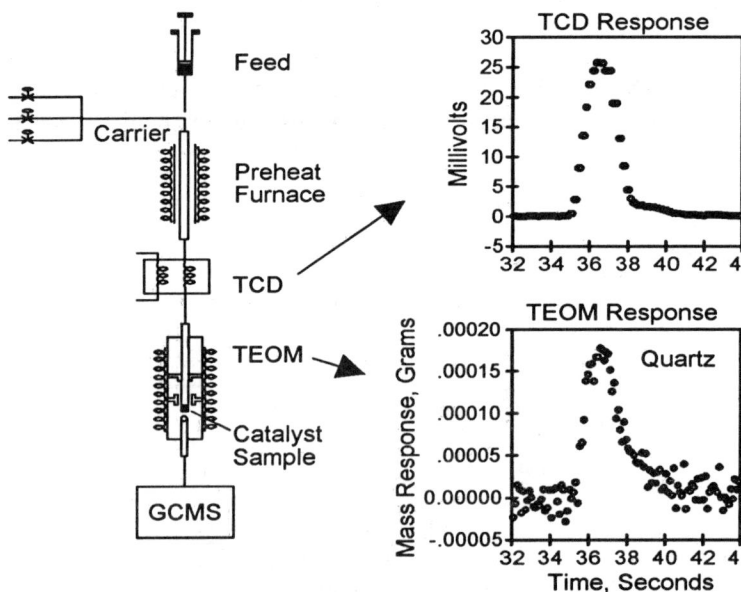

Figure 3. Response in the TCD and TEOM to 5μL Decane pulse fed to fused quartz bed at 300°C, 180 kPa, and carrier gas flow of 100 cm^3(STP)/min Helium.

mass response to an injection over non-adsorbing quartz at these conditions. At these conditions, hydrocarbon would enter the catalyst bed at a peak rate of 1.9 mg/s, which compares well to the 1.4 mg/s rate of mass increase in the bed. The zeolite crystals go from empty to near-saturation in about a half a second. The top of the mass response curve is roughly flat, with concentration of 0.8 mg, or 10% on zeolite.   When the decane pulse ends, desorption of decane begins. Desorption occurs over about 10 seconds, leaving a residue of about 0.16% "coke" on zeolite. Conversion, as measured using the FID detector on the GCMS, is 4%.

This contacting environment closely matches commercial FCC contacting in several important ways. Just as in an FCC unit, the catalyst rapidly switches from an environment without hydrocarbon to one with a high vapor pressure of hydrocarbon -- high enough to drive the zeolite rapidly to saturation. The duration of this exposure to high vapor pressure of hydrocarbon is short, two to three seconds in this particular experiment, and adjustable. After contacting with hydrocarbon, the catalyst is stripped with carrier gas for a variable length of time. Although gas phase residence times are necessarily very short (in range of 10 milliseconds), adsorbed phase residence times are longer, in the range of 1 second.

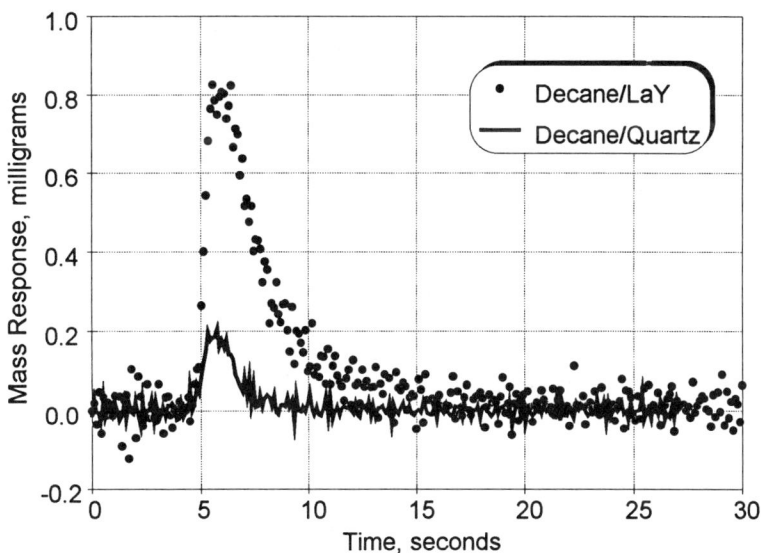

Figure 4.   Mass response to 5μL Decane pulse at 300°C, 180 kPa, and carrier gas flow of 100 cm^3(STP)/min Helium. Individual data points (●) for bed of 8 mg of La^{3+}Y zeolite in 40mg fused quartz; Line is average of 6 pulses over 50 mg of fused quartz alone.

The MPR system's ability to quantify the effects of shape selectivity during reaction are demonstrated in the comparison of cracking of isopropylbenzene to 1,3,5-triisopropylbenzene, as shown in Figure 5. In both cases, 8 mg (dry) of $La^{3+}Y$ is exposed to 5-μL pulses of hydrocarbon at 400°C, 180 kPa, and 100 cm^3(STP)/min of helium carrier gas flow. In both cases mass is also measured over a bed of fused quartz alone (which measures mass in vapor phase). For the isopropylbenzene, the mass measured over zeolite is significantly higher than that measured over fused quartz, indicating adsorption to an extent of about 0.3 mg. For the triisopropylbenzene, there is

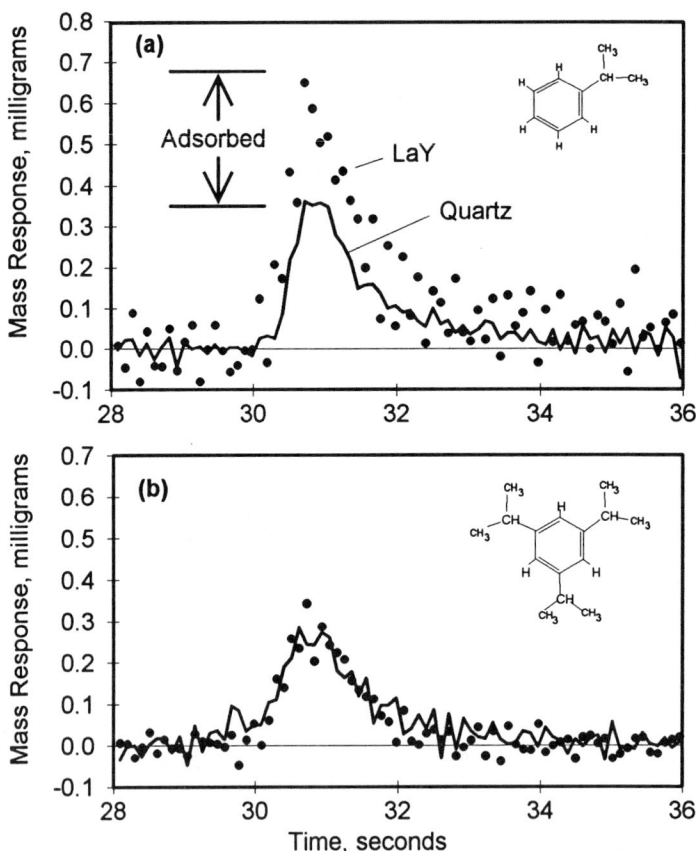

Figure 5.   Mass response to 5μL pulses of (a) isopropylbenzene or (b) triisopropylbenzene at 400°C, 180 kPa, and carrier gas flow of 100 cm^3(STP)/min Helium. Individual data points (●) for bed of 8 mg of $La^{3+}Y$ zeolite in 40mg fused quartz ; Line is for 50 mg of fused quartz alone. All except isopropylbenzene/$La^{3+}Y$ represent averages of 5 pulses.

no discernible difference between zeolite and fused quartz mass uptake, indicating that the zeolite is excluding the triisopropylbenzene from its internal pore structure. Consistent with this difference in adsorption behavior, the isopropylbenzene is converted to the extent of 38%, compared to ~0.1% for the triisopropylbenzene.

Interpretation of Transient Adsorption

In order to calculate adsorption parameters from the transient adsorption curves, we use a regression tool to find the best fit of a numerically-simulated curve to the measured curve, as shown in Figure 6. The numerical simulation uses the measured feed pulse, the measured conversion, unit operation parameters (e.g. flow rate) and adsorption/reaction model parameters to calculate a simulated transient adsorption curve. For the results discussed below, the adsorption/reaction model is based on Langmuir adsorption and first-order kinetics of decane reaction to products and coke. Plug flow in the catalyst bed is approximated as two perfectly stirred reactors in sequence. The adsorption data for this analysis is processed by averaging a number of mass measurements together in regions where the adsorption is changing little (6).

The quality of the fit of this model to a transient adsorption curve is excellent, as shown in Figure 7 for a decane/La^{3+}Y exposure identical to the one shown in Figure 4. Five parameters are regressed in this case: rate

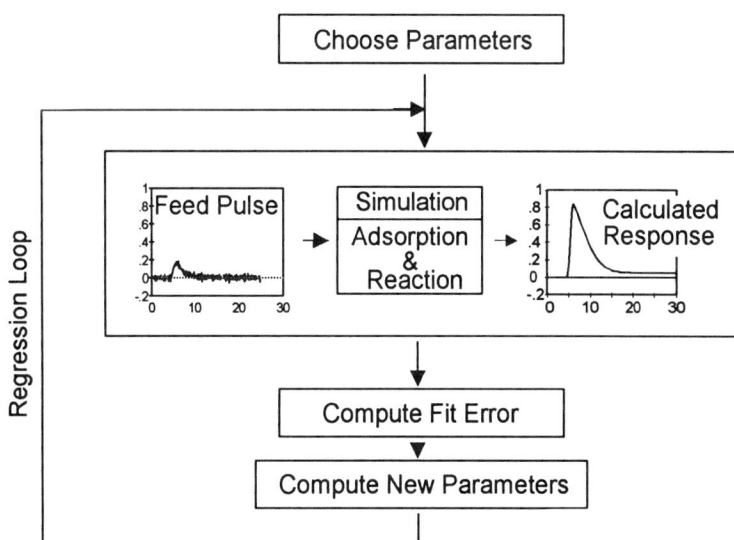

Figure 6. Schematic diagram for the numerical process used to regress adsorption and reaction parameters to fit the transient adsorption data.

constants for the cracking and coking reactions, Langmuir saturation and Henry's law parameters, and a time offset between vapor pulse and adsorption pulse. Fits of this quality are common, in this case with standard deviation of the adsorption data from the simulated response being 0.0192 mg. This particular regression results in Langmuir adsorption parameters of 10.3 g/100g for saturation loading and 469 for the dimensionless Henry's law constant (K of $q = Kc$, where q and c have equivalent units of grams or moles per cm^3, and zeolite concentration, q, is based on crystal volume).

Similar values *cannot* be directly measured at these temperatures by use of conventional means because cracking and coking would be extensive over the time scale of the conventional measurement. However, we can compare these values to extrapolated values. The saturation loading is within 10% of the value of 11.9 that we would extrapolate (using the methods of Dubinin (10)) from the zeolite's low temperature adsorption capacity. The Henry's law constant can be compared to the results reported by Jänchen and Stach (1), and Stach et al. (11), who measured adsorption isotherms of decane on a variety of faujasitic zeolites at temperatures of 0 - 200°C. We have calculated approximate Henry's law constants from those works by taking the ratio of

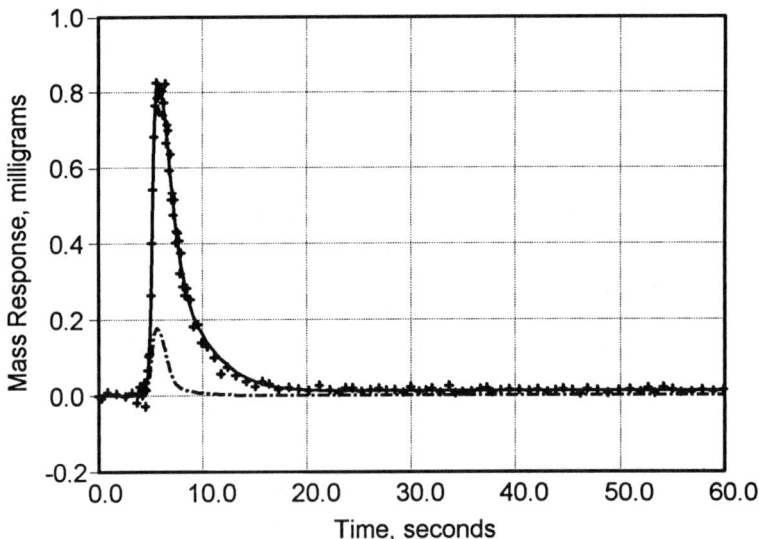

Figure 7. Fit of adsorption/reaction model to transient adsorption data for 5μL Decane pulse at 300°C, 180 kPa, and carrier gas flow of 100 cm^3(STP)/min Helium. Individual data points (✚) are mass measurements (or averages) over 8 mg of $La^{3+}Y$ zeolite in 40mg fused quartz; line (—) is simulated values; broken line (– · –) represents feed pulse shape.

reported vapor and adsorbed phase concentrations at the lowest measured loading. Our regressed, transient-adsorption value of the Henry's law constant is extremely close to the value that we would extrapolate from Stach et al.'s values for decane on a highly dealuminated Y zeolite ("US-Ex), as shown in Figure 8. Such a close fit is fortuitous, as none of the Jänchen and Stach zeolites exactly match our zeolite. The closeness of the fit confirms that what we measure is comparable in magnitude to the expected Henry's law constant.

Deactivation of Zeolites by Coke

In addition to the measurement of adsorption during reaction, the microbalance reactor quantifies of the amount of coke that is left on the zeolite after the reaction. We study the relationship between coke and deactivation by exposing the zeolite to a series of hydrocarbon pulses without intermediate regeneration. A conceptual diagram for this process is shown in Figure 9. Each exposure to hydrocarbon adds an increment of coke. As that coke is added, the activity, adsorption, and diffusion behavior of the zeolite may change, which in turn may result in different measured conversion, and different shape to the transient adsorption curve.

The LZY-82 zeolite is exposed to a set of seven pulses of decane in this manner, under conditions identical to that used for decane on $La^{3+}Y$ ($5\mu L$ decane, 8mg dry zeolite, 100 cm^3(STP)/min He, and 300°C). Zeolite is loaded

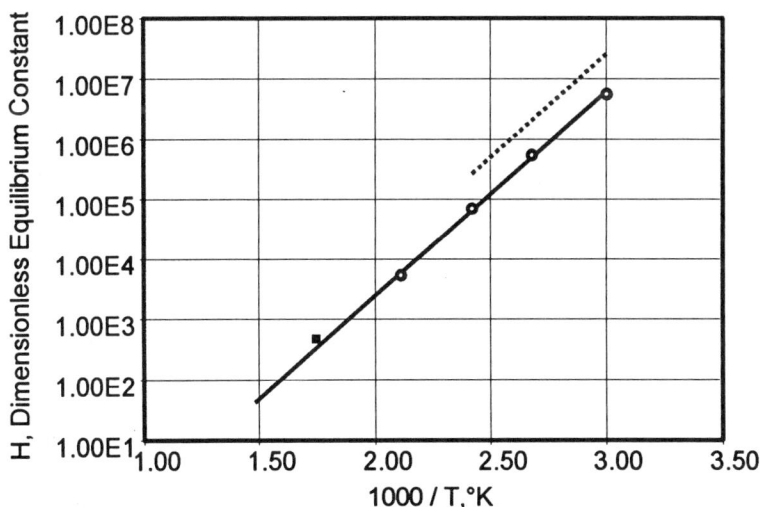

Figure 8. Comparison of the regressed Henry's law value from this work (■) to values approximated from the work of Janchen and Stach (*1*) and Stach et al (*11*) for adsorption of decane on NaY ($\cdots$) and "US-Ex" dealuminated faujasite (–O–).

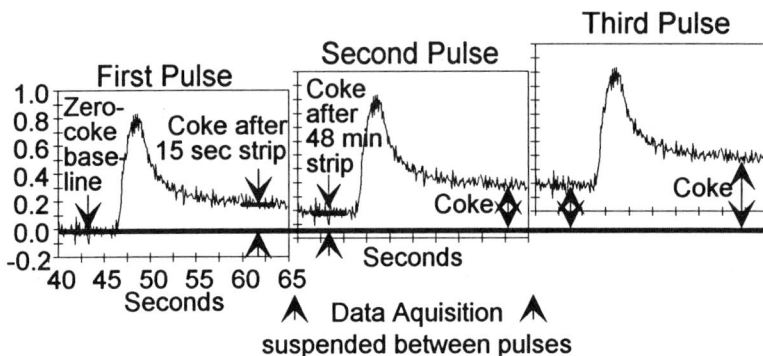

Figure 9.   Schematic diagram of the approach used to measure the effect of coke on subsequent activity, whereby the zeolite is exposed to a sequence of pulses without intermediate regeneration.

as before, and decane is injected at 48 minute intervals without intermediate regeneration. GC analysis for conversion is performed for the first pulse and for every second subsequent pulse. The carrier gas flow of 100 cm^3(STP)/min helium is maintained between pulses. For each pulse, the mass-recording software is run beginning 30 sec before the pulse, and continuing for about two minutes after the pulse. Additional data is not obtained because of the large file sizes that result from longer data acquisition times. The masses within all seven pulses are computed as the mass change relative to the initial mass of zeolite as measured immediately before the first pulse. The mass vs. time data is interpreted using the modeling methodology previously presented. In addition, we tabulate, using 10 second averages, the mass immediately preceding each pulse and 2 minutes following each pulse.

Using these tabulated mass values, we can assign two different measures of delta coke. Within any single pulse's data acquisition, we can define a delta coke based on the 2 minutes of stripping that occurs before the data acquisition is suspended. This is illustrated in Figure 9 using a stripping time of 15 seconds. Computing the mass change from the beginning of one pulse to the beginning of the next results in a delta coke value based on the 48 minutes of stripping that occurs between pulses. The delta coke values decrease slightly as stripping time increases from 2 to 48 minutes, as shown in Figure 10. The coke on zeolite is the difference in mass between the coked zeolite and the initial coke-free zeolite, and is equivalent to the summation of the 48 min delta-coke values. The delta coke based on 2 min stripping can also be summed to a cumulative coke yield. The additional stripping between 2 and 48 min leads to a widening gap between these two cumulative coke yields, also shown in Figure 10. Although coke based on the full stripping time is the theoretically correct measure of coke on zeolite, it relies on the difference

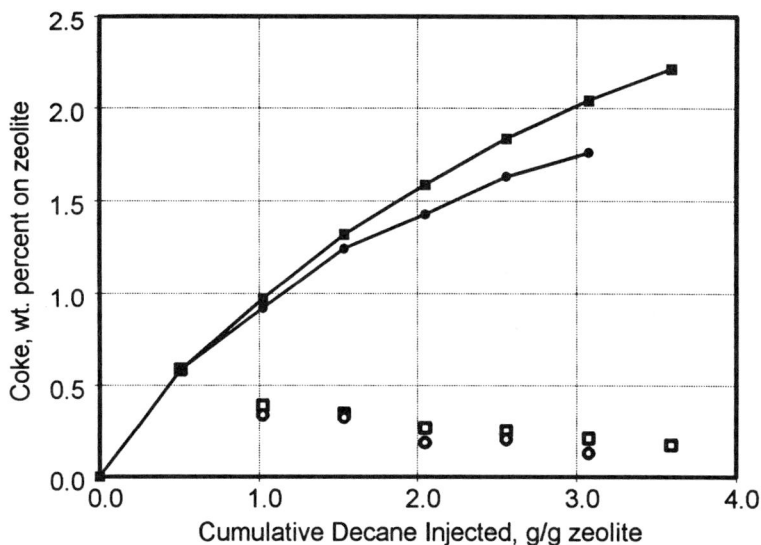

Figure 10. Coke, as wt% on zeolite, versus the cumulative amount of decane injected in a sequence of pulses. Delta-coke is represented by open points: □ for 2-min stripping and o for 48-min. Cumulative coke is represented by closed points: ■ for 2-min stripping and ● for 48-min.

between mass measurements that are made at times many hours apart. Thus, an extremely steady microbalance baseline is required in order to obtain the quality of results reported here. We use this coke (sum of 48 min delta coke) when reporting the coke on catalyst prior to each pulse. However, because of the scatter introduced by even small amounts of baseline drift, we use the 2 min delta coke as the measure of delta-coke for any *single* pulse.

It can be seen, in Figure 10, that the coke yield is highest for the first pulse and decreases as the zeolite is exposed to additional pulses. The resulting shape of the curve for total coke is not sigmoidal, as reported by Dadyburjor and Liu (*12*) for the cracking of hexadecane over similar zeolites at 500°C in a microbalance pulse reactor based on conventional (force-of-gravity-based) microbalance technology. Our results show little of the autocatalytic behavior observed in that work. We believe that this difference is related to the very different contacting between the two systems, in that the TEOM permits us to apply high concentrations on zeolite for short contact times. However, there are also significant differences in feed and temperature.

Each pulse has a shape that is very similar to that shown in Figures 4 and 7 for decane on $La^{3+}Y$ zeolite. The transient adsorption data are interpreted using the modeling methodology discussed previously. Because only four of the seven pulses have measured conversion values, the conversion values used as model input for all seven pulses are approximated using a linear regression of the measured conversion vs coke data, as shown in Figure 11. An initial set of regressions has been performed for all four adsorption/reaction parameters. In these regressions, the saturation loading (which is defined to include coke) is found to be relatively constant ($\sigma = 0.4$ g/100g) and to have no significant increase or decrease with increasing coke. A second set of regressions is then performed holding saturation loading constant at the average value from the first regression. This approach eliminates that scatter in the regressed Henry's law constant that originates with the cross-correlation of the two adsorption parameters.

The effect of increasing coke is primarily on parameters of zeolite reactivity, as shown in Figure 11. In this figure the coke yield (delta coke for 2-min strip), conversion, and Henry's law adsorption parameter are plotted against the coke on zeolite prior to the pulse. These parameters are expressed as a percent of the initial (zero-coke) value, with the initial values defined as the zero-coke intercepts for linear regressions of the parameters vs. coke on zeolite. These linear regressions are also shown in Figure 11.

The zeolite's activity toward coke formation decreases most rapidly with increasing coke. Activity toward decane conversion decreases almost as

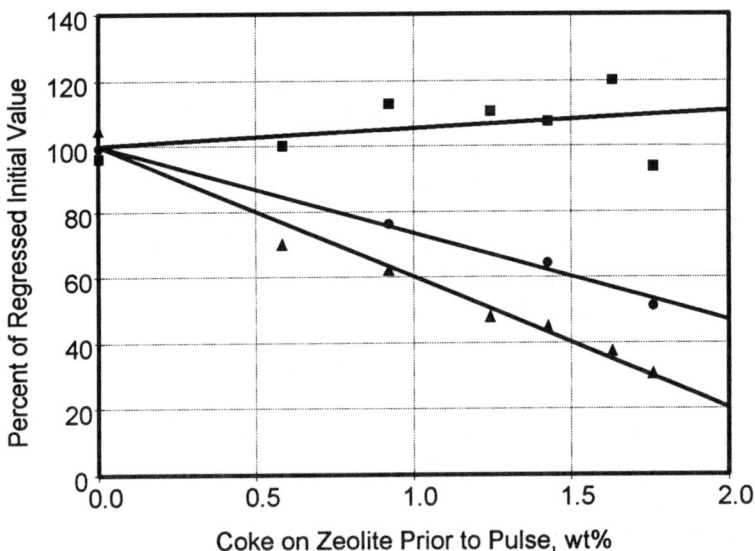

Figure 11. Trend of activity and adsorption parameters with increased coke on zeolite. Delta-coke yield (▲), conversion (●), and Henry's law constant (■) are expressed as a percent of their regressed, zero-coke value, and plotted with the regression lines.

rapidly. Under these conditions, both measures of activity appear to be declining linearly with increasing coke on zeolite. Extrapolation would predict that activity toward coke formation would reach zero at about 2.5 wt.% coke, compared to 3.8 wt.% coke for zero decane conversion. Thus, these two chemistries are not deactivated in exactly the same manner. With a saturation loading of 10 wt.% these zero-activity intercepts represent 25 and 38% of saturation loading, respectively. Alternately, one could say that each volume of coke is eliminating the activity associated with three to four average volumes of zeolite adsorption capacity. Defined as such, these deactivation ratio's are remarkably close to the ratio measured by Beyerlein et al. (*13*) of three framework alumina sites poisoned per sodium atom for LZY-82.

This activity decline is not related to any major change in hydrocarbon access to the zeolite, as indicated by the adsorption parameters. The constant saturation loading indicates that decane penetrates the entire coke-free volume as coke level increases. The Henry's law constant is also relatively constant with increasing coke on zeolite, as shown in Figure 11. The regression shows a slight, but unconvincing increase in K with coke. If real, this increase could mean that the coked zeolite has stronger adsorptive attraction to the decane. Alternatively, it could mean that the desorption of decane is somewhat encumbered by the presence of the coke, causing a slightly slower desorption. The adsorption/reaction model would represent this slower desorption as a larger Henry's law constant.

Conclusions

Characterization of the functioning of zeolitic catalysts, particularly for the reaction conditions of catalytic cracking, has been encumbered by the unavailability of methods to measure key performance features such as adsorption on the catalyst and the coke produced by reaction under realistic reaction conditions of short contact times, high hydrocarbon loading, and high reaction temperatures. The present apparatus and method fill this need by providing a means to measure, *in-situ*, this adsorption and coke. Use of an inertial microbalance in a pulse reactor permits the measurement of the time-variation of the amount of hydrocarbon adsorbed on catalyst during exposure to short-duration, high-concentration pulses of hydrocarbon. This transient mass measurement can be interpreted using simulation and regression tools to yield adsorption parameters that are comparable to values extrapolated from conventional measurements.

The direct measurement of coke permits the use of pulse sequences to study of how catalyst function changes as coke is deposited. In the specific example presented here (decane over LZY-82 at 300°C), activity toward coke and decane conversion both decrease linearly with increased coke on zeolite. The slope of this decline is comparable to a deactivation rate that would have each volume of coke deactivate 3 to 4 volumes of zeolite adsorption capacity. Using the simultaneous measurement of adsorption during reaction, we show that this deactivation is not related to a change in zeolite adsorption strength or amount as the coke is deposited.

Acknowledgment

The authors wish to acknowledge our appreciation of R&P Co.'s contribution in adapting their technology to our application. We are particularly grateful for the help of Dave Hassel and Harvey Patashnick, who personally undertook the challenge and made our balance a reality.

Literature Cited

1. Jänchen, J.; Stach, H. *Zeolites* **1985**, *5*, 57-59.
2. Ruthven, D. M. *ACS Symp. Ser.* **1983**, *218*, 345.
3. Yasuda, Y.; Yamamoto A. *J. Catalysis* **1985,** *93*, 176-181.
4. Nace, D. M. *Ind. Eng. Chem., Prod. Res. Dev.* **1970**, *9*(2), 203-209.
5. Magnoux, P.; Cartraud, P.; Mignard, S.; Guisnet, M. *J. Catalysis* **1987**, *106*, 235-241.
6. Hershkowitz, F; Madiara. P. D. *Ind. Eng. Chem. Res.* **1993**, *32*(12), 2969-2974.
7. Lin, C.; Park, S. W.; Hatcher, W. J. Jr. *Ind. Eng. Chem., Process Des. Dev.* **1983**, *22*(4), 609-614.
8. Dean, J. W.; Dadyburjor, D. B. *Ind. Eng. Chem. Res.* **1988**, *27*(10), 1754-1759.
9. Patashnick, H.; Rupprecht, G.; Wang, J. C. F. *Preprints, ACS Div. Petr. Chem.* **1980**, *25*, 188-193.
10. Dubinin, M. M. *Chem Rev.* **1960**, *60*, 235.
11. Stach, H.; Lohse, U.; Thamm, H.; Schirmer, W. *Zeolites* **1986**, *6*, 74-90.
12. Dadyburjor, D. B.; Liu, Z. *Chem. Eng. Science* **1992**, *47*(3), 645-651.
13. Beyerlein, R. A.; McVicker, G. B.; Yacullo, L. N.; Ziemiak, J. J. *Preprints, ACS Div. Petr. Chem.* **1986**, *31*(1), 190-197.

RECEIVED June 17, 1994

Chapter 16

Effect of Support Texture on Pt Dispersion and CO-Oxidation Catalyst in Fluid Catalytic Cracking

T. Zalioubovskaia[1], A. Gédéon[2], J. Fraissard[2], E. Radchenko[1], and B. Nefedov[1]

[1]All-Russia Research Institute of Oil Refining (VNIIP), 6 Aviamotornaja ul., 111116 Moscow, Russia
[2]Laboratoire de Chimie des Surfaces, Unité de Recherche Associé au Centre National de la Recherche Scientifique 1428, Université Pierre et Marie Curie, 4 place Jussieu, tour 55, 75252 Paris Cedex 05, France

Electron microscopy and hydrogen adsorption have been used to study the active component dispersion on alumina-based, commercial, CO oxidation catalysts. CO oxidation activity and the attrition resistance of the samples were studied as a function of the suppprt structure. The effect of Pt content on the support, and in the total system, of the CO conversion has been also studied. These investigations have shown that the texture characteristics of the alumina support (surface area, pore volume) and the phase composition exert a considerable effect on its physico-mechanical properties, active component (Pt) dispersion and CO oxidation catalyst activity. We have shown that the deactivation of the CO combustion catalysts in the cracking process occurs due to platinum dispersion decrease.

The maximum supported metal dispersion is achieved on the support with surface area 120 - 180 m^2/g, pore volume 0.35 - 0.45 cm^3/g for industrial catalysts, optimum Pt content on the support is 400 - 700 ppm. Our results permitted us to improve the method of the support preparation and to find the optimum procedure for achieving high Pt dispersion and high CO oxidation activity of the catalyst in FCC.

0097–6156/94/0571–0193$08.00/0

CO oxidation catalysts are successfully used in commercial FCC units as separate additives to cracking catalysts. Such an additive promotes the composition of a CO to CO_2 in the dense phase of the regenerator. It prevents afterburning (uncontrolled CO oxidation) in the dilute phase, which can cause metallurgical damage to equipment. Maximizing CO oxidation also minimizes flue gas emissions.

In CO oxidation catalysts, alumina plays the role of a support on which the active component (mainly Pt) is deposited by impregnation (1,2). It is possible to obtain different distribution patterns of the active component in the catalyst as its quantity is less than required for the support surface monolayer cover.

We report here some fragmentary results on the effect of texture and phase composition of the commercial alumina support on active component dispersion and CO oxidation catalyst activity under conditions simulating commercial FCC operation. We have also studied the influence of platinum content on the support on the CO conversion.

Experimental conditions

Preparation of the catalysts. Commercial, precipitated, microspheric alumina (MOA) and technical grade alumina, calcined at 1,000-1,200° C, (G-00) were used as a support for the oxidation catalysts. The microspheric alumina MOA was prepared by precipitation. Alumina support (G-00) was prepared by calcination of alumina hydrate at 1,000 - 1,200° C. The samples KO-9, KO-9M, KO-11 are based on MOA, while the G-00 support was used for the OGR-1 catalyst. Platinum was deposited by impregnation of these supports with appropriate amount of H_2PtCl_6, then dried at 1,200° C. For the oxidation reaction, the platinum content ranged from 500 to 1,000 ppm. A sample with 1 ppm of Pt on MOA was also prepared. These samples were compared with a commercial one named Competitor.

In these samples, the Pt concentration was too low for determining the Pt dispersion and the size distribution of the metal particles by hydrogen adsorption and electron microscopy. As a result, we have prepared some samples with 5,000 ppm of

Pt. For this purpose, we have also prepared the impregnated bimetallic Pt-Re catalyst based on MOA support (5,000 ppm Pt and 5,000 ppm Re). Physico-chemical properties of the samples are presented in tables 1 and 3. MOA based samples had 0.1 wt. % Na_2O, 0.10 wt. % Feas impurity. SO_4^{--} was no more than 0.4 wt. % in MOA, and practically absent in G-00.

The different surface area of MOA can be attributed to the different conditions of its preparation at the commercial unit.

Pt dispersion and the size distribution of the supported Pt. Pt dispersion, on the samples containing 5,000 ppm of Pt and the Pt-Re sample, was measured by the volumetric method of hydrogen adsorption (3). To measure the Pt dispersion, each sample was placed in a vessel and then:

1) slowly heated in oxygen flow to 400° C at a rate of 12° C/hr, then cooled to room temperature and flushed with nitrogen (or argon) for 30 minutes;

2) reduced in hydrogen flow using temperature programmed heating at a rate of 12° C/hr. The samples were maintained at different temperatures: 300 or 500° C for 12hrs, then cooled to room temperaure.

Then the samples were evacuated at 400° C to remove the adsorbed hydrogen and cooled to room temperature.

The quantity of chemisorbed hydrogen was determined as the difference between "total" and "reversible" isotherms. This quantity corresponds to a monolayer coverage of the Pt surface by hydrogen atoms (3).

Electron microscopy. The size distribution of the supported Pt was investigated by high-resolution transmission electron microscope " JEOL 100CX" with electron energy of 100 KeV. Before testing, the samples were oxidized in oxygen flow at 400° C and then reduced in hydrogen flow at 300 or 500° C.

Each sample was manually ground in an agate mortar for one minute. The powder was transfered to 100 cm^3 tetrachloromethane flask, mixed in an ultrasonic vibrator for one minute and, by use of a dropper, several drops of the suspension

Table 1. Physico-chemical properties of the supports

Supports	Preparation	Phase composition	impurities Wt%	Surface area m²/g	Pore volume cm³/g	mean pore nm
MOA	precipitation	γ-Al$_2$O$_3$	Na$_2$O ~ 0.1 Fe ~ 0.1 SO$_4^-$ ~ 0.4	120-180	0.36-0.45	3.7 - 4.0
G - 00	Calcination of alumina hydrate 1000°C	γ-Al$_2$O$_3$-45% θ-Al$_2$O$_3$-30% α-Al$_2$O$_3$-25%	Na$_2$O < 0.1 Fe < 0.1 No SO$_4^-$	40	0.20	15.0

were transfered onto a grid coated with holey carbon. After evaporation of the liquid at room temperature, the sample on its grid was inseted into the microscope.
The electron microscope was operated at magnifications in the range of 230,000 to 470,000. The average diameter of metal particles was determined by the formula:

$$d_{a\,v} = \Sigma\, n_i d_i / \Sigma\, n_i$$

where n_i is the number of the metal particles with diameter d_i and the sum is over all particles.
Many regions were selected from many micrographs to obtain a result as representative as possible of the sample as a whole. Finally, a statistically significant number of particles was counted.

CO oxidation activity. The catalytic activity of the samples was determined in the oxidative regeneration micropilot plant using a reactor with fluidized catalyst bed. The samples were tested as follows: A mixture of 5 vol. % CO and 95 vol. % air was passed through a layer of a mixture (total mass 1g) of a cracking catalyst and CO oxidation catalyst at 600° C, at 0.105 MPa and a gas mixture space velocity of 2.1 cm^3/hr.
For comparing the influence of support, we have always used a mixture of 0.01% CO oxidation catalyst (alumina + Pt) added to FCC catalyst.

The attrition resistance. The attrition resistance of the samples was studied by simulating the operating conditions of the catalyst in a fluidized bed according to the reference (4). After testing the samples, the Pt content in the attrition products collected on the filter was determined.

Results and discussions

The effect of the structure properties of the alumina support on active component dispersion. The catalytic activity of the supported metal is related to the particle size since it depends on the number of the available atoms. The analysis of the derived data (Table 2) shows that the decrease in Pt dispersion, with the increase of the reduction temperature from 300 to 500° C, occurs to different extents for samples with different porosity and phase composition.

For low surface area OGR-1 catalyst, Pt dispersion decreases by 80%, while in the case of higher surface area KO-9M catalyst, Pt dispersion decreases by 53%.

Normally, there is a connection between dispersion of metal particles and their size as determined by electron microscopy. Figures 1 and 2 show particle size distributions depending on the support nature and reduction temperature of the sample. The average diameters of metal particles are presented in Table 2.

The histogram in Figure 1a shows that the predominating platinum particle size on γ- Al_2O_3 (MOA) with high surface area is less than 1 nm (d_{av}=0.7 nm). For the spherical particle model there are 3-4 atoms across the circular base if the Pt particle diameter is 0.7 nm (6).

For sample with a low support surface area and 25 wt. % α-Al_2O_3, the diameters of the metal particles range from 0.5 to 1.8 nm and d_{av} = 1.1nm (5 atoms across the circular base, Figure 2a). Pt dispersion is 80%. The dispersion decrease is apparently connected to the texture and phase composition of the support.

Particles with diameter below 0.5 nm could be attributed to experimental error. At the same time, 95% Pt dispersion for KO-9M and 80% dispersion for OGR-1 determined by hydrogen adsorption indicates that small Pt particles, i.e., consisting of several atoms, are present. So, we think that particles with diameter below 0.5 nm detected by electron microscopy are not experimental error.

At the higher reduction temperature, the average diameter of Pt particles for KO-9M increases to 1.2 nm and dispersion decreases by about 2 times; Pt particles with d = 1.5 - 2.1 nm appear (5-7 atoms across the circular base, Figure 1b). The average diameter of Pt particles for this sample is not in agreement with their dispersion. This might be due to a few larger particles which we have not taken into account, but which decrease the dispersion value.

For OGR-1 (low support surface area), after reduction at 500° C

Table 2. The effect of the structure characteristics of the alumina support on Pt dispersion

Properties	Catalyst		
	OGR-1	KO-9M	KO-11
Metal content, ppm	5000 Pt	5000 Pt	5000 Pt 5000 Re
Reduction temperature, °C	300 500	300 500	300 500
Pt dispersion, %	80 16	95 45	- -
d_{av} of metal particles, nm	1.1 2.4	0.7 1.2	0.7 0.8

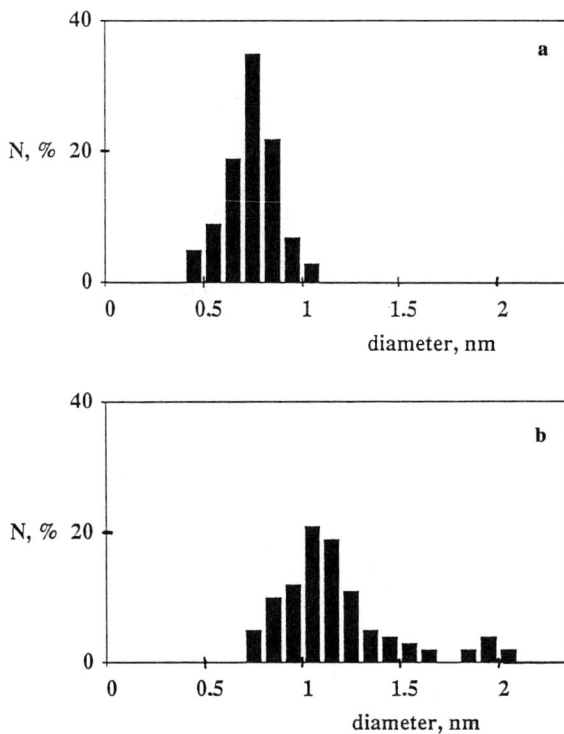

Figure 1 : Platinum particle size distributions on Pt/MOA reduced at:
a) 300° C and b) 500° C.

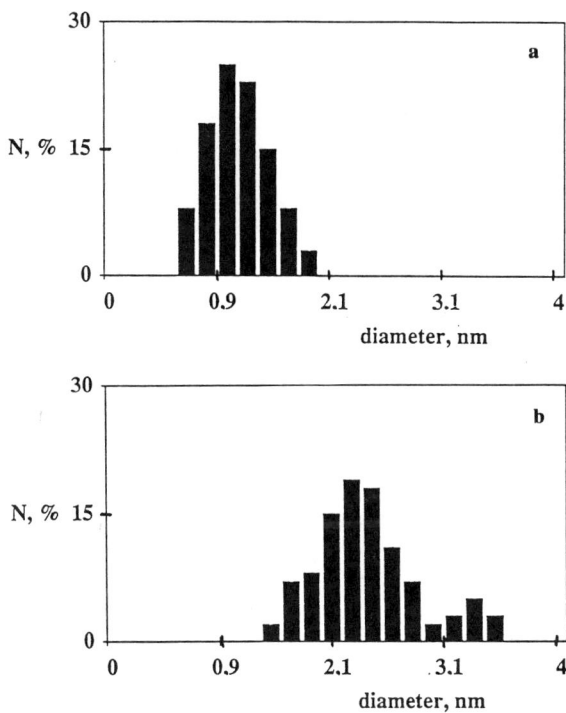

Figure 2 : Platinum particle size distributions on Pt/G-00 reduced at:
a) 300° C and b) 500° C.

(Figure 2b), the diameters of the platinum range from 1.4 to 3.7 nm, d_{av} = 2.4 nm, the Pt clusters appear (d = 3.5 nm, more than 10 atoms across the circular base) and Pt dispersion decreases to 16%. Figure 3b shows Pt clusters and non-uniform particle distribution on the low surface area alumina. The observed effect of Pt particles agglomeration can be attributed to the fact that the surface Pt concentration is much higher in the low surface area and low pore volume sample. The Pt particle size increases with temperature due to the migration process.

The mean pore radius of the G-00 and MOA support is 15 nm and 4 nm, respectively. By comparing the average Pt particle diameters for the above samples, reduced at 500° C, it can be seen that on OGR-1 sample with large pores the average Pt particle diameter is 2.2 times larger (2.4 nm vs 1.1 nm for KO-9M). This is in agreement with the reference (5) showing that Pt particle size depends on the support pore size.

For the bimetallic Pt-Re- catalyst, the size distribution of the metal particles ranges from less than 0.5 to 1.1 nm and from less than 0.5 to 1.2 nm after reduction at 300 and 500° C, respectively (Figure 4). From this histogram It appears that this sample is rather monodispersed and similar to the Pt particles distribution on the same support (after reduction at 300° C). No change in the metal particle size distribution of the sample occured after reduction at 500° C, contrary to the previous sample (Figure 1b). This means that Re can prevent Pt particle agglomeration.

CO Oxidation activity.

Influence of the support structure. From the data presented in Table 3, it can be seen that for equal Pt content on the support the oxidation activity of the samples depends on the support structure; this, apparently, being attributed to different Pt dispersion on different alumina. Whatever the Pt content (500 or 1,000 ppm), the CO oxidation activity of the OGR-1 catalyst based on low surface area alumina is by 1.3 - 1.5 times lower than in high surface area samples. According to the results obtained for samples with 5,000 ppm of Pt, we attribute this difference in CO oxidation activity to the different alumina.

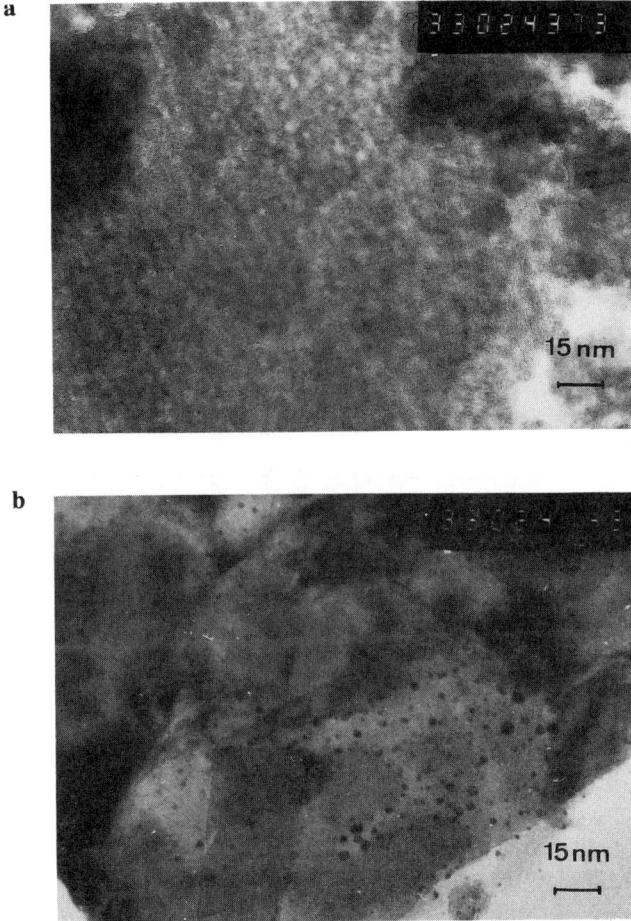

Figure 3 : Microelctron photos of Pt particles reduced at 500° C on:
a) MOA and b) G-00.

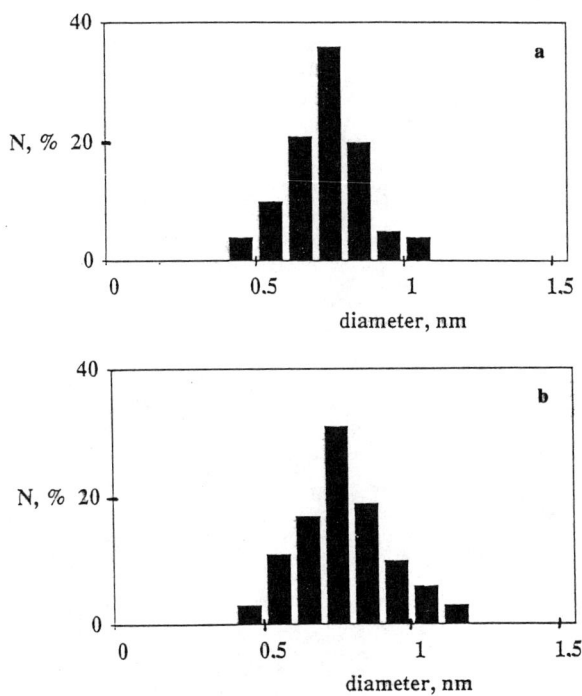

Figure 4 :   Platinum - Rhenium particles size distributions
on Pt - Re/MOA reduced at :
a) 300° C and b) 500° C.

Table 3. Catalytic and physico-mechanical properties of the samples

Sample	Pt content on a support ppm	CO conversion % (0.01 wt. % of CO oxidation catal. added to FCC Catalyst)	Attrition % 1hr	4hrs	Pt content in attrition products ppm
Competitor	800	49	17.9	34.7	380
KO-9	500 / 1000	40 / 60	2.6	12.1	210
KO-9M	500 / 1000	45 / 62	3.9	4.7	100
OGR-1	500 / 1000	30 / 41	2.7	57.1	420

Influence of the Pt concentration. It is known that the Pt dispersion depends not only on temperature but also on supported metal concentration (7). We have investigated samples with different Pt content on the MOA and studied, for each one, the influence of the Pt content in the system by varying the concentration of CO oxidation catalyst and cracking catalyst.

Figure 5 shows that for the same Pt concentration in the system, the lower the Pt content on the support, the higher the sample activity in CO oxidation. In this case, the number of the oxidation catalyst particles increases with a decrease of Pt content on the support. Also, a more homogeneous distribution is achieved in the process of mixing with the cracking catalyst.

As an example for 0.1 ppm Pt content in the system, the CO conversions are 95%, 62% and 60% for samples with 1,500 and 1,000 ppm of Pt/MOA, respectively. From these data, it seems that 1 ppm Pt/MOA is the best. However, in the industry it is difficult to prepare large amount of catalyst (about 100 kg simultaneously) with 1 ppm of Pt, mainly for the lack of homogeneity of Pt distribution. Moreover, it is necessary to use much more alumina support. Taking into account all parameters, including an economical point of view, for the same total Pt concentration in the mixture of FCC and CO oxidation catalysts, the best range of Pt content on the support is from 400 to 700 ppm.

Attrition resistance. Table 3 shows that the mechanical strength of the samples and Pt content in the attrition products during the first 4 hrs depends on the method of catalyst preparation and its structure properties and phase composition.

For OGR-1 catalyst, the Pt content in the attrition products is 84% based on the initial content on the catalyst. So, it can be supposed that while impregnating Pt on the support with low macroporous surface area it is distributed mainly on the external surface of the support particle and is lost in form of dust during surface layer erosion.

On the samples with high microporous surface area, a more uniform Pt distribution in the microsphere volume is achieved since, in this case, Pt is not only on the external surface but also in the pore. Pt particles in the pores are protected from attrition, thus causing less active component loss in the process of operation.

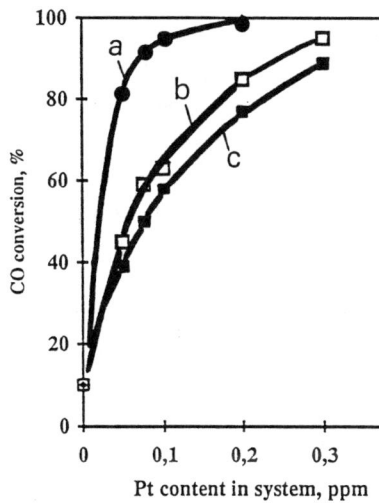

Figure 5 : Effect of the platinum concentration in the system (mixture of FCC and CO oxidation catalysts) on CO conversion for different Pt contents on the alumina support:
a) 1 ppm; b) 500 ppm; c) 1,000 ppm.

Conclusion

The investigations conducted here have shown that the texture characteristics of the alumina support (surface area, pore volume) and phase composition exert an important effect on its physico-mechanical properties, active component (Pt) dispersion and CO oxidation catalyst activity.

The deactivation of the CO combustion catalysts in the cracking process occurs due to a decrease of Pt dispersion and depends on the support structure properties and phase composition. Our results show that the maximum supported metal dispersion is achieved on the support with surface area 120 - 180 m^2/g and pore volume 0.35 - 0.45 cm^3/g. This permits efficient utilization of the entire oxidation catalyst surface. In this case, Pt particles are not only on the external support surface, but also in the pores, so it results in fewer active component losses in the process of layer erosion. The CO oxidation activity of the catalyst based on such alumina is 1.3-1.5 times higher than in low surface area OGR-1 catalyst. We have shown that, for a given concentration of Pt in the mixture, the best range of Pt content on the suport is 400 -700 ppm if we take into account all parameters, including an economical point of view.

Our resuts permitted us to improve the method of support preparation and allow optimization of procedures which yield high Pt dispersion and high CO oxidation activity in FCC units.

Literature Cited

1.   Zalioubovskaia T., Radchenko E., Melik-Akhnazarov, T. Aliev R. *J. Chem. and Tech. of Fuels* **1 9 8 7**,2, 39

2.   Melik-Akhnazarov T., Zalioubovskaia T., Aliev R. *Review CO Oxidation Cat in FCC units*.. Moscow. NIITEKhim **1 9 8 9**,80

3   Benson J.E., Boudart M. *J. Catal.* **1 9 6 5**,4 ,704

4.   Shkarin A.V., Kundo L.P. *Phys.of Commercial Catalysts and test methods* Moscow. NIITEKhim **1 9 8 9**, 92

5.   Anderson J.R.- Structure of metallic *Catalysts Ac. Press- N.Y.* **1 9 7 5**, 319

6.   Geus I.W., Wells P. B; *Appl. Catal.* **1 9 8 5**, 18, 231

7.   Rodionov A.V., Duplyakin V.K. *J. Chem. Kinetics and Catalysis* **1 9 7 9**, 157

RECEIVED June 21, 1994

Chapter 17

Scanning Electron Microscopy—Energy-Dispersive X-ray Analysis of Nickel Migration in Model Fluid Cracking Catalysts

Yan-Fei Shen[1], Hui Cao[1], Steven L. Suib[1,2,4], and Mario L. Occelli[3]

[1]Department of Chemistry and [2]Department of Chemical Engineering, University of Connecticut, Storrs, CT 06269
[3]Zeolites and Clays Program, Georgia Tech Research Institute, Georgia Institute of Technology, Atlanta, GA 30332

SEM/EDX (scanning electron microscopy/energy dispersive x-ray analysis) has been used to probe interparticle migration of nickel between an Eu^{3+}-exchanged zeolite Y and AAA-alumina in model fluid cracking catalysts (FCC). Migration depends on the nature of nickel precursors and their loadings.

No migration can be observed between the two components during calcination or steaming when nickel naphthenate is used and only 0.5 wt % Ni is loaded. When the loading is increased to 1 wt %, Ni does not migrate from the matrix to the zeolite or vice versa during calcination. Steaming induces about 20 to 30 % of the initially-deposited nickel to move from the zeolite to the matrix, but only trace migration is observed from the matrix to the zeolite.

When 1.0 wt % Ni is loaded using nickel porphyrin, no migration from matrix to the zeolite occurs, as is the case for vanadyl porphyrin. Steaming allows some of the nickel that migrated to the matrix during calcination to move back to the zeolite. However, only a trace of Ni can migrate from the initially-deposited matrix to the zeolite during steaming. These results demonstrate that nickel preferentially remains in the matrix during steaming.

When the matrix is not intimately contacted with the zeolite, nickel porphyrin molecules are removed from the zeolite to the gas phase during calcination. A particle contact mechanism prevails, however, when the two components are intimately contacted.

---

[4]Corresponding author

0097–6156/94/0571–0209$08.18/0

The deleterious effects of nickel and vanadium on equilibrium fluid cracking catalysts (FCC) have stimulated considerable interest in fundamental research (1-12). It is well known that nickel contaminants cause overcracking of gasoline, leading to the generation of excess hydrogen and coke. Nickel does not reduce cracking activity even at the 2 to 3 % level (13-18). Nickel usually produces 2 to 5 times more hydrogen and coke than vanadium (2,13), which instead destructively reacts with the zeolite present in the FCC (3). The poisoning activity of nickel is known to depend on whether this metal is deposited on the zeolite or on the FCC nonzeolitic components. Larger amounts of hydrogen and coke are generated during gas oil cracking when nickel is deposited on zeolite-containing particles (2). Because of different performance of a nonzeolitic component in terms of interactions with metal contaminants (5,8,12), a nonzeolitic component or passivating agent has been widely mixed with a zeolite to form an FCC with the desired metal-resistant properties.

A considerable volume of research work has been recently reported describing the behavior of Ni and V in FCC and metals migration between the catalyst components (3-12, 19-20). Vanadium can migrate from zeolite to matrix or vice versa depending largely on the nature of the matrix or passivator (5, 7-12) as well as the vanadium precursors used (12). In the case of interparticle nickel migration, little or no migration has been reported in many cases (3-5). However, surface migration of Ni is not uncommon, and luminescence spectroscopy results have indicated the exchange of nickel migration from a Eu^{3+}-exchanged zeolite Y (of a model FCC doped with 2 % Ni) to an aluminosilicate matrix (6). Furthermore, nickel was found to go into the bulk of an FCC after high-temperature calcination (8). Steaming dramatically enhances nickel dispersion on a sepiolite surface in dual function FCC (DFCC) mixtures (8).

The objective of the present study is to elucidate: (i) nickel migration between the Eu^{3+}-exchanged zeolite Y and an AAA aluminosilicate, (ii) effects of nickel precursors on Ni migration, and (iii) migration mechanisms during calcination and steaming. SEM/EDX is used to track nickel migration by spot analysis on either zeolite Y or matrix particles, characterized by considerable difference in morphology.

Experimental

Sample Preparation. An Eu^{3+}-exchanged zeolite Y (EuY) was used as the cracking component and an AAA aluminosilicate (AAA alumina) was used as the matrix component of an FCC. The preparation of EuY has been described previously (12). EuY and AAA were loaded with nickel naphthenate and 5, 10, 15, 20-tetraphenyl-21H, 23H-porphine nickel oxide (Ni TPP) in the following ways. In the case of nickel naphthenate, 0.5 and 1 wt % nickel were loaded by incipient wetness impregnation with a nickel naphthenate-benzene solution and then the samples were dried at 80°C. Since Ni TPP is not as soluble in benzene as Ni naphthenate, Ni TPP in benzene was boiled before EuY or AAA powder was added to the boiling Ni TPP benzene solution. Benzene was then evaporated and the Ni TPP (1 wt % Ni) loaded powders were dried at 80°C. The Ni (Ni TPP or Ni naphthenate)-loaded solid was then mixed with pure AAA or EuY, respectively, either by benzene or by mechanical grinding, at a ratio of 1:1 by weight. In some cases, Ni-loaded EuY or AAA was calcined, before mixing with the metal-free component.

After mixing, samples were calcined and steam-aged. Calcination was performed in a vertical quartz reactor with flowing air at 540°C for 10 hours to decompose the organic compounds. Steam-aging was conducted at 740°C for 10 hours in a flowing 95 % H_2O/5% N_2 mixture with a water flow rate of 4 ml/hr.

Nomenclature of Samples. The nomenclature used to describe samples and sample treatments has been outlined in detail elsewhere (8). In this study, EuYNi(p)AAACS corresponds to EuY loaded with 1.0 wt% of nickel porphyrin (p), then mixed with the AAA-aluminosilicate (AAA), and finally pretreated with calcination (C) and subsequent steaming (S). EuYNi(AAA)C indicates an EuYNi sample which was calcined together with the AAA-aluminosilicate without contacting each other. EuYNi(n)AAAC represents a sample prepared by initially loading nickel naphthenate (n) onto the zeolite at 1 wt % Ni (EuYNi(n), mixing EuYNi(n) with the matrix (AAA), and finally calcining the mixture (C); while EuYNi(0.5, n)AAA indicates a sample containing 0.5 wt % Ni (nickel naphthenate) on the zeolite.

SEM/EDX Measurements. Sample powders were coated on a carbon coated aluminum sample holder. After drying, the samples were evacuated and analyzed with an AMRAY 1810 scanning electron microscope and a Philips PV9800 EDAX spectrometer. Care was taken to find a pure matrix or zeolite spot, and different spots were analyzed for accuracy. Quantitative calculations were done with the Super Quant program with a ZAF correction to obtain relative concentrations. Because the instrument is not sensitive to oxygen, data presented in this paper do not include oxygen composition.

Results

A. Nickel Naphthenate as Contaminants

Most literature reports have focused on catalysts artificially contaminated with solutions of Ni or V naphthenate in benzene or toluene (1, 21-22). Ni naphthenate was used for migration studies for this reason.

It was found that mixing of the two components with benzene results in interparticle migration of nickel naphthenate. Because of this, mechanical mixing is chosen, since it avoids migration during the mixing process.

SEM/EDX results of EuYNi(0.5, n)AAAC and AAANi(0.5, n)EuYC are shown in Fig. 1. Five Eu L's peaks are observed at 5.85, 6.46, 6.84, 7.48, and 5.18 KeV, which are assigned to Lα1, Lβ1, Lβ2, Lγ1 and LL and have a relative intensity of 100, 56, 20, 7 and 3, respectively.

Nickel $K\alpha1$ appears at 7.48 KeV. As seen in Fig. 1, no Ni $K\alpha1$ peak is observed in the matrix of EuYNi(0.5,n)AAAC (Fig. 1a), clearly indicating the absence of migration from the zeolite to the matrix. For the zeolite part of AAANi(0.5,n)EuYC (Fig. 1b), care needs to be taken when interpreting SEM/EDX spectra, because the Ni $K\alpha1$ and the Eu $L\gamma1$ (7.48 KeV) peaks overlap and small zeolite particles are dispersed on the matrix. One way to solve this problem is to compare the intensity of the peak at 7.48 KeV in the observed spectrum to that of Eu $L\gamma1$ peak (7.48 KeV) in the fitted spectrum. If the observed peak has the same intensity as the fitted peak, no Ni migration occurs. If the observed peak has stronger intensity than the fitted peak, there is Ni migration. As shown in Figs. 1b and 2b, no Ni migration takes place from the matrix to the zeolite during calcination (Fig. 1b) or during steaming (Fig. 2b). Thus, when 0.5 wt % Ni (from nickel naphthenate) is loaded, nickel migration between the two components does not occur during either calcination or steaming. This conclusion agrees with some literature reports (2).

Results of 1 wt % Ni (nickel naphthenate) being loaded are presented in Figs. 3-6 and Table 1. Fig. 3 shows no Ni $K\alpha1$ peak in the matrix of EuYNi(n)AAAC (Fig. 3a), revealing that Ni migration from the zeolite to the matrix during calcination did not occur.

Steaming EuYNi(n)AAAC induced some Ni migration from the zeolite to the matrix, as revealed by the Ni peak in the matrix in Fig. 4a. Data in Table 1 show that either the Ni concentration or Ni/Eu molar ratio decreases by about 27 % in the zeolite part and increases proportionally in the matrix part, as compared to the corresponding values of EuYNi(n)AAAC. These results demonstrate that Ni migrates from the zeolite to the matrix during steaming. This observation is in accord with fluorescence studies of a similar sample with 2 wt % Ni initially deposited onto the zeolite (6).

Fig. 5 shows a strong Ni $K\alpha1$ peak in the matrix of AAANi(n)EuYC (Fig. 5a). There is no significant increase in the intensity of the peak at 7.48 KeV in the zeolite component (Fig. 5b). This result demonstrates that little (trace) Ni moves to the zeolite from the matrix during calcination.

After steaming AAANi(n)EuYC, the intensity of the peak at 7.48 KeV in the observed spectrum was stronger than that of the fitted peak (Fig. 6b), revealing some Ni migration from the matrix to the zeolite during steaming. Quantitative data in Table 1 show that the Ni concentration or Ni/Eu molar ratio somewhat increases in the zeolite, while Ni concentration decreases in the matrix, as compared to corresponding values of AAANi(n)EuYC. Therefore, steaming induces a small migration of Ni from the matrix to the zeolite.

In summary, in the presence of 0.50 wt % Ni (from nickel naphthenate), there was no nickel migration between the two model components either during calcination or during steaming. Likewise, no migration was found during calcination when Ni loadings increase to 1.0 wt %. At 1.0 wt % Ni loading, however, steaming induced about 27 % of the Ni on the zeolite to move to the matrix and 9 % of the Ni on the matrix to move to the zeolite.

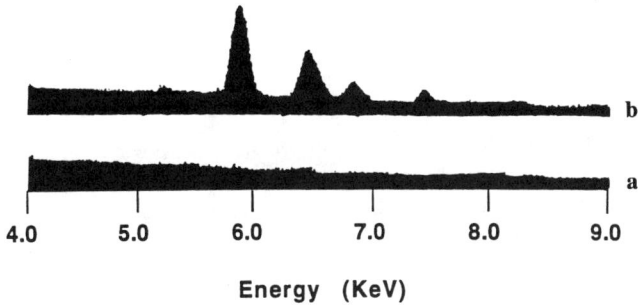

Figure 1. SEM/EDX spectra of (a) the matrix of EuYNi(0.5,n)AAAC and (b) the zeolite of AAANi(0.5,n)EuYC.

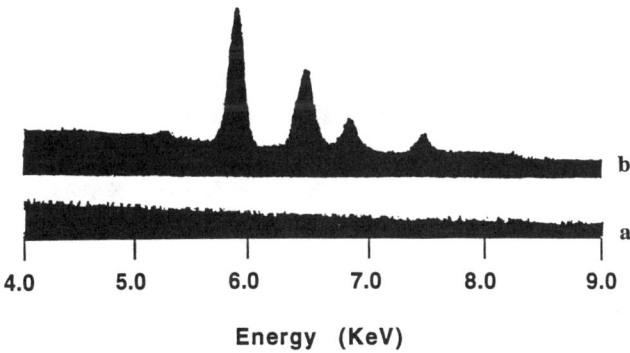

Figure 2. SEM/EDX spectra of (a) the matrix of EuYNi(0.5,n)AAACS and (b) the zeolite of AAANi(0.5,n)EuYCS.

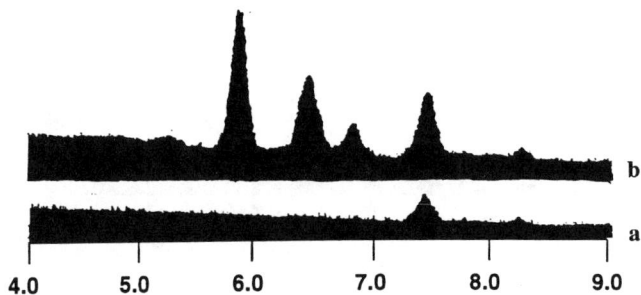

Figure 3. SEM/EDX spectra of EuYNi(n)AAAC: (a) the matrix and (b) the zeolite.

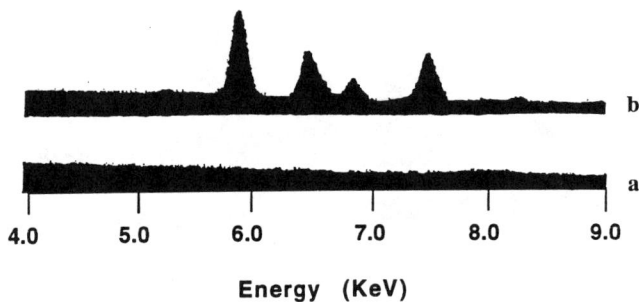

Energy (KeV)

Figure 4. SEM/EDX spectra of EuYNi(n)AAACS: (a) the matrix and (b) the zeolite.

Table 1. SEM/EDAX data of model FCC catalysts doped with 1 wt % Ni from Ni naphthenate

Sample	elemental concentration (wt %)											
	EuY						AAA-alumina					
	Al	Si	Ni	Eu	Si/Al (molar)	Ni/Eu (molar)	Al	Si	Ni	Eu	Si/Al (molar)	Ni/Eu (molar)
EuYNi(n)AAAC	16.17	67.60	3.45	12.78	4.01	0.70	17.86	81.54	0.11	0.49	4.38	0.31
EuYNi(n)AAACS	16.23	68.67	2.40	12.71	4.06	0.49	18.57	78.98	1.05	1.32	4.08	2.12
AAANi(n)EuYC	15.96	69.44	0.42	14.19	4.18	0.08.	18.62	77.41	3.45	0.52	3.99	17.17
AAANi(n)EuYCS	15.58	67.72	0.98	15.72	4.17	0.16	18.56	76.52	3.48	1.44	3.96	5.80

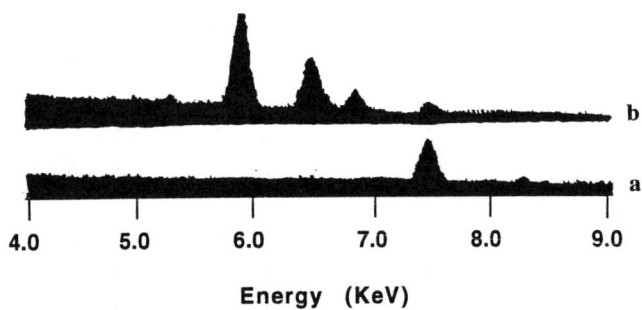

Figure 5. SEM/EDX spectra of AAANi(n)EuYC: (a) the matrix and (b) the zeolite.

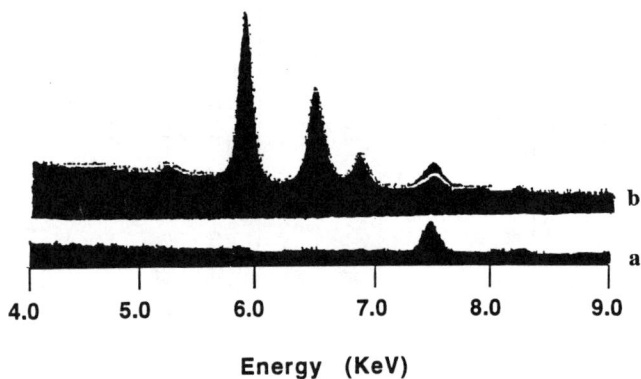

Figure 6. SEM/EDX spectra of AAANi(n)EuYCS: (a) the matrix and (b) the zeolite.

B. Nickel Porphyrins as Contaminants

As already reported (*12*), vanadyl porphyrins exhibit significant differences in migration behavior from vanadyl naphthenate. Therefore, Ni TPP, which is contained in crude oil (*23-24*), has also been tested as a source of Ni-contamination.

Effects of calcination and steaming treatments on migration are shown in Figs. 7-10 and Table 2. Like the case of nickel naphthenate, a strong Ni Kα1 peak is observed in the matrix of AAANi(p)EuYC (Fig. 7a) and almost no Ni peak in the zeolite (Fig. 6b). This indicates little migration from the matrix to the zeolite during calcination.

After steaming AAANi(p)EuYC, the Ni Kα1 peak is not significantly increased in the zeolite portion of AAANi(p)EuYCS (Fig. 8b). Data in Table 2 show that either the Ni concentration or the Ni/Eu ratio increases slightly in the zeolite part of AAANi(p)EuYCS, as compared to corresponding values for AAANi(p)EuYC. This indicates that only a trace of Ni can move from the matrix to the zeolite during steaming.

In the case of EuYNi(p)AAAC, Fig. 9 shows a strong Ni peak in the matrix component (Fig. 9a) and a very small Ni peak in the zeolite (Fig. 9b), revealing a pronounced migration from the zeolite to the matrix during calcination, as is the case for vanadyl porphyrins (*12*). Data in Table 2 demonstrate that most of the Ni initially loaded on the zeolite moves to the matrix.

After steaming EuYNi(p)AAAC, Fig. 10b shows that the peak at 7.48 KeV is stronger than the peak at 6.84 KeV. Compared to Fig. 9b, where the peak at 7.48 KeV is weaker than the peak at 6.84 KeV, Fig. 10b indicates an increase in Ni concentration in the zeolite after steaming EuYNi(p)AAAC. Data in Table 2 show that steaming decreases the Ni concentration or Ni/Eu ratio in the matrix and increases these values in the zeolite component. More precisely, an analysis of Ni concentration in the matrix of EuYNi(p)AAAC and EuYNi(p)AAACS reveals that about 20 % of the Ni that migrated to the matrix during calcination moves back to the zeolite during steaming. This observation is identical to the one reported when using vanadyl porphyrin (*12*), indicating a similar interaction between the porphyrins and the matrix or zeolite surface.

C. Migration Mechanisms

As was the case for vanadium systems (*12*), the following experiments have been performed to obtain insight into migration mechanisms. The two components, such as EuYNi(p) and AAA or EuYNi(p)C and AAA, were put respectively in the upstream and downstream portions of a vertical reactor and separated either by treated glass wool for calcination or by treated asbestos for steaming. These separated components were then pretreated under the same conditions as above.

It is found that Ni TPP is not dispersed on the zeolite as homogeneously as nickel naphthenate. This conclusion requires an

Table 2.   SEM/EDAX data of model FCC catalysts doped with 1 wt % Ni from nickel porphyrin

Sample	elemental concentration (wt %)											
	EuY						AAA-alumina					
	Al	Si	Ni	Eu	Si/Al (molar)	Ni/Eu (molar)	Al	Si	Ni	Eu	Si/Al (molar)	Ni/Eu (molar)
EuYNi(p)AAAC	15.80	69.92	0.24	14.04	4.25	.044	19.08	76.84	3.69	0.39	3.87	24.58
EuYNi(p)AAACS	16.91	68.54	1.00	13.54	3.89	.192	20.39	76.03	2.84	0.74	3.58	9.95
AAANi(p)EuYC	16.79	68.20	0.39	14.63	3.90	.070	18.70	77.60	3.38	0.34	3.98	25.72
AAANi(p)EuYCS	16.43	67.89	0.61	14.97	3.97	.106	19.18	77.13	3.32	0.37	3.80	23.23

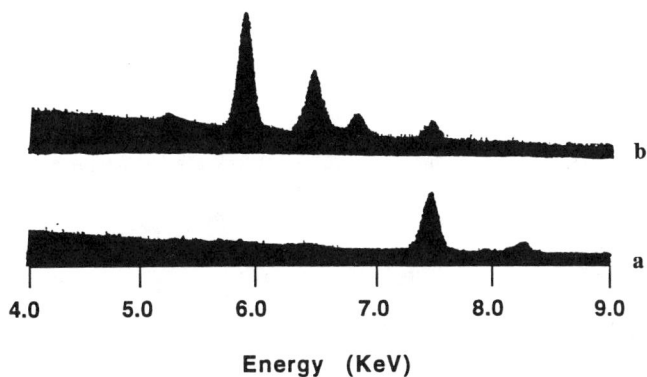

Figure 7. SEM/EDX spectra of AAANi(p)EuYC: (a) the matrix and (b) the zeolite.

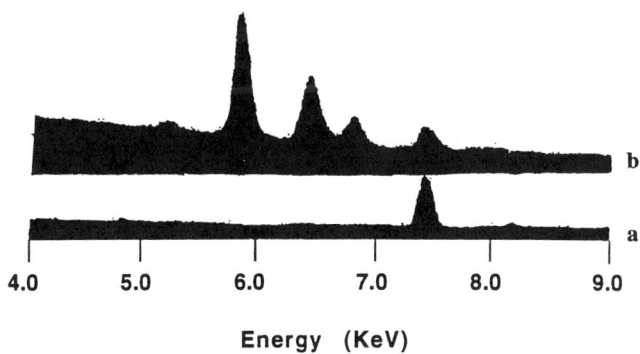

Figure 8. SEM/EDX spectra of AAANi(p)EuYCS: (a) the matrix and (b) the zeolite.

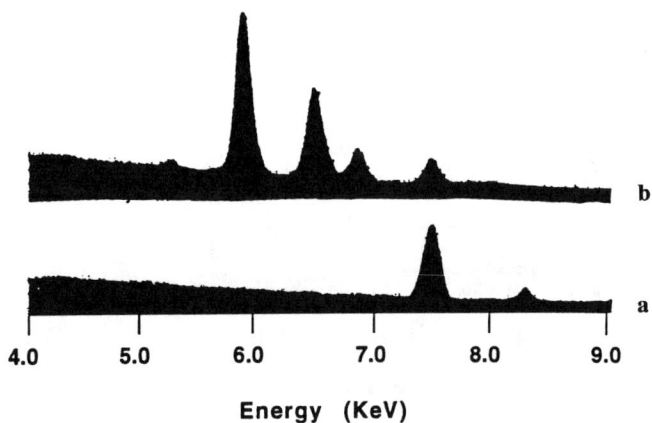

Figure 9. SEM/EDX spectra of EuYNi(p)AAAC: (a) the matrix and (b) the zeolite.

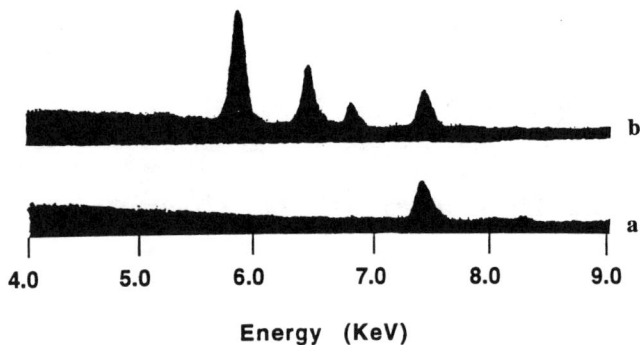

Figure 10. SEM/EDX spectra of EuYNi(p)AAACS: (a) the matrix and (b) the zeolite.

analysis of as many zeolite particles as possible for EuYNi(p) to look for regions of inhomogeneity. Data in Table 3 are average values taken from different zeolite particles, though nickel is dispersed much more homogeneously after calcination or steaming.

The results for calcination are shown in Fig. 11. No nickel is detected either in the matrix (Fig. 12a) or in the glass wool (Fig. 12b), indicating no Ni deposition to the matrix or to the glass wool during calcination. Compared to Fig. 9, which reveals a significant migration from the zeolite to the matrix during calcination, this observation seems to suggest that nickel migration occurs mainly through particle contact interactions. A quantitative analysis (Table 3) shows that pure EuYNi(p) has 40-60 wt % higher Ni concentration than EuYNi(p)C as well as the zeolite component of EuYNi(p)(...)AAAC and EuYNiC(...)AAAS. These results demonstrate that without intimate contact with the matrix, calcination also causes a significant loss of nickel from the zeolite. During calcination, the lost Ni cannot be trapped by either the matrix or the glass wool. Comparison of this result to that in Fig. 9 appears to indicate that an intimate contact of the zeolite with the matrix is necessary for Ni to migrate to the matrix. Without intimate contact, however, nickel may be removed via gas phase reactions; that is, migration mechanisms may be different, depending on whether or not the zeolite is intimately contacted with the matrix component (the gel).

To confirm a gas-phase migration mechanism, the following experiment was performed. EuYNi(p) was calcined at the same condition as above. However, the matrix was not loaded in the same reactor, but was kept at room temperature in another glass tube which was connected to the outlet of the reactor. The matrix was thus expected to trap gaseous products formed during calcination. It is found that the color of the matrix starts to turn dark brown when the calcination temperature increases to about 300 to 400°C, indicating that some species, Ni TPP or organic compounds, were removed from the zeolite and deposited onto the matrix. An SEM/EDX analysis shows a strong Ni peak in the matrix (Fig. 12a), revealing significant removal of Ni from the zeolite via gas-phase transport during calcination. When EuYNi(p)AAA, rather than EuYNi(p), is calcined, however, no color change and no obvious Ni peak are found in the matrix (Fig. 12b), indicating that intimate contact of the zeolite with the matrix is needed for Ni migration (Fig. 9) from the zeolite to the matrix to occur via a particle contact mechanism.

SEM/EDX data during steaming are shown in Fig. 13. Likewise, no nickel can be observed in the matrix (Fig. 13a) or in the asbestos (Fig. 13b). Quantitative data (Table 3) show that the Ni/Eu ratio [Ni concentration in the zeolite of EuYNi(p)C(...)AAAS] is slightly smaller than that of EuYNi(p)C and close to the corresponding value of EuYNi(p)AAACS (Table 2). These results indicate that steam-aging removes small amounts of Ni from the zeolite via a gas-phase mechanism (steam stripping).

Table 3.   SEM/EDAX data of model FCC catalysts doped with 1 wt % Ni from nickel porphyrin; EuY and AAA-matrix are separated

Sample	elemental concentration (wt %)										
	EuY						AAA-alumina				
	Al	Si	Ni	Eu	Si/Al (molar)	Ni/Eu (molar)	Al	Si	Ni	Eu	Si/Al (molar)
EuYNi(p)	17.08	65.40	2.74	14.78	3.68	0.41					
EuYNi(n)C	16.93	65.95	1.61	15.51	3.74	0.28					
EuYNi(p)(…)AAAC	16.51	66.97	1.19	15.33	3.89	0.21	19.80	80.2	0	0	3.89
EuYNi(n)C(…)AAAS	16.67	66.75	1.28	15.30	3.84	0.21	20.16	79.81	0	0	3.80

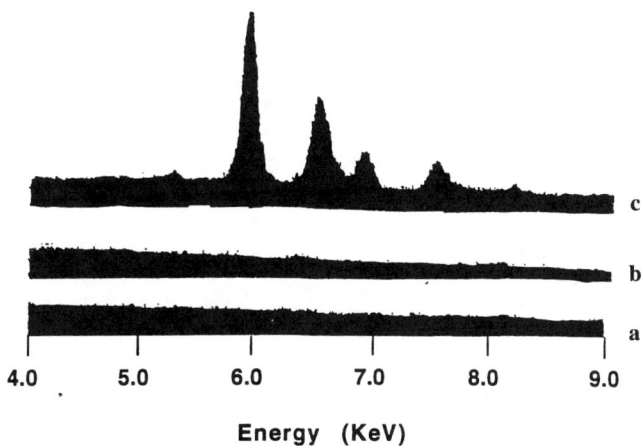

Figure 11. SEM/EDX spectra of EuYNi(p)(...)AAAC: (a) the matrix, (b) glass wool and (c) the zeolite.

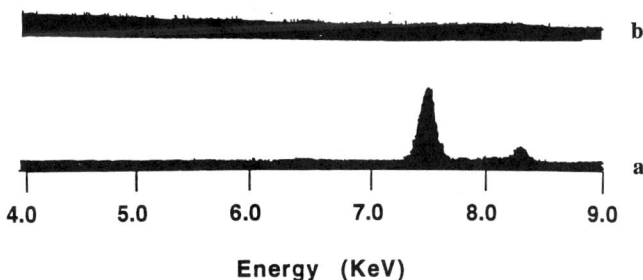

Figure 12. SEM/EDX spectra of the matrix that was kept in a glass tube at 25°C and exposed to the effluent of a calcination reactor during calcination of (a) EuYNi(p) and (b) EuYNi(p)AAA.

Discussion

A. Effects of Ni loadings

Effects of Ni loading on Ni migration appear to indicate heterogeneous sites on the zeolite and the matrix. At 0.5 wt % Ni, most of the loaded Ni species might be strongly bound on the surface of the matrix or zeolite or react with the surface to form a stable compound. Thus, no migration can be observed, as revealed in Figs. 1-2. At 1.0 wt % Ni, however, some Ni species are adsorbed on weak sites or weakly react with the surface, so that they can migrate from one component to the other during steaming (Figs. 4, 6 and Table 1). The above results show preferential migration from the zeolite to the matrix than vice versa (Figs. 4, 6 and Table 1) possibly due to the formation of more stable Ni-Al compounds on the matrix than on the zeolite.

B. Effects of Ni Precursors

It has been reported that vanadium (V) precursors show considerable differences in V migration (12) and in V-species formation on the surface (25). As shown above, Ni naphthenate and Ni TPP also show a significant difference in migration behavior. Ni naphthenate does not migrate from one component to the other during calcination (Figs. 3 and 5), while Ni TPP markedly moves from the zeolite to the matrix during calcination (Fig. 9). This difference may be due to the size of the two molecules and the nature of the interactions between the surface and the molecules. Ni TPP is much larger than Ni naphthenate. Ni naphthenate might be able to go into the supercage of the zeolite, but Ni TPP can only stay on the external surface. As will be described below, 1.0 wt % Ni (Ni TPP) molecules cannot totally disperse on the external surface of the zeolite. Some Ni TPP molecules need to be adsorbed physically and can migrate from the zeolite surface during calcination. This size difference can be used to explain the different migration behavior of the two precursors exhibited in the calcination process. After calcination, the location and nature of Ni species may be different for the two precursors, as is the case for V systems (25), so that migration is different during steam-aging (Figs. 4, 6, 8, 10 and Tables 1 and 2).

C. Migration Mechanisms

The above results clearly indicate that there are two migration mechanisms for Ni porphyrin systems: (1) a gas-phase mechanism for catalysts without an intimate contact of the zeolite and the matrix (Scheme I) and (2) a particle-contact mechanism for catalysts in which an intimate contact of the zeolite with the matrix exist (Scheme II).

Without an intimate contact between the zeolite and the matrix, Ni TPP molecules that are weakly adsorbed on the zeolite may be evaporated into the gas phase before decomposing during heating, as shown in Scheme I. Once Ni TPP or related Ni species are removed from the zeolite to the gas phase, gaseous Ni species do not deposit on either the matrix or glass wool at high temperature (400 to 500°C), as

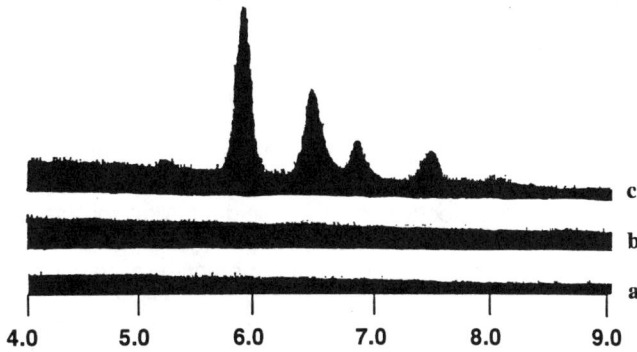

Figure 13. SEM/EDX spectra of EuYNi(p)C(...)AAAS: (a) the matrix, (b) asbestos and (c) the zeolite.

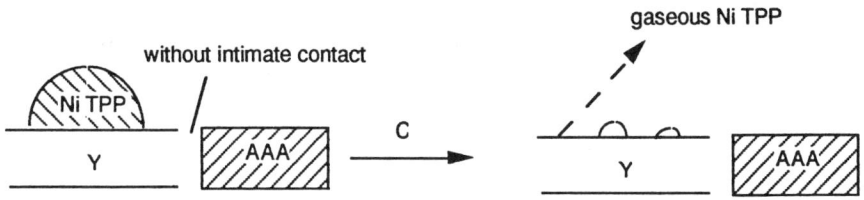

Scheme I. A gas-phase migration mechanism when EuYNi(p) is not intimately contacted with AAA matrix.

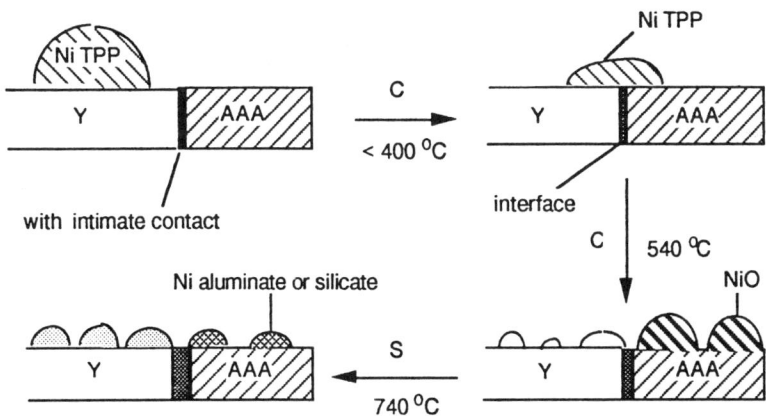

Scheme II. A particle-contact migration mechanism between the two intimately-contacted components

revealed in Fig. 11, where the zeolite, the matrix and glass wool are not intimately contacted. However, they can be trapped by the matrix kept at room temperature, as shown in Fig. 12a. The dark brown color of deposited Ni species probably indicates that the gaseous Ni species is Ni TPP.

With an intimate contact of the zeolite with the matrix, as shown in Scheme II, however, Ni TPP might preferentially react with the matrix through donation of π electron to Lewis acid sites on the matrix (26-28) before it is removed to the gas phase, as illustrated in Model I. Our previous results (indicating a significant interaction of vanadyl TPP with alumina and no interaction of vanadyl TPP with silica (12)) support the role of Lewis acid sites in trapping porphyrin molecules on the surface of the matrix. An EPR signal due to a porphyrin cation radical, that is found in Ni/Al_2O_3-SiO_2 systems and ascribed to an electron transfer from the porphyrin ring to the oxide (29), also supports this model.

The driving force for Ni migration to the matrix from the zeolite should be due to the difference in nature of the two components. Ni TPP molecules are too large in size to go into the supercage of EuY, and can only stay on the external surface. It is reported that the porphyrin plane of Ni TPP or VOTPP is parallel to the support surface (29).

Our calculation shows that one Ni TPP molecule occupies at least an area of 200 $Å^2$. This means that at least a surface area of 210 m^2/g is required for 1.0 wt % Ni of Ni TPP molecules to disperse. Thus, the external surface of the zeolite, which has an area of about 60 m^2/g, is not large enough for the Ni TPP molecules to be homogeneously dispersed and to be tightly bonded on its surface. Some of the doped Ni TPP molecules are adsorbed physically, which may be removed molecularly upon heating to 300 to 400°C, as revealed in Fig. 12a. Only tightly bound Ni TPP molecules remain on the zeolite until thermal decomposition occurs at about 450°C (Fig. 14a).

The AAA-matrix has pores in the 100 to 1000 Å range and the surface area of this matrix (394 m^2/g) is much larger than the external surface area of the zeolite. Thus, all the loaded Ni TPP can be dispersed on or into the matrix. Moreover, the number and strength distribution of Lewis acid sites are much larger in the AAA-matrix than on the external surface of the zeolite. Both these differences might contribute to more Ni TPP being bound on the matrix than on the zeolite. This explanation is further supported by the much broader DSC exotherm for Ni TPP on the matrix than for Ni TPP on the zeolite (Fig. 14).

As far as the nature of migrating species is concerned, molecular Ni TPP might be removed during calcination of EuYNi(p), since Ni species deposited from the gas phase to the matrix (Fig. 12a) have a dark brown color. In the case of EuYNi(p)AAA, there is no strong evidence concerning the nature of the migrating species. However, it is reasonable to assume that molecular Ni TPP is also responsible for Ni migration from the zeolite to the matrix during calcination. After calcination, Ni has been reported to be present on the matrix mainly as a NiO-like phase (18), as shown in Scheme II. But, no evidence has been given regarding the nature of Ni on or in the zeolite. Since Ni

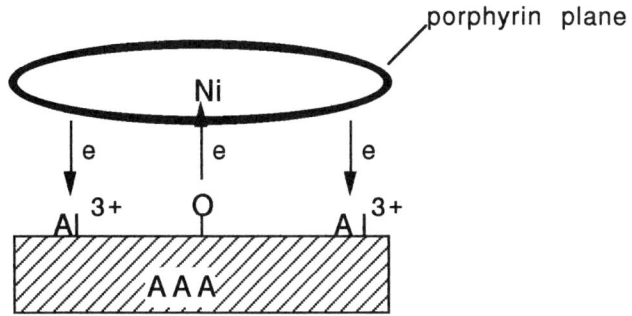

Model I.   Schematic representation of Ni TPP/AAA interaction

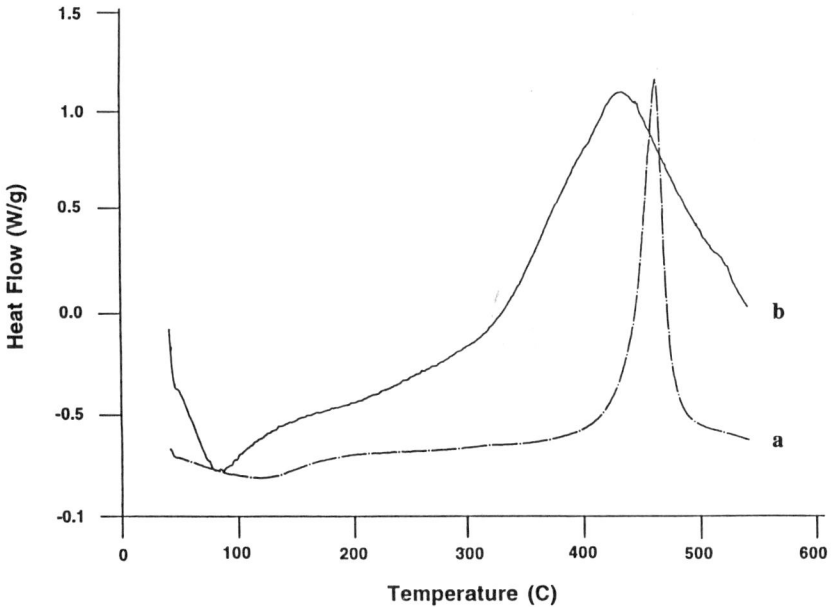

Figure 14.   DSC profiles of (a) AAANi(p) and (b) EuYNi(p).

does not destroy the zeolite structure, it may be adsorbed on the zeolite surface also as a NiO-like species. During steam-aging, the NiO-like phase may be hydrated to some extent and in the meantime may interact with the matrix to form a $NiAl_2O_4$- and/or $NiSiO_3$-like phase, as reported elsewhere (18).

A reviewer has suggested that it is unlikely that Ni TPP does not react with the matrix. It might be possible that Ni TPP reacts to form a species that is not very labile before it migrates or that H_2O vapor desorbed from the matrix causes reaction of Ni TPP with subsequent migration of the reacted species.

Conclusions

Benzene used for mixing the zeolite and the matrix in conventional methods has been found capable of dissolving metal precursors that are already doped on either the zeolite or the matrix component and hence causing migration of the precursors from one component to another. Mechanical mixing was found to induce little migration in most cases and was therefore used in the preparation of all the samples examined in the present study.

Ni migration is found to depend on loading and on the type of precursors used. At 0.5 wt % Ni (with nickel naphthenate), no migration is observed during calcination or steaming. At 1.0 wt % Ni, however, steaming induces some migration from the zeolite to the matrix or vice versa, though calcination does not cause observable migration. When 1.0 wt % Ni (nickel porphyrin) is loaded, a significant migration is found to occur from the zeolite to the matrix during calcination, but not vice versa. Steaming appears to induce a part of the Ni that had migrated to the matrix during calcination to move back to the zeolite. It seems that nickel porphyrin prefers to stay on the matrix and migrates more readily than nickel naphthenate.

Nickel migration occurs through a particle contact mechanism if the two components are intimately contacted. However, nickel can also be removed during pretreatment via a gas-phase mechanism, if the two components are not contacted. It must be remembered however that transport experiments at microactivity test conditions as well as examination of equilibrium FCC from a U.S. refinery did not reveal evidence of interparticle Ni transport (4,5).

Acknowledgments

Support for this research from the National Science Foundation under the Kinetics and Catalysis Program through grant CBT-08814974 is gratefully acknowledged.

Literature Cited

1.   Occelli, M. L. *Catal. Rev.* **1991**, *33* (3-4), 241.
2.   Schubert, P. F.; Altomare, C. A. In *Fluid Catalytic Cracking --- Role in Modern Refining*: M. L. Occelli, Ed.,**1988**, p.182.

3.   Jaeras,S. *Applied Catal.* **1982**, *2*, 207.
4.   Kugler E. L.; Leta, D. P. *J. Catal.* **1988**, *109*, 387.
5.   Occelli, M. L. In *Fluid Catalytic Cracking --- Role in Modern Refining*: M. L. Occelli, Ed., **1988**, p.162.
6.   Occelli, M. L.; Psaras, D.; Suib, S. L. *J. Catal.* **1985**, *96*, 363.
7.   Andersson, S. L. T; Lundin, S. T.; Jaeras, S.; Otterstedt, J. E. *Applied Catal.*,**1984**, *9*, 317.
8.   Occelli, M. L.; Stencel, J. M. In *Fluid Catalytic Cracking --- Role in Modern Refining*: M. L. Occelli, Ed., **1988**, p.195.
9.   Anderson, M. W.; Suib, S. L.; Occelli, M. L. *J. Catal.* **1989**, *118*, 31.
10. Woolery, G. L.; Chin, A. A.; Kirker, G. W.; Huss Jr., A. *Physica B*, **1989**, *158*, 213; In *Fluid Catalytic Cracking --- Role in Modern Refining*: M. L. Occelli, Ed., **1988**, p. 215.
11. Maselli, J. A.; Peters,A. W. *Catal. Rev. Sci. Eng.* **1984**, *26*, 525.
12. Shen, Y. -F.; Suib, S. L.; Occelli, M. L. (a) *Passivation and migration of contaminant metals in model FCC catalysts*, 202nd ACS meeting, New York, Aug. 25-30, **1991**; (b) *SEM/EDAX study of vanadium migration in model FCC catalysts*, In *Selectivity in Catalysis*: M.E. Davis and S.L. Suib, Eds., **1993**,p.185.
13. Cimbalo, R. N. et al., *Oil and Gas J.* **1972**, *70*, 112.
14. Meisenheimer, R. G. *J.Catal.* **1962**, *1*, 356.
15. Magee, J. S.; Ritter, R. E.; Rheume, L. *Hydrocarbon Processing* **1979**,*123*.
16. Occelli, M. L.; Psaras, D.; Suib, S. L. *J. Catal.* **1986**, *96*, 336.
17. Occelli, M. L. ; Psaras, D.; Suib, S. L. In *Proc. Int. Symp. Zeo. Catal.*: Siofok, Hungary, **1985**, p.423.
18. Occelli, M. L.; Stencel, J. M. In *Proc. 9th Int. Congr. Catal.*: M.J. Phillips and M. Ternan, Eds., Vol. 1, **1988**, p.214.
19. Occelli, M. L.; Kennedy, J. V. U.S. Patent 4 465 588, 1984.
20. Occelli, M. L.; Swift, H. E. U.S. Patent 4 466 884, 1984.
21. Mitchell, B. R. *Ind. Chem. Prod. Res. Dev.* **1980**, *19*, 209.
22. Chester,A. W. *Ind. Eng. Chem. Res.* **1987**, *26* (5), 863.
23. Mitchell, P. C. H.; Scott, C. E. *Catal. Today* **1990**, *7*, 467.
24. Ledoux, M. J.; Hantzer, S. *Catal. Today* **1990**, *7*, 479.
25. Kurihara, L. K.; Occelli, M. L; Suib, S. L. In *Fluid Catalytic Cracking II Concepts in Catalytic Design*: M.L. Occelli, Ed., Washington, DC, **1991**, p. 224.
26. Knözinger, H; Cordischi, D.; Vielhaber, B. *Catal. Today* **1990**, *7*, 447.
27. Mitchell, P. C. H.; Scott, C. E. *Polyhedron.* **1986**, *5*, 237.
28. Cordischi, D.; Vielhaber, B.; Knözinger, H. *Applied Catal.* **1987**, *30*, 265.
29. Mitchell, P. C. H.; Scott, C. E. *Catal. Today* **1990**, *7*, 467.

RECEIVED August 3, 1994

Chapter 18

Coking and Regeneration of Fluid Cracking Catalysts by ^{129}Xe and ^{27}Al Nuclear Magnetic Resonance Spectroscopy

J. L. Bonardet, M. C. Barrage, and J. Fraissard

Laboratoire de Chimie des Surfaces, Unité de Recherche Associé au Centre National de la Recherche Scientifique 1428, Université Pierre et Marie Curie, 4 place Jussieu, tour 55, 75252 Paris Cedex 05, France

129-Xe and 27-Al NMR have been used to study the regeneration of a partially dealuminated Y zeolite which has been coked by cracking of hexane or orthoxylene. We show that alumininium atoms of the framework take part in the coking process. A complete oxidation of the carbonaceous residus , even under mild conditions, induces structure defects and/or partial amorphization leading to a signifiant loss of internal microporous volume of the catalyst.

Because of their high catalytic activity and their form stereospecificity, zeolites are catalysts frequently employed in industrial processes, particularly cracking. Unfortunately, the corollary of their high catalytic activity is the ease with which some of them are deactivated by formation of carbonaceous residues (paraffinic, olefinic, aromatic and polyaromatic), called "coke", which poison or block access to the active centers. Thus, Y zeolites lose more than half their activity in heptane cracking when the amount of coke formed is only 3% (1). Along with numerous physical and chemical techniques, 129-Xe NMR has already enabled us to obtain very interesting complementary information about the mode of formation and the location of the carbonaceous residues in FCC catalysts and to describe in detail the role of the Al atoms (3-5). In the present article, we define by means of Xe and Al NMR the mode of regeneration of industrial samples coked by cracking hexane and ortho-xylene.

0097–6156/94/0571–0230$08.00/0

Experimental conditions.

The samples studied were derived from an ultrastable industrial HY zeolite partially dealuminated by steaming and followed by acid wash (FY-SW samples). Coke was produced by cracking n-hexane or ortho-xylene at 673K. The highest coke content was 8.6% for n-hexane (FSW8.6H) and 8.7% for ortho-xylene (FY-SW8.7X). Coke was partially or totally eliminated either by pyrolysis at 673K (FY-SW HorX$_{p\ y}$ (partial) or in a stream of pure oxygen (8 l.h^{-1}) at 573 or 623K (partial and total). The reoxidized samples are denoted FY-SW8.6Hox$_y$ or FY-SW8.7Xox$_y$ for re-oxidized samples where y represents the percentage of residual coke and FY-SW8.6H$_{p\ y}$ or FY-SW8.7X$_{p\ y}$ for the samples partially regenerated by pyrolysis.

Xenon is adsorbed in a volumetric apparatus at 300K and 212K to obtain saturation. Adsorption equilibrium is attained after 30 minutes. Before adsorption and NMR measurements, all the samples are outgassed under vacuum (P<10^{-5} torr) at 473K. Under these conditions, even "light coke" remains in the solids.

The 129-Xe NMR spectra are recorded at the same temperature on a Bruker CXP 100 spectrometer at 24.9 MHz. The duration of recycle delay is 0.5s and the number of scans (NS) between 1,000 and 20,000. The 27-Al NMR spectra are obtained on a Bruker MSL spectrometer operating at 94 or 130 MHz with high speed rotation (>10^4 Hz) at the magic angle. The duration of recycle delay is 4s and NS = 160.

The principle of 129-Xe NMR.

The high polarizability of the xenon atom makes it very sensitive to its environment. Small variations in physical interactions cause marked perturbations of the electronic cloud which result in large chemical shifts. Consequently, 129-Xe NMR is a convenient tool for studying microporosity in zeolite systems.

Fraissard, et al. (6) have shown that the chemical shift δ of adsorbed xenon is the sum of several terms corresponding to the various perturbations it suffers:

$$\delta = \delta_{ref} + \delta_s + \delta_{Xe-Xe} + \delta_{SAS}$$

where $\delta_{ref} = 0$ (chemical shift of the gas at 0 pressure). δ_s depends on the Xe -wall interactions, δ_{Xe-Xe} is the contribution of Xe-Xe collisions and δ_{SAS} expresses the contribution due to the existence

of strong adsorption sites. These SAS are often charged and
sometimes paramagnetic. In practice, the information provided by
Xe-NMR is contained in the form and the expression of the $\delta = f(n_{Xe})$ curve where n_{Xe} is the number of xenon atoms adsorbed by
one gram of anhydrous zeolite. In the absence of SAS, δ_s can be
obtained by extrapolation of the δ-curve to zero pressure and
related to the mean free path of xenon imposed by the zeolite
structure (7). The slope, $d\delta/dn$, of the linear section of the curve is
inversely proportional to the void volume of the cavities
accessible to the xenon (8).

Results and discussion.

-Xe adsorption isotherms. First, let us consider the samples
FY-SW coked during ortho-xylene cracking, then partially or
totally regenerated. Figure 1 shows the 300K xenon adsorption
isotherms for these different samples. It can be seen that for a
given P (whatever it is, below 1,000 torr), the number of xenon
atoms increases with the amount of residual coke. This result
shows that the Xe-wall interactions increase with the coke level
even though the internal volume of the zeolite diminishes, either
the "coke molecules" form adsorption centers stronger than the
internal surface of the zeolite or the Van der Waals interactions
relative to the curvature of the cages increases when their
diameter decreases (9). On the other hand, the amount adsorbed
at saturation (212K) naturally depends on the residual volume
(Table 1), increasing when the coke level decreases, first of all
slightly (1.75×10^{21} atom.g^{-1} for FY-SW 8.7X-Ox$_{4.8}$), then more
significantly(2.3×10^{21} atom.g^{-1} for FY-SW 8.7X-Ox$_{1.6}$). This result
shows that the first oxidation step almost exclusively concerns
the external coke, as is confirmed by the 129-Xe NMR results.
     If one now compares the initial sample (FY-SW) with that
coked then totally reoxidized (FY-SW 8.7X-Ox$_0$), it can been seen
that whatever the given P (P<1,000 torr) the amount of xenon
adsorbed is twice as small for the reoxidized sample (for example,
5.2×10^{18} and 10^{19} atom.g^{-1} for FY-SW 8.7X-Ox$_0$ and FY-SW,
respectively, at 20 torr). This difference appears again in the
values of the amounts adsorbed at saturation (Table 1): 1.3×10^{21}
and 2.5×10^{21} atom.g^{-1}.
     Roughly the same observations can be made for the samples
coked during hexane cracking (FY-SW 8.6H). However, the
differences between the initial sample and that which has been
completely reoxidized are much smaller : the 300K xenon

adsorption isotherms (Figure 2) show a difference of only 20% between the amounts adsorbed whatever P (P < 1,000 torr); for example, 10^{19} and 8×10^{18} atom.g^{-1} at 20 torr for FY-SW and FY-SW 8.6H-Ox_0, respectively.

All these results therefore seem to show very qualitatively that the elimination of carbonaceous residues formed during hydrocarbon cracking does not allow one to recover the initial qualities of the original catalyst, at least under our experimental conditions. Elimination of the last "coke molecules" leads to the appearance of structure defects or of a short-range amorphization of the lattice which blocks access to certain cavities. This phenomenon is distinctly more pronounced when an aromatic molecule (ortho-xylene) is cracked. This result is doubtless a consequence of the nature of the coke, essentially aromatic and polyaromatic as is shown by the 13-C NMR (10).

Table 1:Maximal number of xenon atoms adsorbed at saturation

Samples	n(Xe) atom g^{-1}
FY-SW	2.50×10^{21}
FY-SW-8.6H	1.50×10^{21}
(FY-SW-8.6H-Ox)$_{4.7}$	1.7×10^{21}
(FY-SW-8.6H-Ox)$_{1.7}$	2.1×10^{21}
(FY-SW-8.6H-Ox)$_0$	2.2×10^{21}
FY-SW-8.7X	1.70×10^{21}
(FY-SW-8.7X-Ox)$_{4.8}$	1.75×10^{21}
(FY-SW-8.7X-Ox)$_{1.6}$	2.30×10^{21}
(FY-SW-8.7X-Ox)$_0$	1.30×10^{21}

129-Xe NMR results. All the spectra show only one symmetrical line resonance which broadens when the coke content increases.

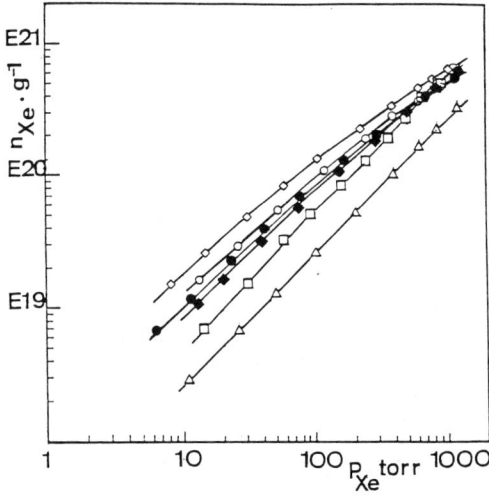

Figure 1 : <u>Xenon adsorption isotherms at 300K:</u> □ FY-SW;
◇ FY-SW8.7X; ● FY-SW8.7X$_{p y}$; ○ FY-SW8.7XOx$_{4.8}$;
◆ FY-SW8.7XOx$_{1.6}$;   △ FY-SW 8.7XOx$_0$.

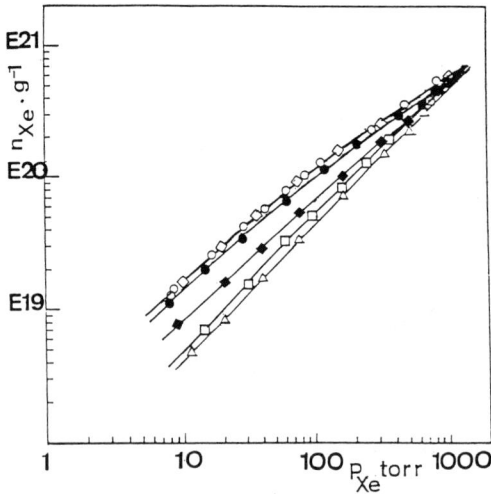

Figure 2 : <u>Xenon adsorption isotherms at 300K:</u>   □ FY-SW,;
◇ FY-SW8.6H,; ● FY-SW8.6H$_{p y}$; ○ FY-SW8.6HOx$_{4.7}$;
◆ FY-SW8.6HOx$_{1.7}$; △ FY-SW 8.6HOx$_0$.

Let us consider the samples coked with ortho-xylene (FY-SW 8.7X-Ox$_y$). Figure 3 shows the $\delta = f(n_{Xe})$ variations and Table 2 gives the values of δ_s and dδ/dn for the different samples studied. In the case of the initial product, FY-SW, this has at low concentrations a curvature towards high chemical shifts characteristic of SAS (stronger than Na$^+$ or H$^+$ at least) corresponding to more or less complexed extraframework aluminated species, as has been shown by Barrage, et al. (3). After coking (FY-SW8.7X), the $\delta = f(n_{Xe})$ variation is linear over the entire concentration range; the disappearance of the curvature indicates that the coke forms first of all on or near the extraframework aluminated species, thus masking the effect of the SAS. If we consider that coke attacks the strong acid sites of the zeolite first, this result confirms that extraframework aluminiums are implied in the formation of these strong acid sites (certainly by an interplay between framework and non framework aluminiums as it is now generally accepted). The ratio of the slopes of FY-SW8.7X (5.6x10^{-20} ppm.atom^{-1}) and FY-SW (3.85x10^{-20}) is 1.45, which expresses a loss of internal volume of about 30%. This result is in perfect agreement with the inverse ratio of the amount of Xe adsorbed at saturation at 212K (2.5x10^{21} and 1.7x10^{21} atom.g^{-1}, respectively) which is equal to 1.47. A large fraction of the coke is therefore at the external surface of the catalyst. On the other hand, xenon diffusion in the internal volume is strongly restricted, since δ_s goes from 60 ppm (FY-SW) to 85 ppm (FY-SW8.7X); the internal coke is therefore situated mainly at the windows between the supercages.

For the samples partially regenerated by pyrolysis or under oxygen (residual coke content >4.8%), the $\delta = f(n_{Xe})$ plots are straight lines parallel to that of FY-SW8.7X; the slopes are therefore the same but δ_s decreases slightly. This result shows that the first steps of regeneration eliminate the external coke blocking the pore openings. Under these conditions the internal pore volume is practically unchanged, as is confirmed by the amount of xenon adsorbed at saturation at 212K (1.75x10^{21} atom.g^{-1}). For a smaller residual coke load (FY-SW8.7Xox$_{1.6}$), there is both a decrease in the slope (3.85x10^{-20} ppm/atom.g^{-1}), which returns to that corresponding to the non-coked sample, and in δ_s (75 ppm). However, the δ-variation remains linear over the whole concentration range. These results show that the internal coke affecting the supercages is to a large extent eliminated; the amount of xenon adsorbed at saturation (2.3x10^{21} atom.g^{-1})

confirms this result. The absence of curvature indicates that the residual coke that is most difficult to eliminate is located mainly at the supercage windows on the extraframework aluminiums of the catalyst.

After total elimination of the carbonaceous residues (FY-SW8.7Xox$_0$), the shape of the $\delta = f(n_{Xe})$ curve is roughly the same as that of the non-coked sample, but:

-the curvature at low xenon concentration is accentuated : complete oxidation of the coke leads to an increase in the dealumination of the lattice;

- the slope of the linear straight section is greatly increased (x2.3): the internal free volume must therefore be divided by the same factor. This is what is in fact observed by measuring the amount of xenon adsorbed at saturation (1.3×10^{21} and 2.5×10^{21} atom.g^{-1} for FY-SW8.7Xox$_0$ and FY-SW, respectively).

Complete elimination of coke under our regeneration conditions leads therefore to the appearance of structure defects and/or short-range amorphization of the zeolite lattice.

For the samples coked by hexane (FY-SW 8.6H), the $\delta = f(n)$ variations as a function of the residual coke content (Figure 4) are fairly similar to those observed with FY-SW 8.7X, but if one compares the totally reoxidized sample (FY-SW 8.6H-Ox$_0$) with the initial sample (FY-SW), one finds:

- that the curvature at low xenon concentration is hardly more pronounced: complete reoxidation leads to little further dealumination.

- that the $d\delta/dn$ slope for the regenerated sample is admittedly higher, but the factor is only 1.2 (4.6×10^{-20} (FY-SW) and 3.85×10^{-20} atom.g^{-1} (FY-SW 8.6H-Ox$_0$), respectively). The ratio of the slopes indicates a loss of internal volume of about 15%, which is confirmed by the amounts of xenon adsorbed at saturation (Table 2).

Total oxidation of a sample coked by hexane cracking leads to fewer modifications of the structure or loss of crystallinity than that of a sample coked by ortho-xylene cracking. 129-Xe NMR confirms clearly and more quantitatively that the carbonaceous residues formed by cracking an aromatic hydrocarbon are more difficult to eliminate because of the higher proportion of polyaromatic coke, as is shown by the 13-C NMR.

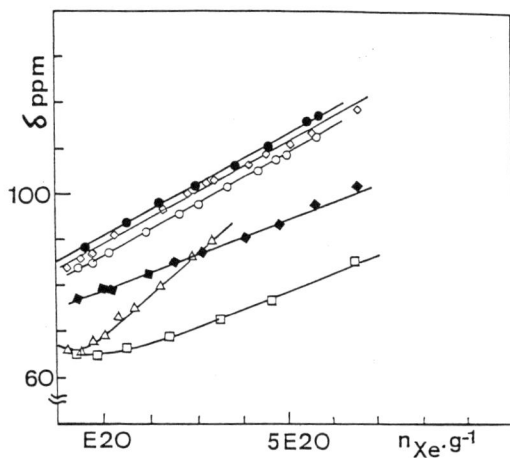

Figure 3 : $\delta = f(n_{Xe})$ at 300K: □ FY-SW; ● FY-SW8.7X;
◇ FY-SW8.7X$_{p\ y}$; ○ FY-SW8.7XOx$_{4.8}$; ◆ FY-SW8.7XOx$_{1.6}$;
△ FY-SW8.7XOx$_0$.

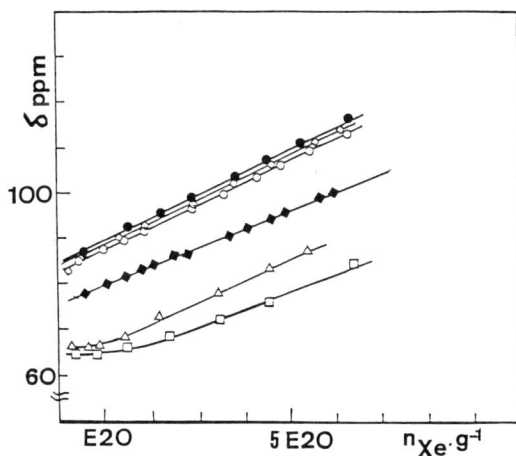

Figure 4 : $\delta = f(n_{Xe})$ at 300K: □ FY-SW, ● FY-SW8.6H,
◇ FY-SW8.6H$_{p\ y}$; ○ FY-SW8.6HOx$_{4.7}$; ◆ FY-SW8.6HOx$_{1.7}$;
△ FY-SW 8.6HOx$_0$.

Table 2: NMR parameters from the $\delta = f(n_{Xe})$ curves

300K	$\delta_{as \to 0}$ (ppm)	slope $d\delta/dn$ (ppm/at.g^{-1}) x 10^{20}
FY-SW	59.5	3.85
FY-SW-8.6H	85.0	5.10
FY-SW-8.6H pyrolyzed 673K	84.2	5.00
(FY-SW-8.6H-Ox)$_{4.7}$	83.0	5.00
(FY-SW-8.6H-Ox)$_{1.7}$	75.7	4.25
(FY-SW-8.6H-Ox)$_0$	62.5	4.60
FY-SW-8.7X	85.0	5.60
FY-SW-8.7X pyrolyzed 623K	83.5	5.60
(FY-SW-8.7X-Ox)$_{4.8}$	80.8	5.60
(FY-SW-8.7X-Ox)$_{1.6}$	75.0	3.85
(FY-SW-8.7X-Ox)$_0$	60.0	8.85

The interpretation of our preceding results is confirmed by the 27-Al NMR spectra. Figure 5 shows the spectra obtained after coking and decoking the FY-SW samples. For the non-coked sample, spectrum (2-a) shows two lines at 60 ppm (A, tetracoordinated Al of the lattice) and at 0 ppm (B, hexacoordinated extraframework Al). Coking gives rise to an additional signal (C), at about 30 ppm, whose intensity increases with the amount of coke (spectra 2-b and 2-c). After total elimination of the coke, the relative intensity of signal C decreases and that of signal B increases (Figure 5-d). We attribute the signal C to lattice aluminiums distorted by the coke and therefore weakened. It is reasonable to think that the elimination of this coke causes the extraction of this weakened Al from the lattice, leading, as we said before, to structure defects and partial amorphization of the lattice.

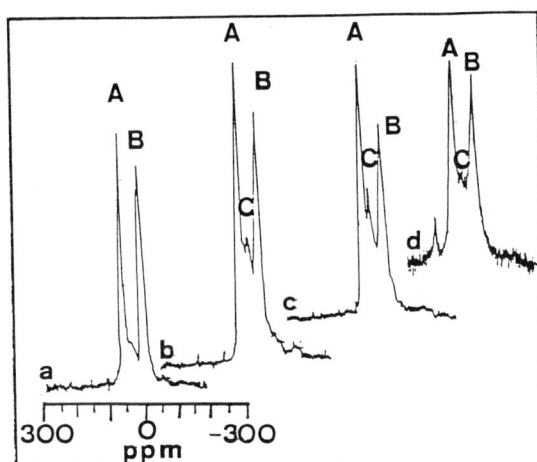

Figure 5: <u>Al-NMR spectra</u> : a-FY-SW; b-FY-SW4.7X; c-FY-SW8.7X
d-FY-SW8.7X-Ox$_0$

Conclusion

Coupled with 27-Al NMR, which allows us to specify the role of certain lattice aluminiums, 129-Xe NMR has proved to be an interesting tool for studying the deactivation of zeolites by coking. Morever, this technique makes it possible to monitor the different stages of regeneration and therefore to specify the experimental conditions required to re-establish the structure and the initial activity of the catalyst. In this way, we have shown unambiguously that a relatively mild oxidation (O_2 8 l.h^{-1} at 573K) nevertheless causes structure defects leading to a significant loss of microporosity. This phenomenon is more pronounced for the samples coked by an aromatic hydrocarbon (ortho-xylene, in our case) because of the large proportion of polyaromatic coke which is more difficult to eliminate.

Literature cited

1.   Magnoux, P; Cartraud, P.; Mignard, S.; and Guisnet, M..*J. of Catal.* **1987**, 106, 235
2.   Ito,T; Bonardet,J.L.; Fraissard J.; Nagy J.B.; André C.; Gabelica Z and Derouane E.G. *Appl. Catal.* **1 9 8 8**, 43, 5
3.   Barrage, M.C.; Bonardet,J.L.; and Fraissard.*J. Catalys. Lett.* **1 9 9 0**, 5, 143 1
4.   Barrage, M.C.; Bauer, F.; Ernst, H.; Fraissard, J., Freude D., Pfeifer, H . *Catalys. Lett.* **1 9 9 0**, 6, 201
5.   Barrage, M.C; Bonardet,J.L; Fraissard, J; Kubelkova, L; Novakova,J.; Ernst, H.,; Freude, D. *Coll. Czech Chem Commun.* **1 9 9 2**, 57, 733
6.   Ito,T.; Fraissard J. In :*Proc. Vth Int. Conf. Zeolites, (Naples)* Rees L.V .Ed.; Heyden: London, Great-Britain,**1980**, pp510 - 5 1 5
7.   Springuel-Huet, M.A.; Demarquay, J.; Ito ,T.; Fraissard, J. In innovation in zeolite materials science; Grobet,P.J. et al Ed.; *Studies.Surf.Sci.andCatal..* Elsevier:Amsterdam, Netherlands, **1 9 8 8**, Vol 37, pp183-189
8.   Fraissard J., and Ito T. *Zeolites* **1 9 8 8**, 8, 350
9.   Derouane E.G; Nagy J.B. *Chem. Phys. Lett.* **1 9 8 7**,137, 4, 3 4 1
10. Barrage M.C. **T h e s i s**, **1 9 9 2**, Paris

RECEIVED June 17, 1994

Chapter 19

Fluid Cracking Catalyst ZSM-5 Additive Performance at Overcracking Mode

L. H. Hsing and R. E. Pratt

Texaco Research and Development, 4545 Savannah at Highway 73, Port Arthur, TX 77641

A series of experiments was conducted on a circulating pilot unit with 4 wt% ZSM-5 additive (one wt% pure ZSM-5 zeolite) present in an overcracking mode to study their combined effects on product yields and qualities in meeting future reformulated gasoline requirements. The changes in product yields and qualities are correlated with the reaction rates of cracking, hydrogen transfer, skeletal isomerization reactions at high temperature and the use of ZSM-5 additive.

The Clean Air Act Amendments of 1990 mandate a significant change in gasoline compositions to meet increasingly stringent environmental emission requirements. It is anticipated that an FCCU in the refining industry will undergo a significant change from a major gasoline producer to an important olefin generation unit to meet future reformalated gasoline needs. Among the options available on the FCCU to produce olefins is operation at high temperature (over 1000F) or in an overcracking mode. The would increase olefin yields for down-stream alkylation, MTBE, and TAME processes and can be easily adapted without committing major investment cost. The use of ZSM-5 additive in an FCCU is another option available to the refinery, and has been practiced in the industry to increase olefin production. Therefore, it is desirable to study the combined effects of ZSM-5 additive and overcracking operation in an FCCU in meeting reformulated gasoline requirements.

Experimental Work

The properties of the feedstock used for this study are given in Table I. A commercial equilibrium catalyst was used to carry out the work. The properties of the equilibrium catalyst are listed in Table II. The ZSM-5 additive used was Z-CAT PLUS from INTERCAT Corporation. According to INTERCAT, the active ZSM-5 content of Z-CAT PLUS additive is 25 weight percent. The additive was steam-deactivated at 1500F for three hours to simulate the equilibrium additive.

0097–6156/94/0571–0241$09.80/0

Table I. Feedstock Test Results

API GRAVITY	21.4
ANILINE POINT, F	163
BROMINE NO	16.6
WATSON AROMATICS, WT%	60.8
X-RAY SULFUR, WT%	2.517
BASIC N2, WPPM	412
TOTAL N2, WPPM	1949
R. I. 70C	1.4974
MICRO CARBON RESIDUE, WT%	0.680

	D1160, F
IBP	546
5	645
10	680
20	723
30	761
40	805
50	834
60	868
70	905
80	950
90	1003
95	1046
EP	1078

Table II. Catalyst Test Results

Catalyst	Equilibrium Catalyst
Metals on catalyst	
Ni, wppm	270
V, wppm	700
Fe, wt%	0.54
La, wt%	1.2
Ce, wt%	0.24
Na2O, wt%	0.47
Compositions	
Al2O3, wt%	35.4
SiO2, wt%	59.1
Surface Area (BET), m^2/gm	
Total	153
Matrix	50
Zeolite	103
Pore Volume, cc/gm	0.36
Bulk Density, gm/cc	0.846
Unit Cell size, A	24.31

The performance data were obtained on a Miniature Riser unit (MRU), which has a catalyst inventory of 3500 grams and feed rate capacity of 500 - 2000 cc/hr. A simplified sketch of the unit is shown in Figure 1. All the runs were carried out in an adiabatic mode at riser outlet temperature of 960 - 1040 F by adjusting the jacket temperature along the riser to simulate a commercial FCCU. The feed rate for all the runs was kept constant at 1000 cc/hr. For runs conducted at riser temperatures of 960 - 1040F, the feed preheat temperature was constant at 500F, and regenerator temperature was varied from 1300 to 1367F to simulate a heat-balanced commercial FCCU. Under these operational conditions, a temperature profile in the riser similar to that of a commercial FCCU was established. The temperature profile in the riser is very important for a pilot plant in matching product distribution of a commercial FCCU. Each run was carried out for 5 hours to collect enough liquid product to obtain sufficient light and heavy naphtha for octane determinations. An on-line wet gas meter and gas chromatograph were used to continuously monitor gas product volume and compositions. The liquid product was collected in a product tank for further fractionation, and analyzed by a GC for compositions.

Results and Discussion

Yields with ZSM-5 Additive in the Overcracking Mode. Conversion and yields in the overcracking mode are plotted in Figures 2-18. As shown in Figure 2, the addition of 4 wt% additive (1 wt% pure ZSM-5) to the base FCC equilibrium catalyst did not change the conversion over the temperature range studied. This agrees with the general consensus that the addition of reasonable amounts of additive to the unit will have only minimum effect on catalyst activity (1). Higher concentrations of additive will lower catalyst activity due to a dilution effect, thus resulting in higher catalyst circulation rate to maintain conversion. The high conversion in the overcracking mode was achieved by changing riser outlet temperature and catalyst/oil ratio simultaneously. In general, commercial ZSM-5 content ranges from 0.25 to 1.5 wt% depending on yield improvement needed. Effectiveness of ZSM-5 additive in increasing octane number and olefin production is a function of feedstock properties, base values of the product, and the ZSM-5 level used. Effectiveness of the ZSM-5 is the highest when the concentration is about 0.25 wt% and declines with increasing ZSM-5 content. 4 wt% ZSM-5 additive (1 wt% pure ZSM-5) is a typical dosage used in the industry to achieve the optimum performance.

As shown in Figure 3, dry gas yields increased with temperature, but no apparent increase at constant conversion was noticed with the addition of 4 wt% additive. As shown in Figures 4 and 5, C3=, N-C4=, and I-C4= yields increased at constant riser temperature with the addition of 4 wt% additive over the temperature range studied. As shown in Figure 6, branched C5= yields increased, but linear C5 olefin yields decreased with the addition of 4 wt% additive. This can be explained by the fact that ZSM-5 additive not only cracks the olefins in the heavy-end gasoline to lighter olefins, but also isomerizes the linear olefins to branched olefins, particularly for C5 olefins. It is known that n-C5 olefins are much easier to isomerize to their respective branched olefins than n-C4 olefins (2,3).

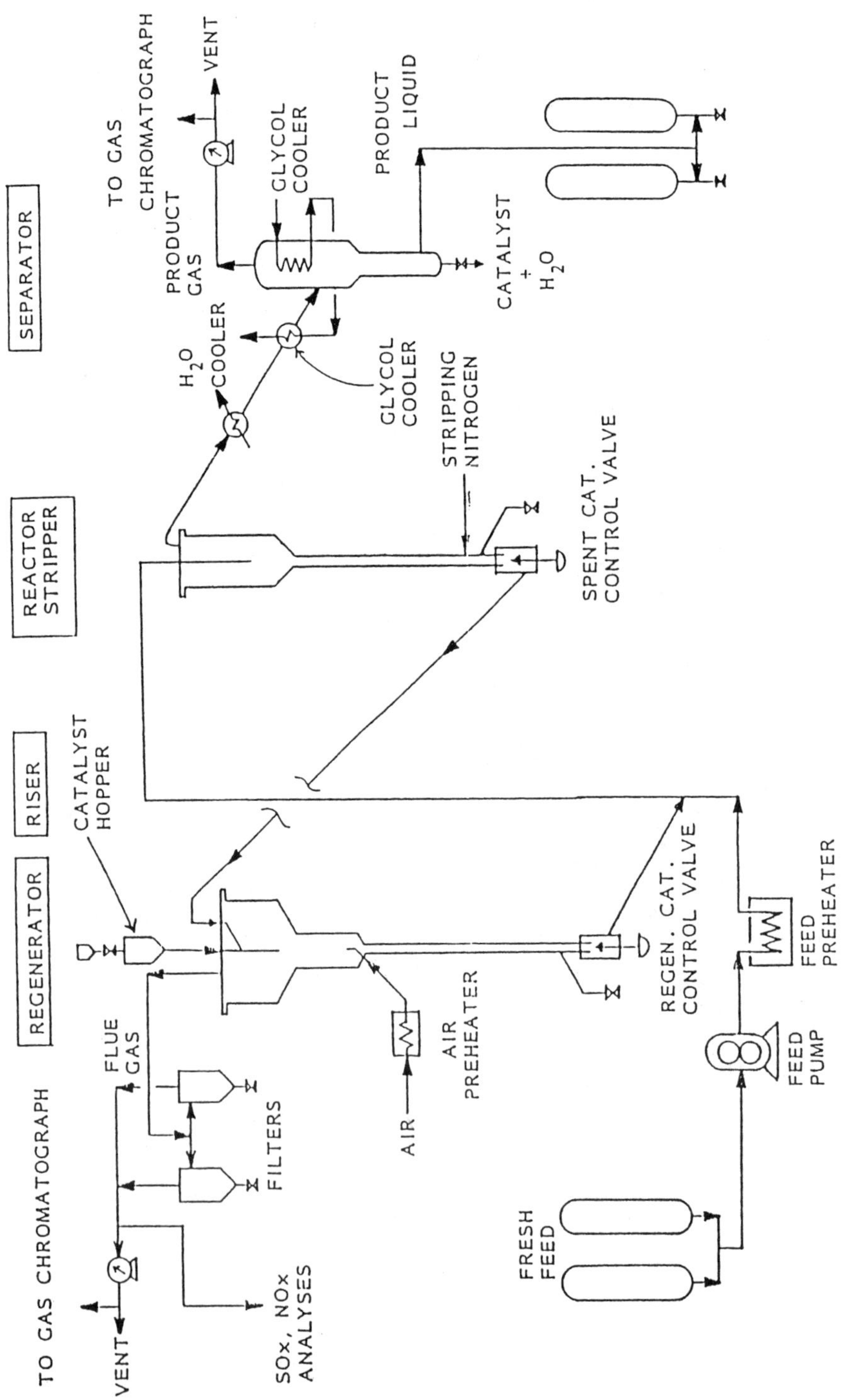

Figure 1. Miniature Riser FCC Pilot Unit.

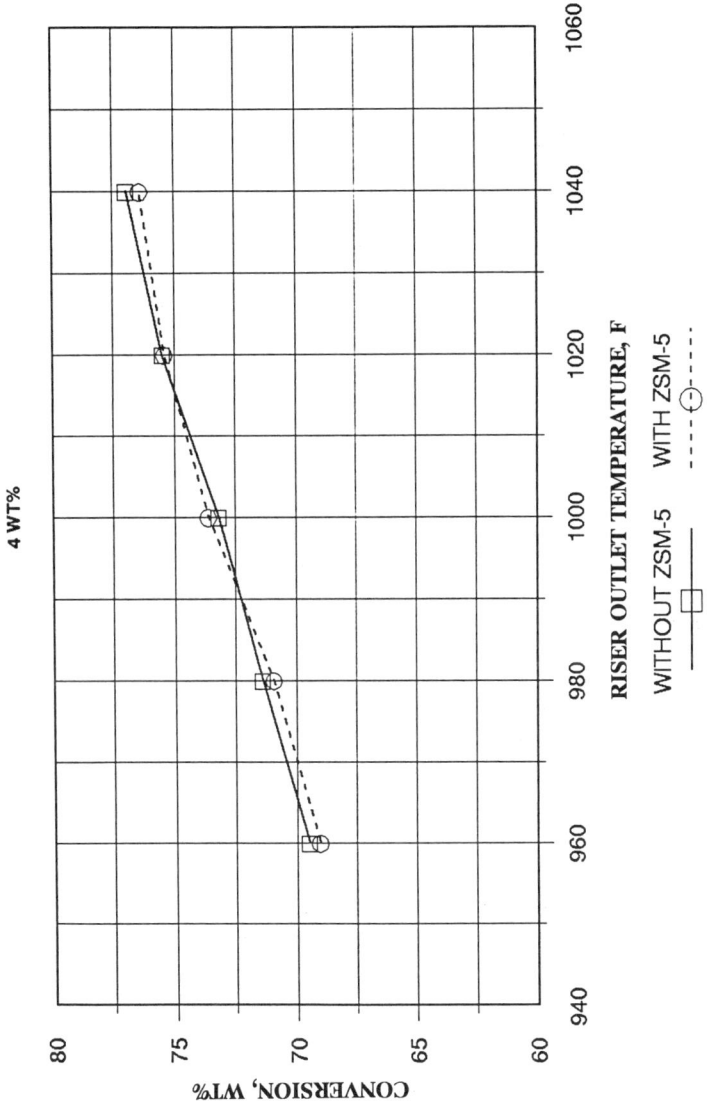

Figure 2. Conversion vs Riser Outlet Temperature.

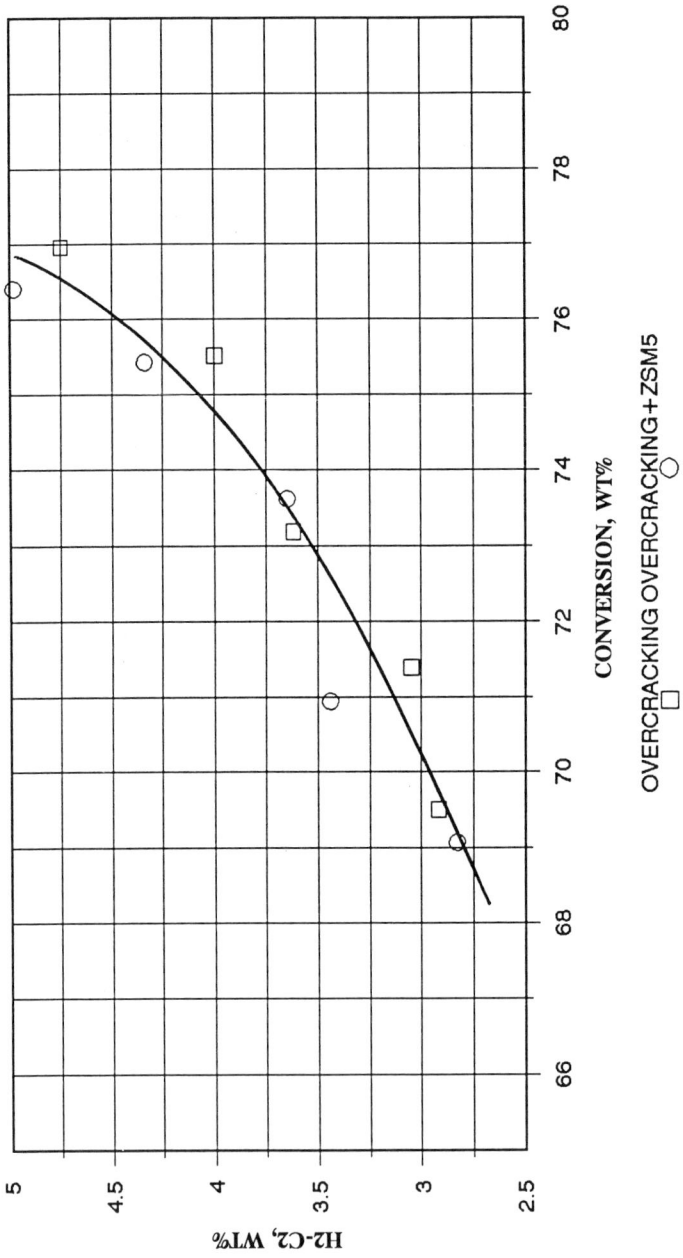

Figure 3. Dry Gas Yields vs Conversion.

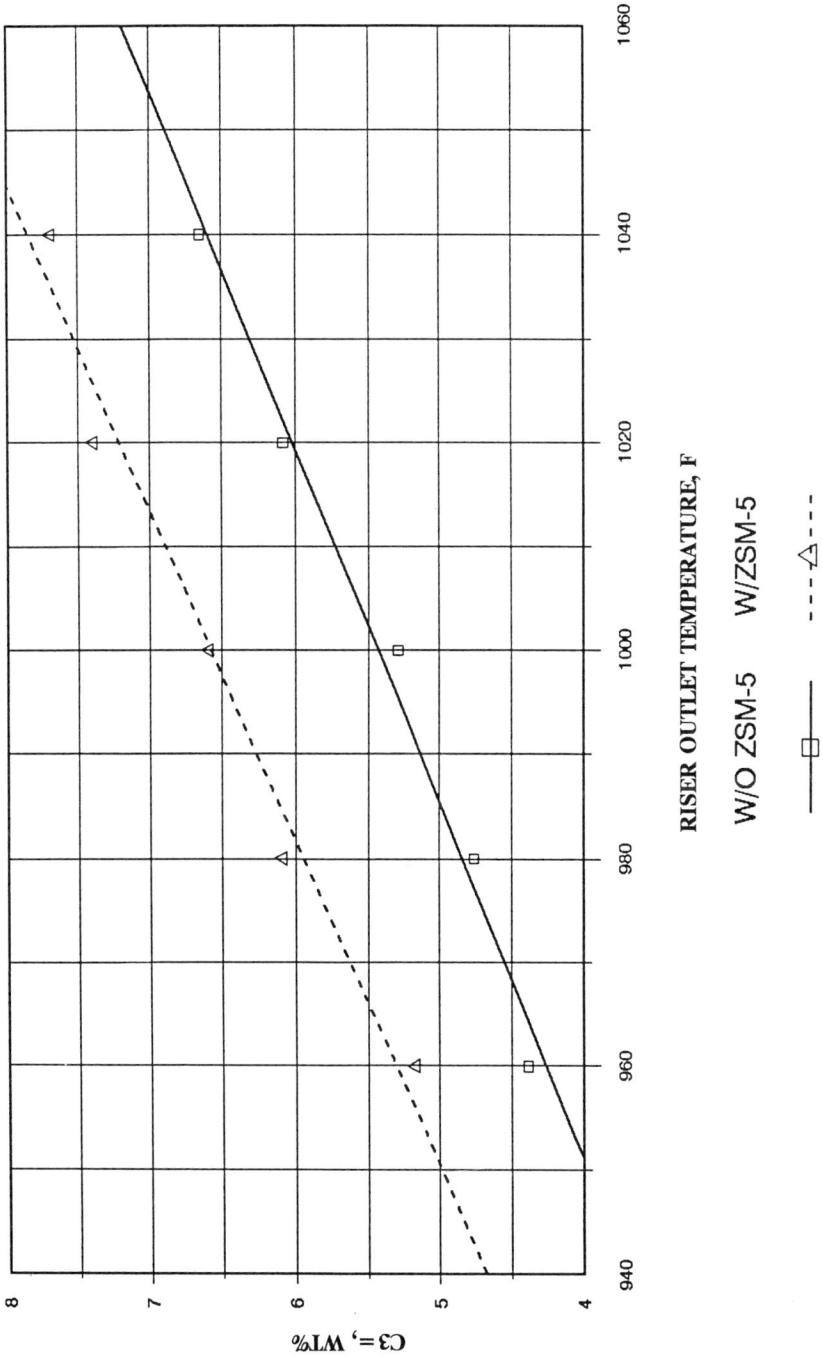

Figure 4. Propylene Yields vs Riser Outlet Temperature.

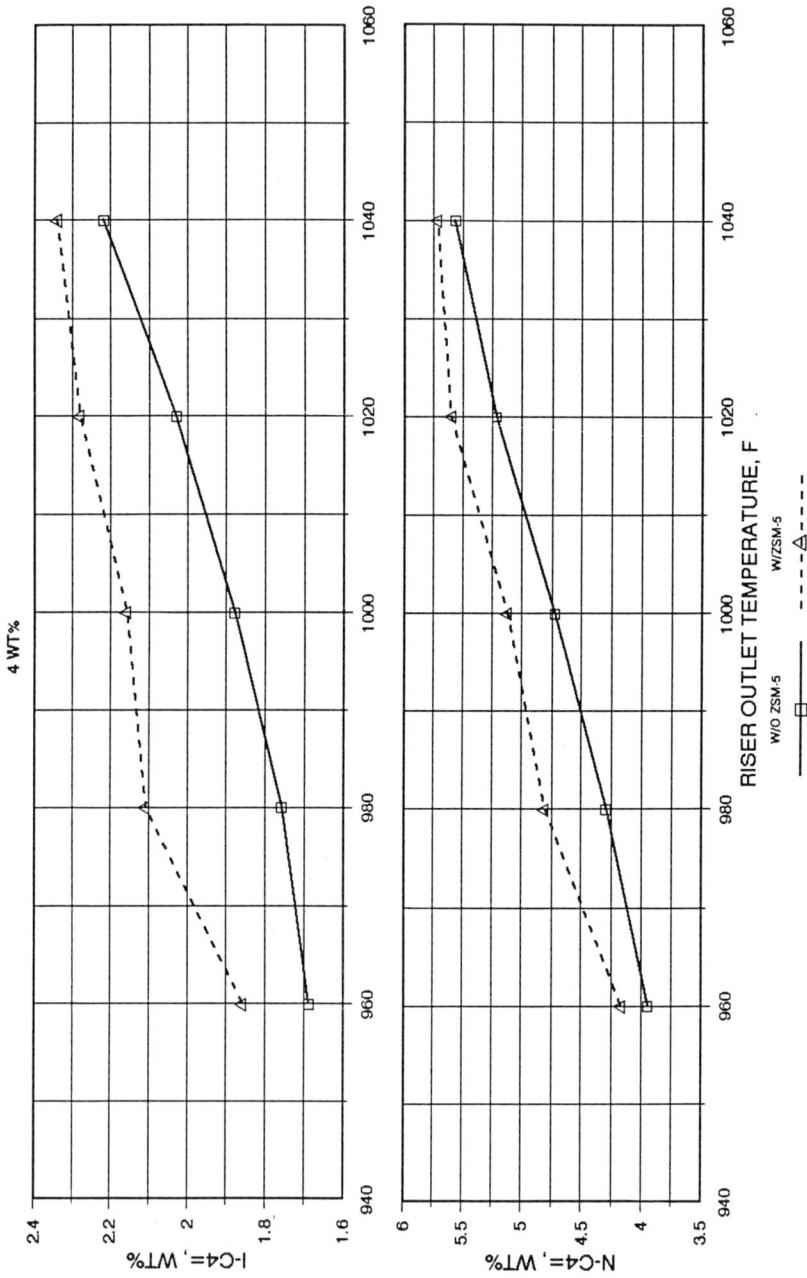

Figure 5. Butylene Yields vs Riser Outlet Temperature.

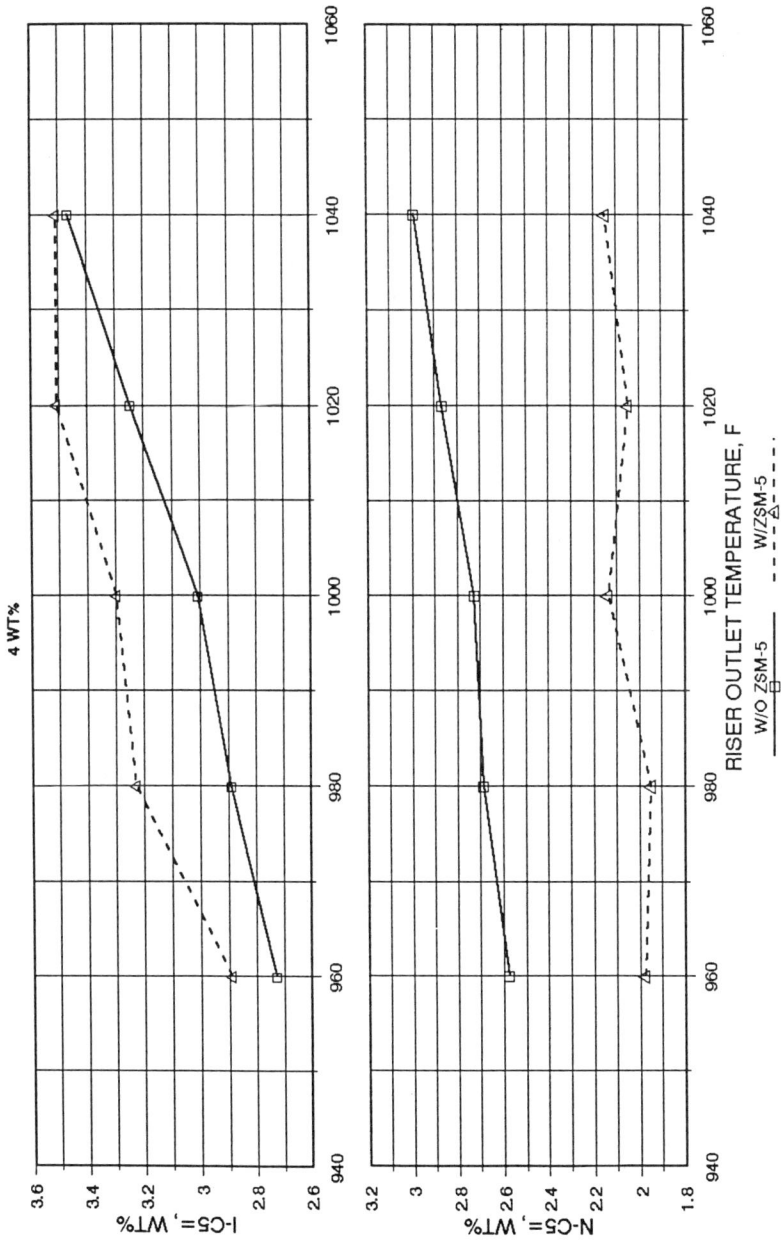

Figure 6. Pentene Yields vs Riser Outlet Temperature.

ZSM-5 additive effect on olefin yield ratio is plotted in Figure 7. Olefin yield ratio was calculated based on yields with additive divided by yields without additive at same temperature for C3, C4, C5 olefins. As shown, olefin yield increase with the addition of ZSM-5 additive was the highest for C3=, followed by C4 and C5 olefins. As a matter of fact, C5 olefin yields decreased with the use of additive. It is noted that the additive performance in promoting olefin production was the most effective when the riser outlet temperature ranged from 980-1000F. In an overcracking mode, effectiveness of ZSM-5 additive in promoting additional olefin generation (at 1040F) is diminished due to the high olefin yield at this temperature. Generally speaking, the higher increase in C3= than C4 and C5 olefins by ZSM-5 additive is not desirable due to C4 and C5 olefins being more valuable than C3= in making MTBE, TAME and alkylate. Therefore, an additive with the capability of preferentially producing C4/C5 olefins, particularly, branched olefin production, will command higher value. This means that an additive with acid sites promoting isomerization more than cracking will be needed to accomplish the desirable performance.

As shown in Figure 8, ZSM-5 additive increased I-C4= yield more than N-C4 olefins over the entire temperature range studied, and was most effective at 980F. A 20 wt% increase in I-C4= can be achieved by using the additive at this temperature. The higher increment in I-C4= than N-C4= results from the skeletal isomerization of N-C4= by the additive.

As shown in Figure 9, ZSM-5 additive increased I-C5= yields significantly more than N-C5=. Although the additive decreased overall C5 olefin production at all temperatures, branched C5= yields were always higher with additive. This indicates that branched C5= were produced at the expense of N-C5= from isomerization. At high temperature (1040F) the additive lost some of its effectiveness in increasing C5 olefin production. This is because C5 olefins at high temperature can be further cracked to lighter components.

As shown in Figure 10, C3-C5 paraffin yields increased with the use of additive except for N-C5. This is particularly true for iso-paraffins. This is true because part of the increased olefins produced by the additive were subsequently converted to paraffins by hydrogen-transfer reactions. The same reasoning can be applied to the decrease in N-C5 yield using the additive being due to lower N-C5= produced in the presence of the additive.

As shown in Figure 11, I-C4=/total C4= yield ratio increased from 0.3 to 0.32 at 960F, and 0.28 to 0.3 at 1040F; I-C5=/total C5= yield ratio increased from 0.5 to 0.57 when ZSM-5 was added to equilibrium FCC catalyst. A greater increase in the yield ratio with the additive was seen for the C5 olefins, since they are easier isomerization than the C4 olefins. This may explain why the C5 yield ratio is closer to the equilibrium limitation of 65 wt%. Nevertheless, both C4 and C5 yields were far below equilibrium limitations of 45 and 65 wt%, respectively.

C5=/total C5 yield ratio, shown in Figure 12, increased for branched C5=, but decreased for N-C5= by the addition of ZSM-5 additive. This also supports the isomerization mechanisms for C5 olefins by ZSM-5 additive.

I-C4/I-C4= and I-C5/I-C5= yield ratios vs conversion in Figures 13 and 14, are approximately the same for catalyst with and without ZSM-5 additive.

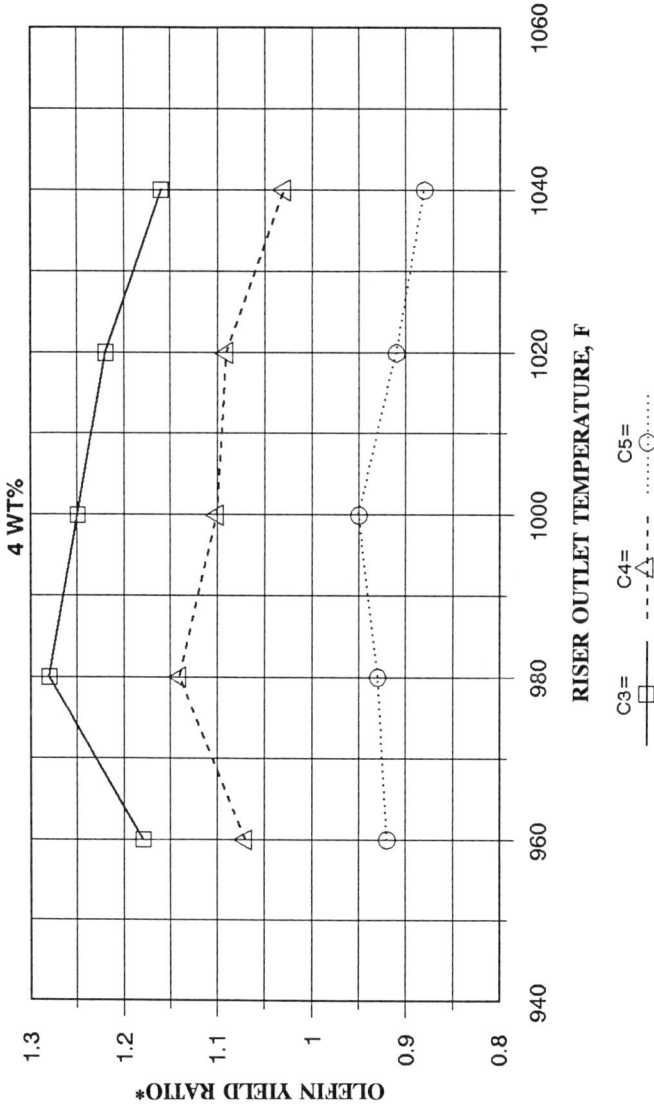

Figure 7. Olefin Yield Ratio vs Riser Outlet Temperature.

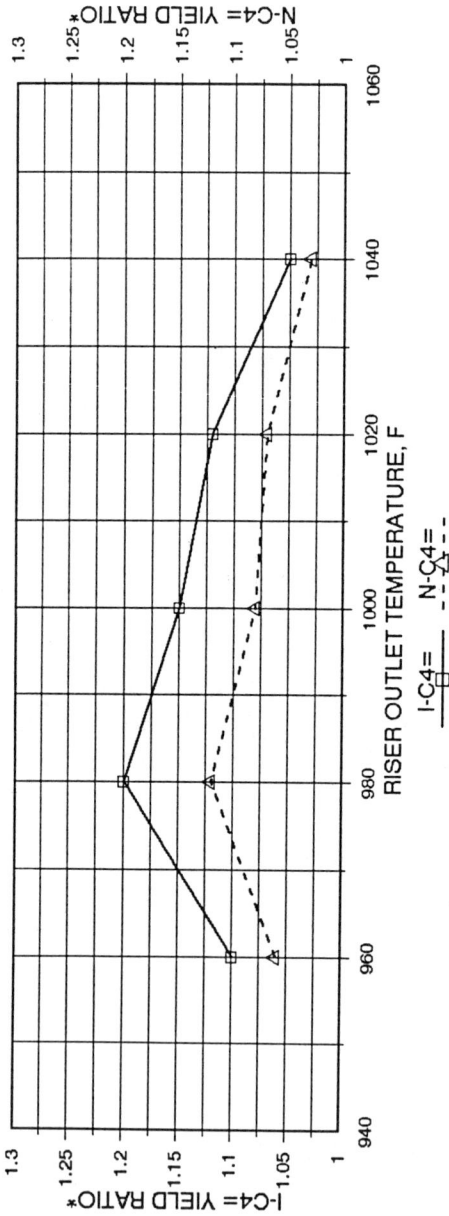

Figure 8. Isobutylene Yield Ratio vs Riser Outlet Temperature.

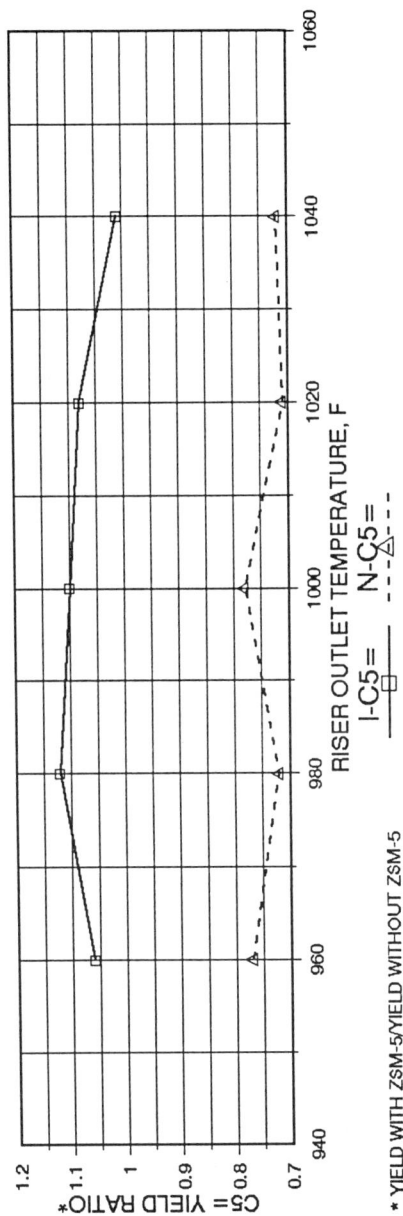

Figure 9. Pentene Yield Ratio vs Riser Outlet Temperature.

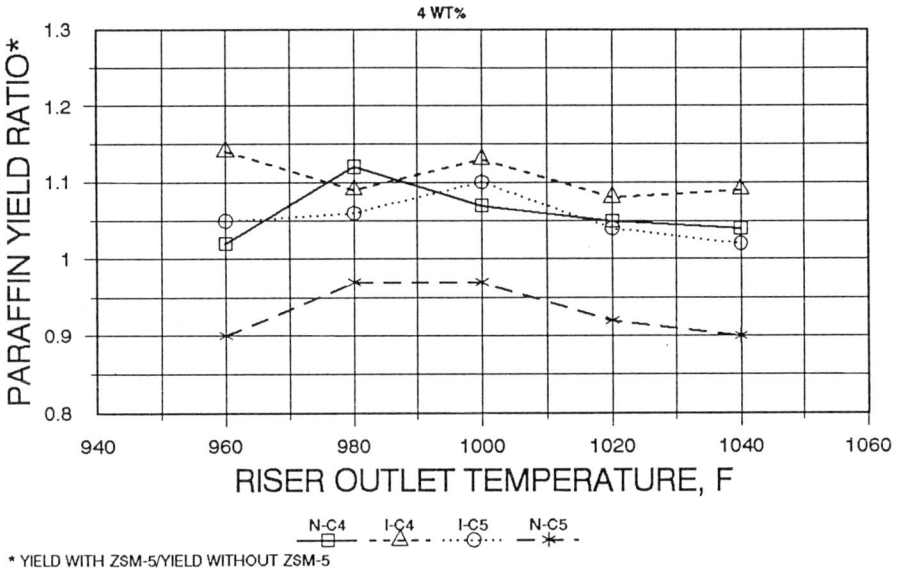

Figure 10. Paraffin Yield Ratio vs Riser Outlet Temperature.

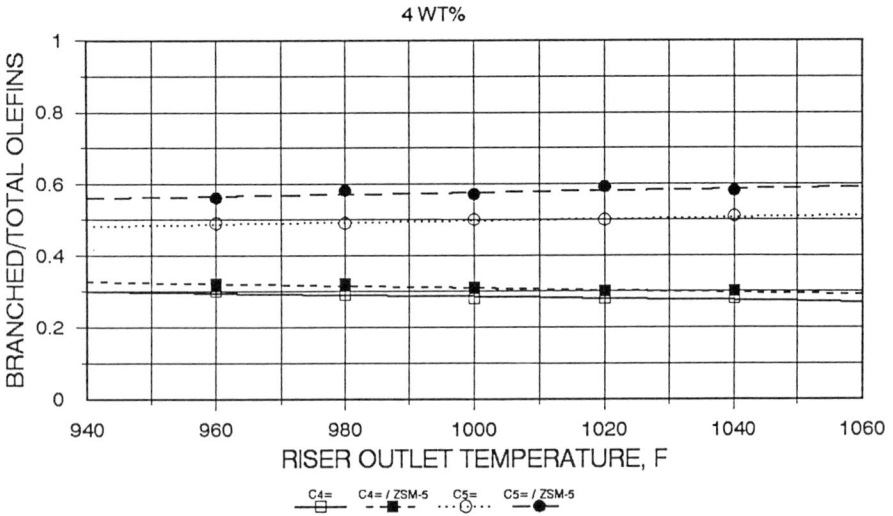

Figure 11. Branched/Total Olefins Yield Ratio vs Riser Outlet Temperature.

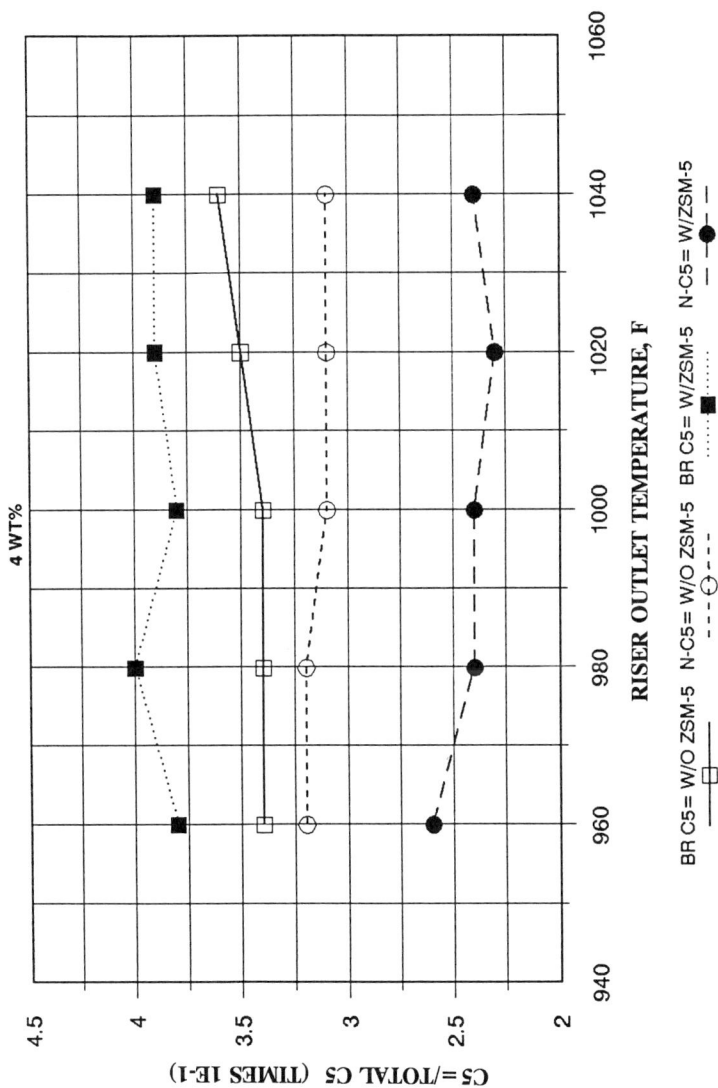

Figure 12. C5=/Total C5 Yield Ratio vs Riser Outlet Temperature.

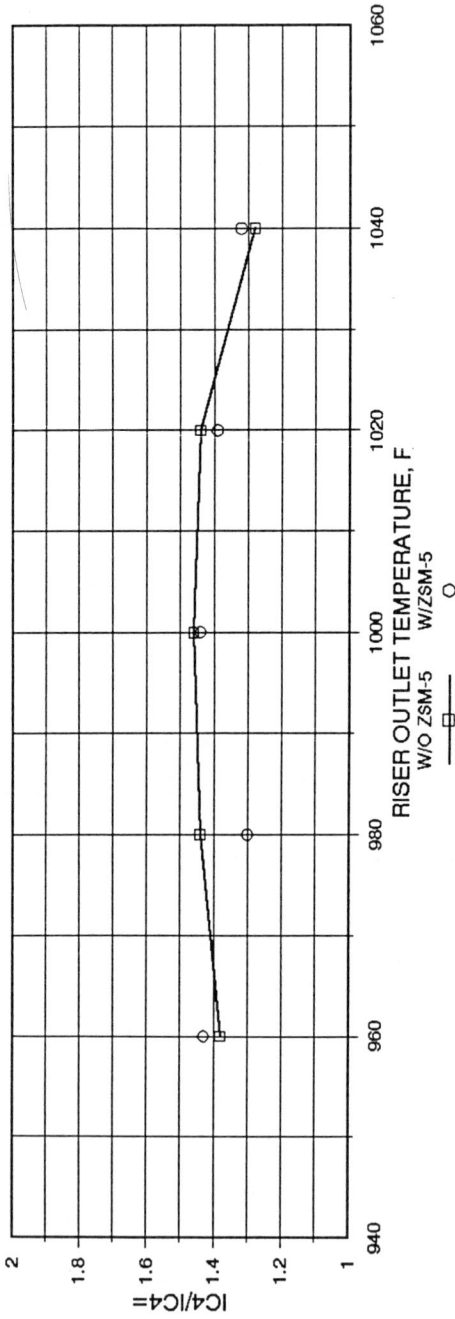

Figure 13. Isobutane/Isobutene Yield Ratio vs Riser Outlet Temperature.

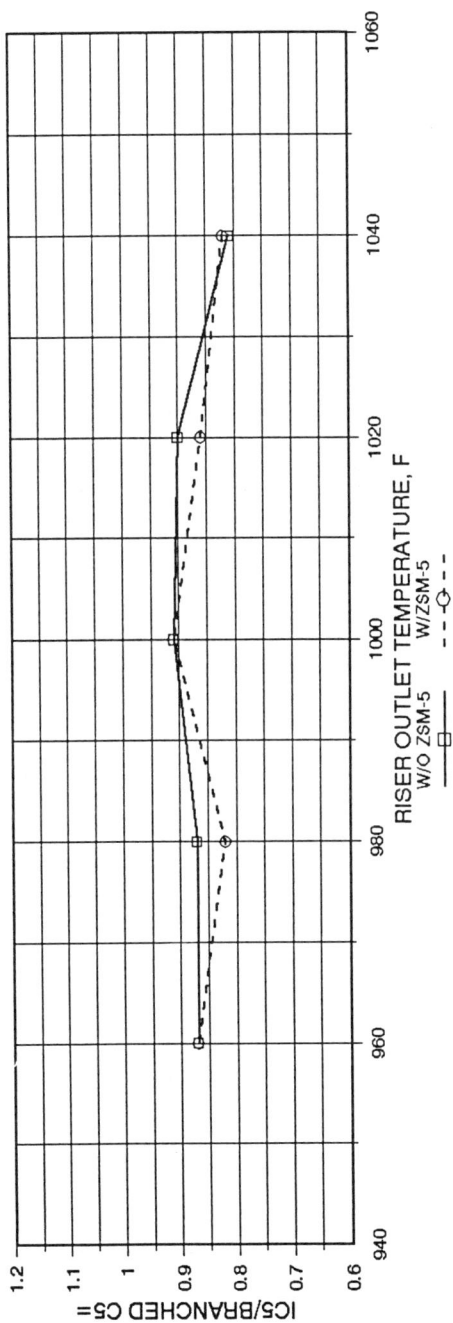

Figure 14.   Isopentane/Branched Pentene Yield Ratio vs Riser Outlet
Temperature.

This indicates that the increase of both saturated and unsaturated branched C4 and C5 components is at about the same rate with the addition of ZSM-5. (2MB1+2MB2)/I-C4= yield ratios (Figure 15) were lower with the use of ZSM-5 additive than without ZSM-5 indicating that the relative amounts of TAME/MTBE feedstock is slightly smaller with the use of ZSM-5 additive than without ZSM-5 at all temperatures tested.

It is noted in Figure 16 that the branched/total yield ratios for C4 and C5 increased with the addition of ZSM-5 and exceed the equilibrium limits of 0.45 and 0.65 for C4 and C5, respectively. This occurs because of ZSM-5's capability in promoting skeletal isomerization, particularly for C5 olefins. FC cracking produces isoolefins and n-olefins, some of which are converted to paraffins by hydrogen transfer (2). Normal olefins are also isomerized to isoolefins. The rate of hydrogen transfer which produces isoparaffins is greater than that which produces n-paraffins. Thus the total isoolefins + isoparaffins to n-olefins + n-paraffins ratio can be higher than the equilibrium produced during cracking.

The data in Figure 17 indicate that the increase of C3-C5 components by ZSM-5 was obtained at the expense of naphtha. This agrees with the consensus that ZSM-5 shape-selectively cracks gasoline range components into lighter products. The naphtha yield ratio shown in Figure 18 indicates that both C5-430 and C6-430F cuts lose 5-7 and 4-8 wt%, respectively, over a temperature range of 960 to 1040F. The most effective performance by the use of ZSM-5 was at temperature range of 980 to 1020F. The naphtha yield ratio is plotted based on yields without ZSM-5 at the same temperature.

Product Quality with ZSM-5 Additive. Product composition with and without ZSM-5 additive is plotted in Figures 19-30. All light naphthas shown in the Figures are depentanized (DP) LT naphthas unless otherwise noted. As shown in Figure 19, aromatics contents in both light and heavy naphthas increased with the addition of ZSM-5 additive due to preferential cracking of gasoline range olefins and paraffins into lighter components. This is supported by the lower gasoline olefin and paraffin contents shown in Figures 20 and 21. It is noted that the loss of gasoline olefins in the presence of ZSM-5 is higher than paraffins indicating the faster reaction rate of gasoline olefins than gasoline paraffins in the presence of an acid catalyst such as ZSM-5. As shown in Figure 22, the gasoline naphthene content when using ZSM-5 was slightly higher than without ZSM-5 since they concentrate when other material crack out of the gasoline range (concentration effect). The increase in the gasoline naphthene content in the presence of ZSM-5 is less than the increase in aromatics indicating that some gasoline naphthenes may crack to lighter components. The higher aromatics content in the gasoline pool when using ZSM-5 may not be realized after compensating for the increased oxygenate precursors produced by ZSM-5. As shown in Figure 23, benzene in Lt naphtha with ZSM-5 is higher due to concentration effect and can be at least partially compensated for by the increased oxygenates production (due to increased oxygenate precursors).

The product composition of the debutanized (DB) naphtha is plotted in Figures 24-28. The aromatics content of the DB naphtha shown in Figure 24 was significantly lower than that of the DP naphtha due to concentration effect, but the addition of ZSM-5 additive still increased the aromatics content of the DB naphtha; and the increase in aromatics content of the DB naphtha with temperature was slower than that of the DP naphtha.

4 WT%

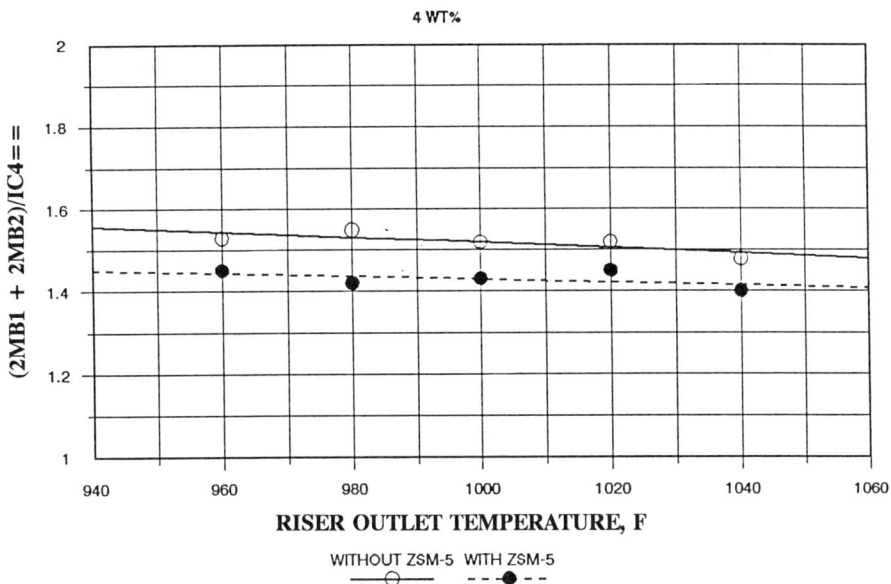

Figure 15.   (2MB1 + 2MB2)/Isobutene Yield Ratio vs Riser Outlet Temperature.

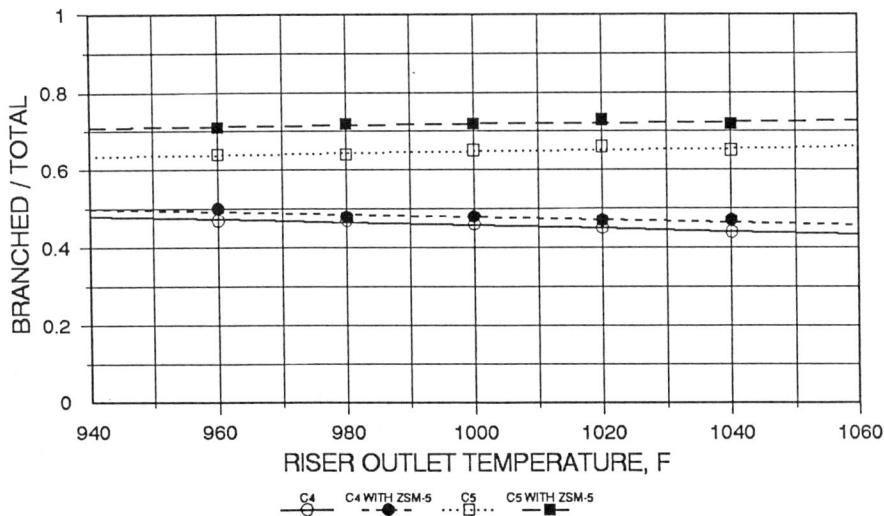

Figure 16. Branched/Total Yield Ratio vs Riser Outlet Temperature.

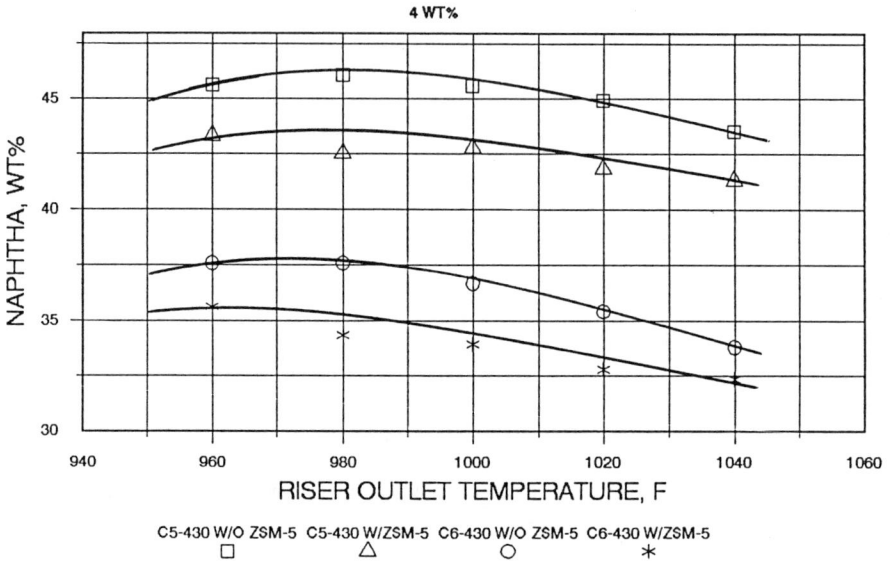

Figure 17. Naphtha Yields vs Riser Outlet Temperature.

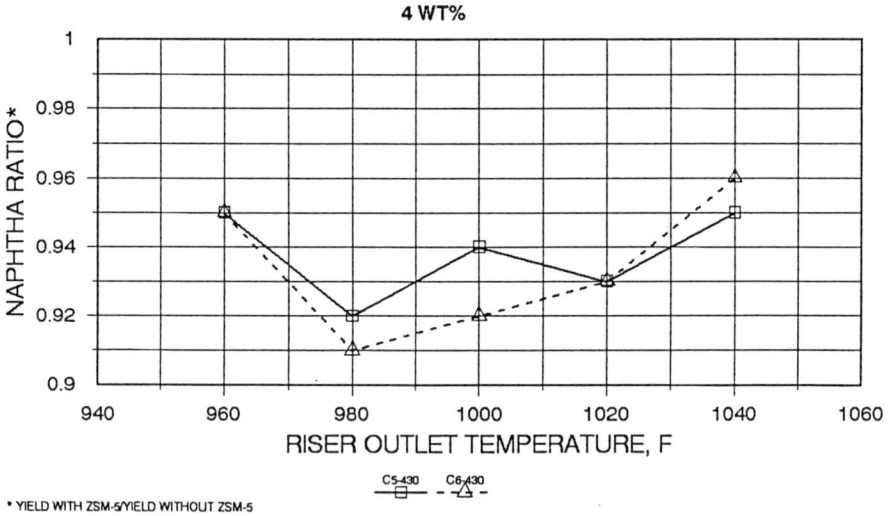

Figure 18. Naphtha Yield Ratio vs Riser Outlet Temperature.

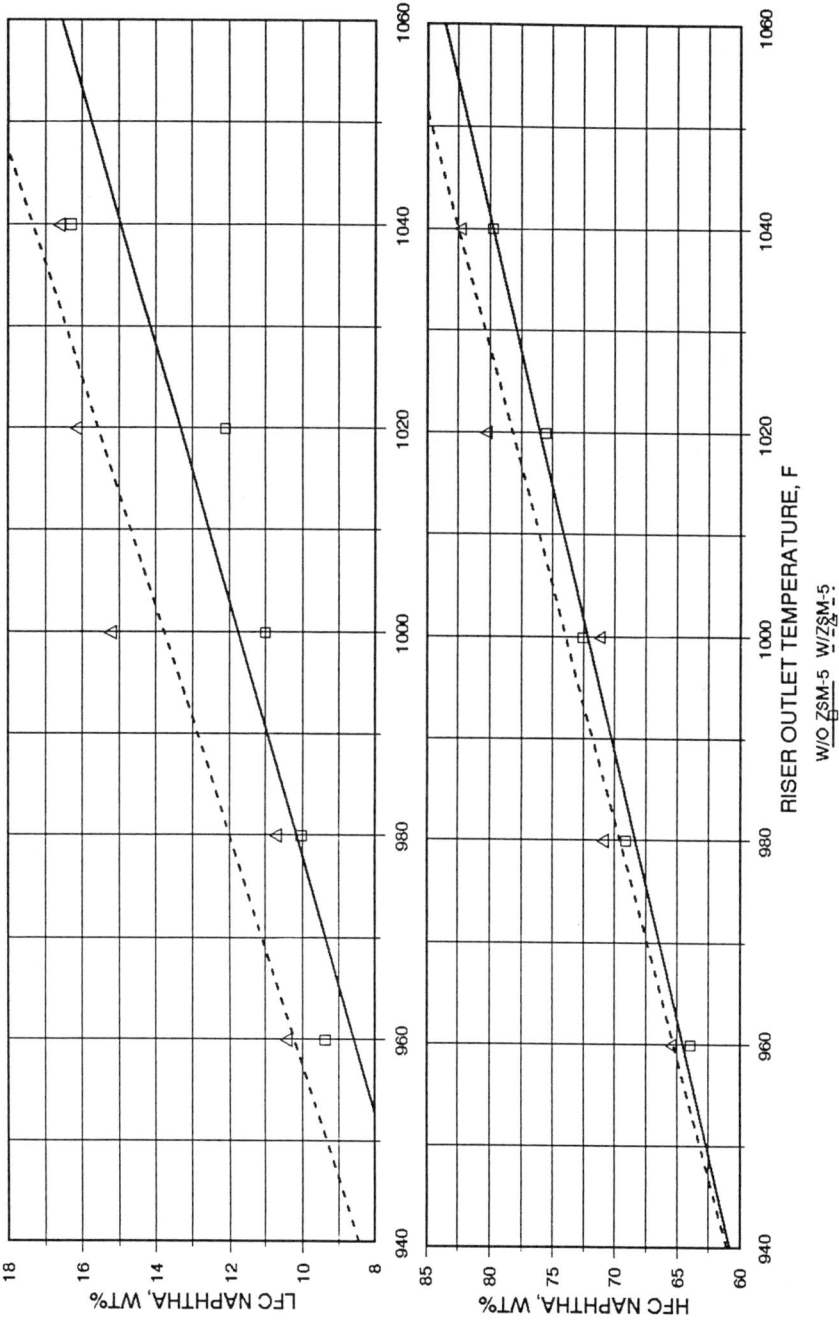

Figure 19. Naphtha Aromatics Content vs Riser Outlet Temperature.

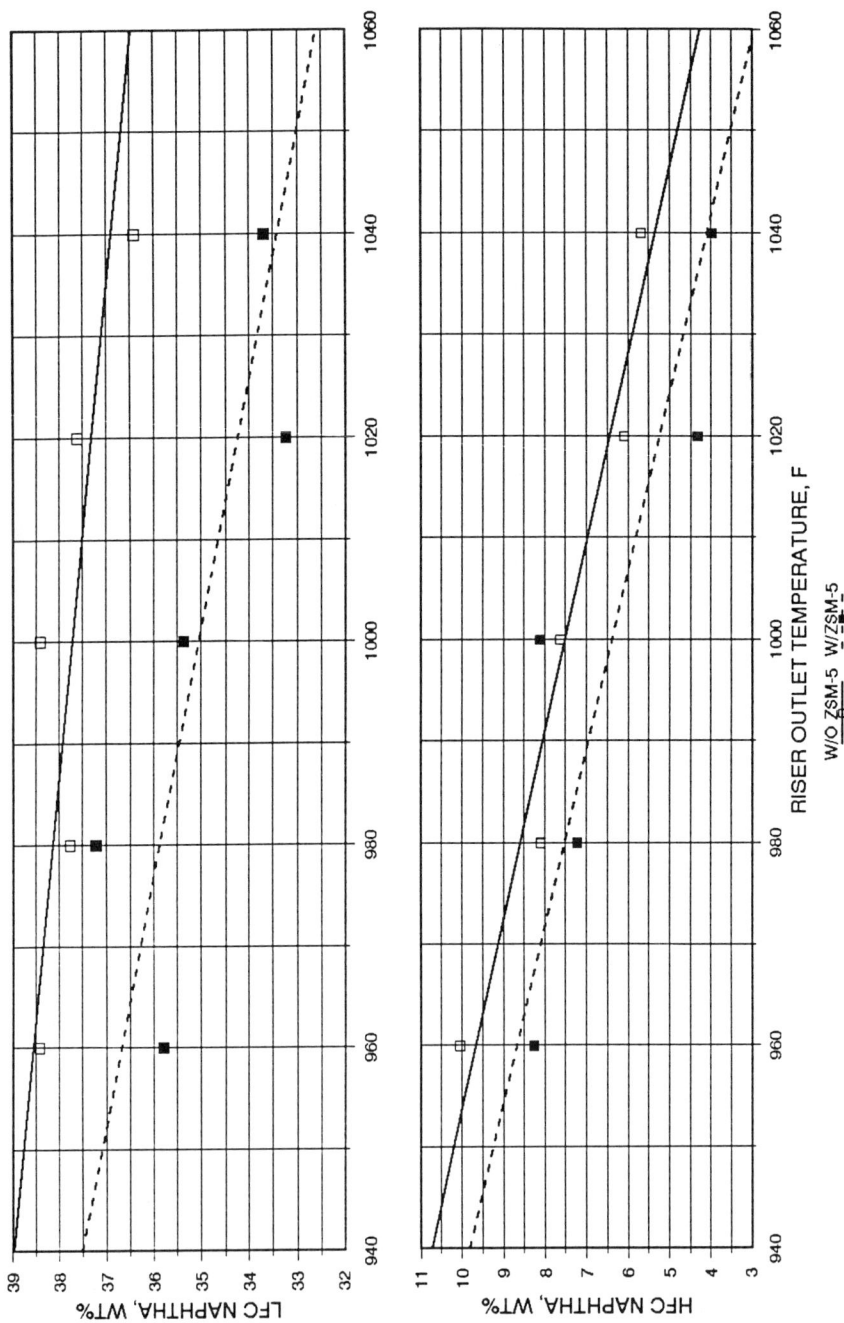

Figure 20. Naphtha Olefin Content vs Riser Outlet Temperature.

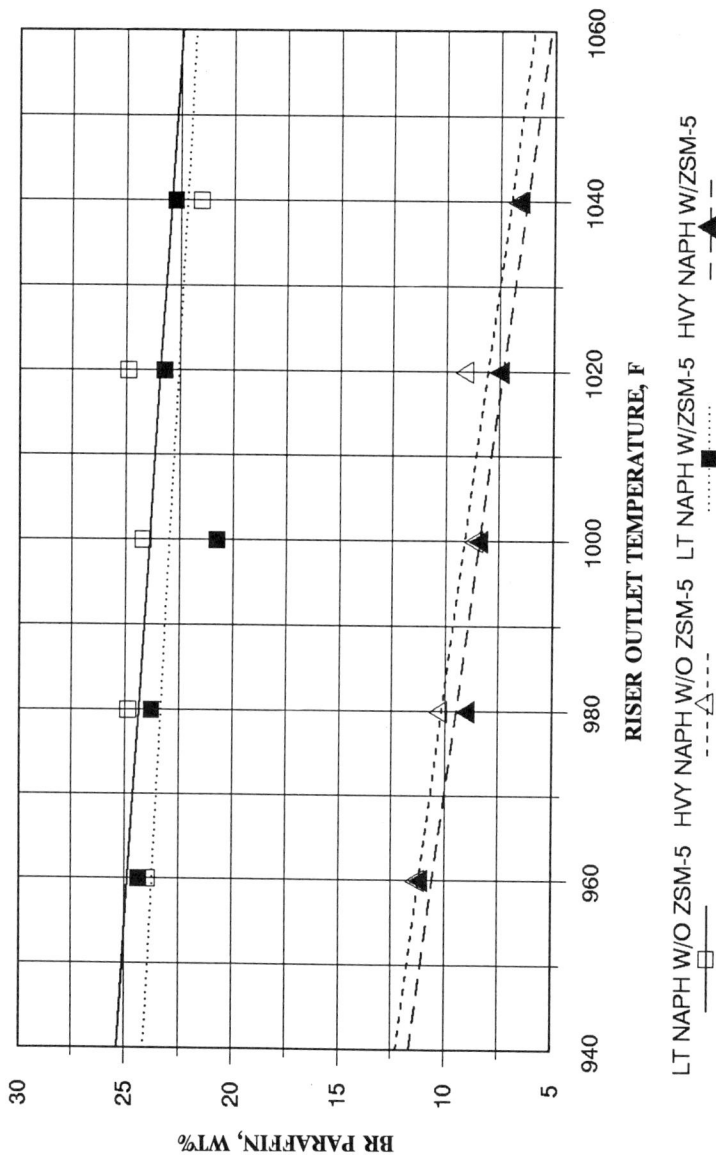

Figure 21. Naphtha Branched Paraffin Content vs Riser Outlet Temperature.

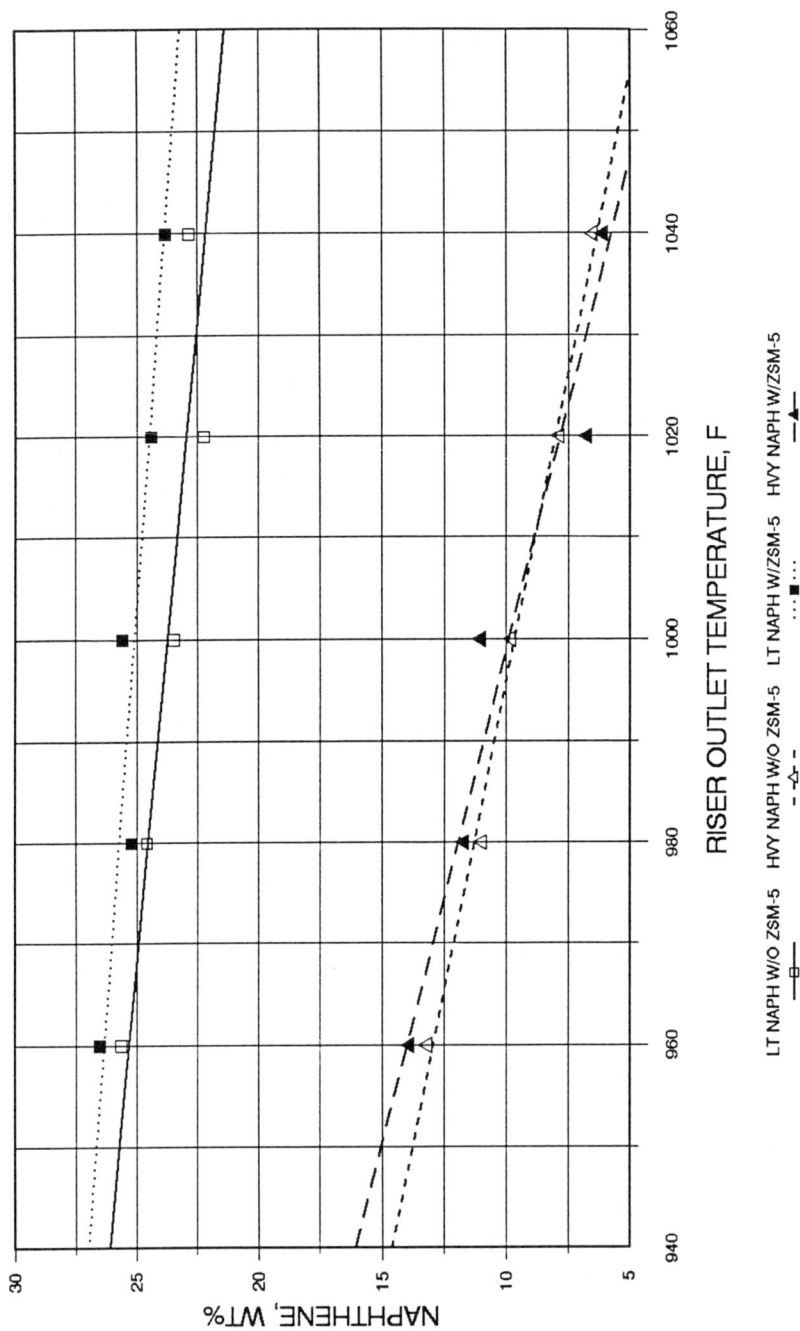

Figure 22. Naphtha Naphthene Content vs Riser Outlet Temperature.

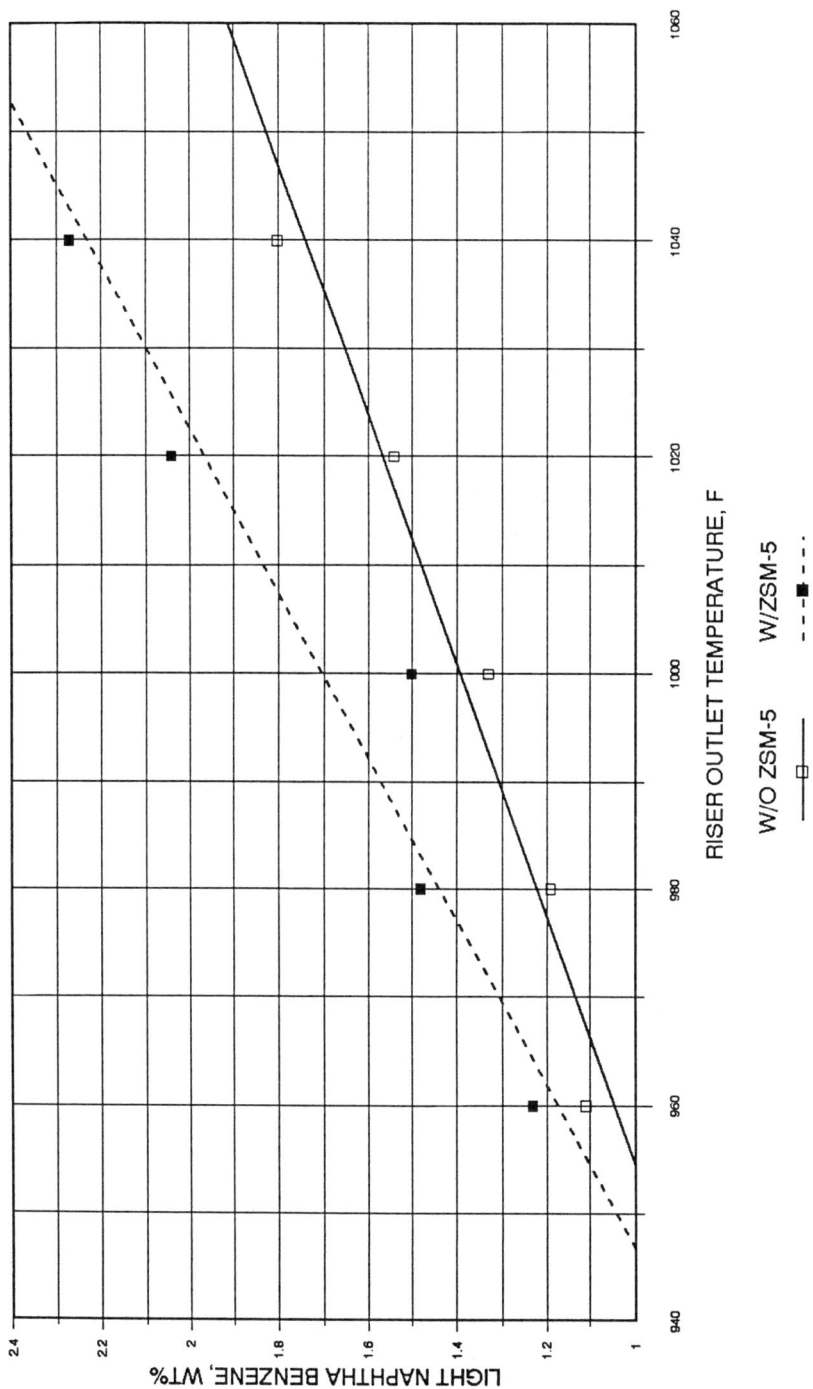

Figure 23. Light Naphtha Benzene Content vs Riser Outlet Temperature.

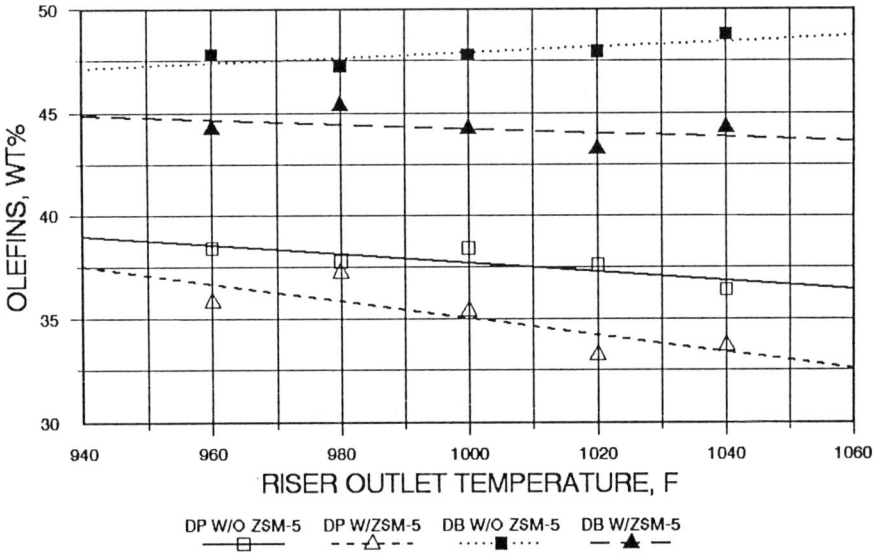

Figure 24. Light Naphtha Aromatics Content vs Riser Outlet Temperature.

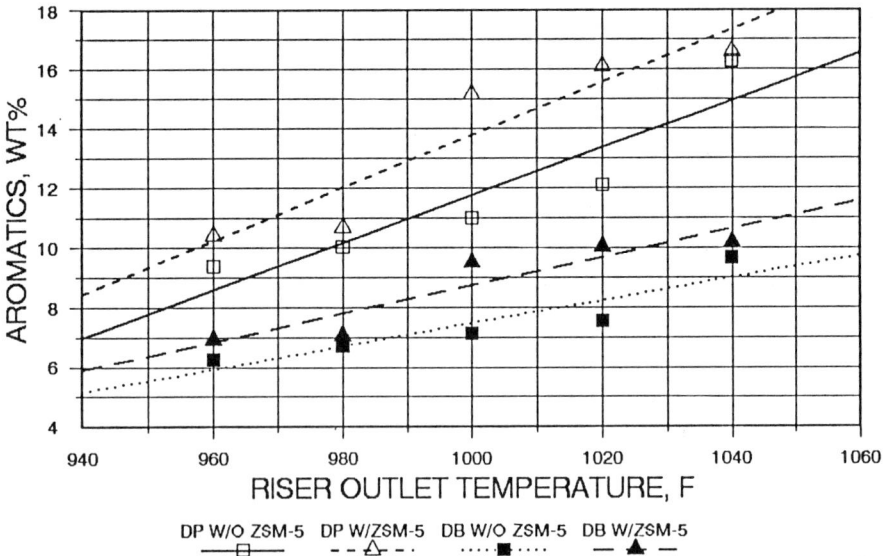

Figure 25. Light Naphtha Olefin Content vs Riser Outlet Temperature.

As shown in Figure 25, the olefin content of the DB naphtha (different from DP naphtha) was fairly flat with temperature when the ZSM-5 additive was present. This is different from the DB naphtha olefin content without ZSM-5 additive, which increased slightly with riser outlet temperature. This indicates that the removal of C5 components from the gasoline pool, when ZSM-5 additive is used, is even more desirable to not only increase TAME production, but also lower gasoline pool vapor pressure and reactive olefin contents. This is particularly true when operating in an overcracking operation.

It is noted that branched paraffin content in the DB naphtha was relatively constant with riser outlet temperature and significantly higher than the DP naphtha paraffin content due to high I-C5 concentration in C5 stream shown in Figure 26. It is also noted that the branched paraffin content in the DB naphtha was higher with ZSM-5 present, while ZSM-5 decreased the branched paraffin content of the DP naphtha. This could mean that high molecular weight gasoline olefins and paraffins were preferentially cracked to lighter components including I-C5.

It is noted in Figure 27 that the naphthene contents in both the DB and the DP naphthas were higher with the addition of ZSM-5 additive due to concentration effect over all temperature range studied. This indicates low cracking of gasoline naphthenes by ZSM-5 additive.

As shown in Figure 28, benzene content in the DB naphtha was significantly lower than in the DP naphtha and the addition of ZSM-5 increased both the DB and the DP naphthas' benzene content due to concentration effect. However, the increased ability to produce oxygenates and alkylates when using ZSM-5 should be sufficient to maintain or decrease the benzene in gasoline from an FCCU.

Heavy naphtha octanes shown in Figure 29 indicate that both RON and MON were increased with the addition of ZSM-5. The increment of increase in MON appears to be higher at high riser temperature. Octane numbers of the DP light naphtha also increased with the use of ZSM-5 but less than for the heavy naphtha. This agrees with the assumption that ZSM-5 primarily cracks olefins and paraffins in the heavy naphtha fraction into lighter components.

Conclusions

1. ZSM-5 increased olefin production in the order of C3= > C4= > C5= over the temperature range studied.

2. ZSM-5 increased branched C4 and C5 olefin yields more than linear olefins over the temperature range studied.

3. ZSM-5 promoted olefin skeletal isomerization, particularly n-C5=.

4. ZSM-5 increased branched C5= but lowered n-C5= yields.

5. ZSM-5 increased C3, C4, C5 paraffin yields, particularly I-C4 and I-C5.

6. ZSM-5 had negligible effect on conversion, coke and dry gas yields.

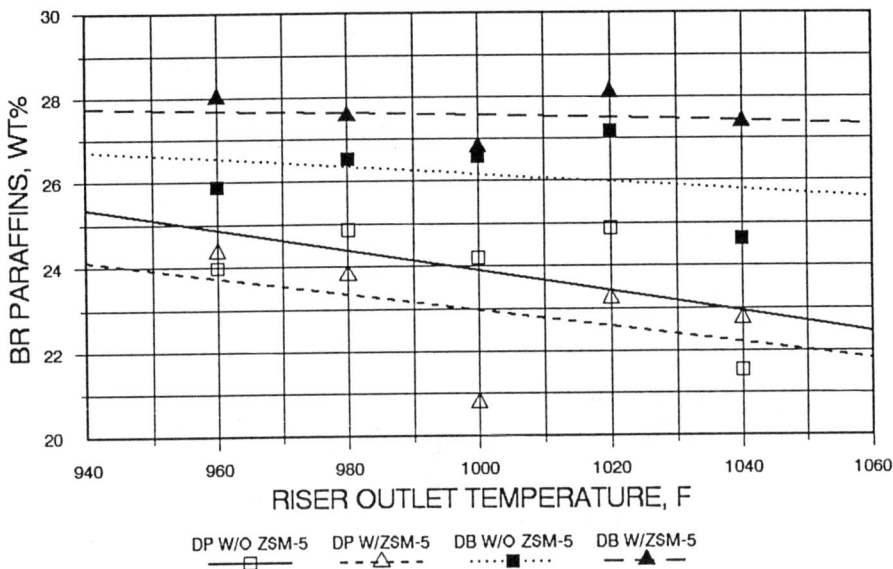

Figure 26. Light Naphtha Branched Paraffin Content vs Riser Outlet
Temperature.

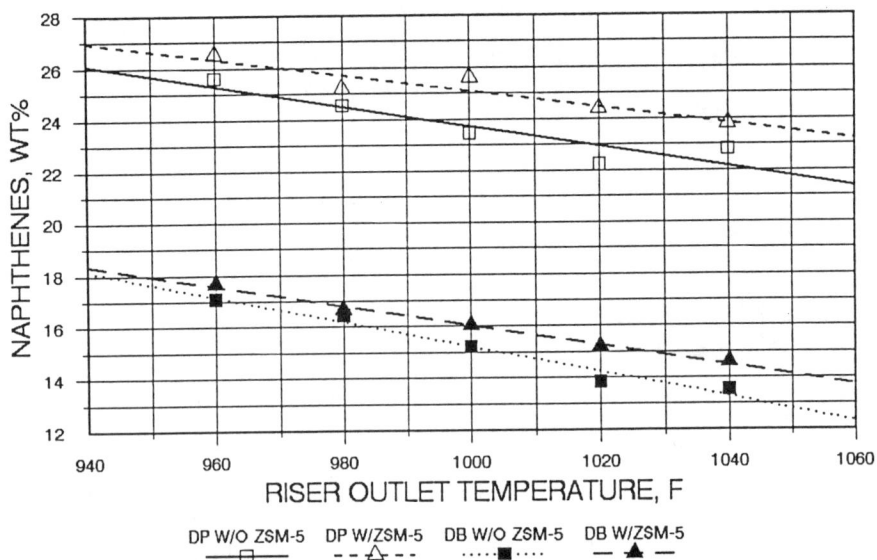

Figure 27. Light Naphtha Naphthene Content vs Riser Outlet Temperature.

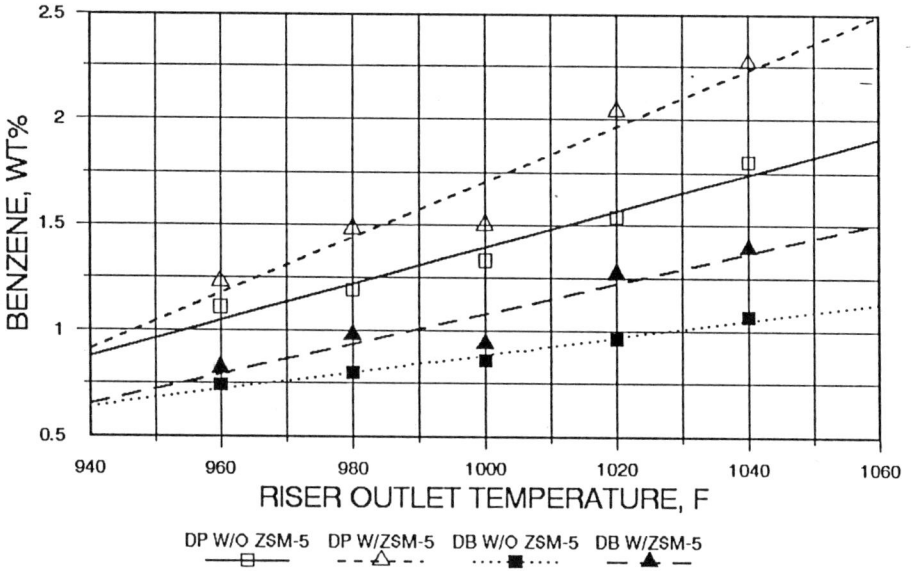

Figure 28. Light Naphtha Benzene Content vs Riser Outlet Temperature.

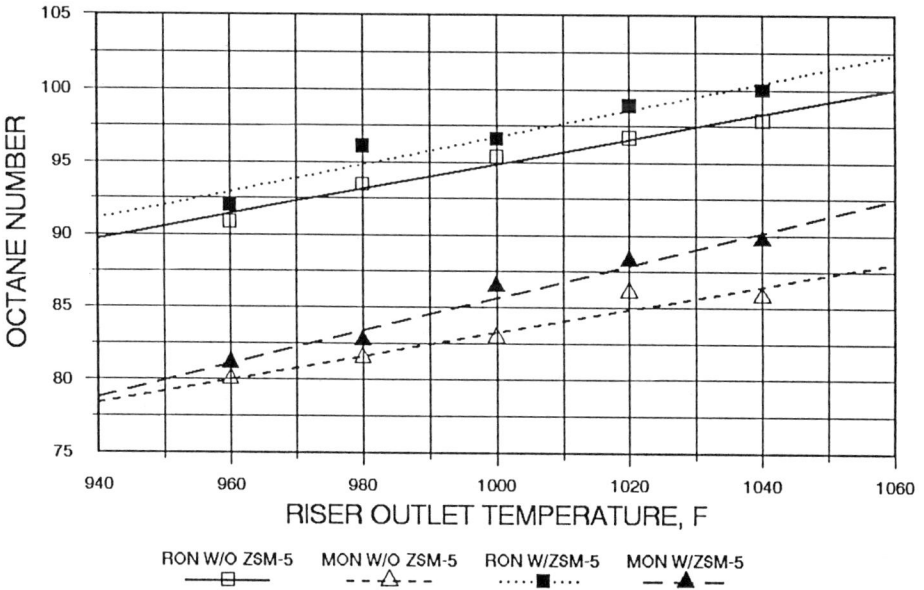

Figure 29. Heavy Naphtha Octanes vs Riser Outlet Temperature.

7. ZSM-5 decreased the naphtha yield and increased the heavy naphtha octane by 2 numbers.

8. Performance effectiveness of ZSM-5 was the highest at riser temperatures ranging from 980 to 1000F. The combination of ZSM-5 and an overcracking operation was more effective in increasing naphtha octane numbers as well as olefin production for oxygenates than the use of ZSM-5 or overcracking operation alone.

9. ZSM-5 increased aromatics and benzene, but lowered olefin and branched paraffin contents of DP light and heavy naphthas.

Acknowledgments

I would like to express my appreciation to Texaco, Inc for allowing the publication of the paper.

Literature Cited

(1)　　Krishna, A. S.; Hsieh, C. R.; Pecoraro, T. A.; Kuehler, C. W., Hydrocarbon Processing November 1991, pp. 59-66.
(2)　　Cheng, W. C.; Suarez, W.; Young, G. W., Presented at AICHE Annual Meeting, Nov. 19, 1991.
(3)　　Young, G. W.; Suarez, W.; Roberie, T. G.; Cheng, W. C., 1991 NPRA Annual Meeting , Paper AM-91-34.

RECEIVED June 17, 1994

Chapter 20

Atomic Force Microscopy Examination of the Topography of a Fluid Cracking Catalyst Surface

Mario L. Occelli[1], S. A. C. Gould[2], and B. Drake[3]

[1]Zeolites and Clays Program, Georgia Tech Research Institute, Georgia Institute of Technology, Atlanta, GA 30332
[2]Claremont Colleges, Claremont, CA 91711
[3]Imaging Services, Santa Barbara, CA 93111

Examination of the topography of a fluid cracking catalyst (FCC) using atomic force microscopy (AFM) has revealed the presence of a unique surface architecture characterized by valleys, ridges, crevices, dislodged plates, and narrow slits 6-9 nm wide. Formation of pits and craters with micrometer dimensions are believed to result from incomplete delamination of clay platelets used in catalyst preparation. Atomic scale resolution images were found consistent with the presence of kaolin and faujasite-type crystals.

When V was added at the 2-4% level, the catalyst surface roughness decreased. Furthermore, AFM images indicate the formation of vanadia islands and that vanadia coat the surface causing blockage of the narrow slits and cracks responsible for most of the catalyst microporosity.

The extent of clay delamination can influence macropores formation on the FCC surface and therefore influence the tendency of the catalyst to occlude hydrocarbons during cracking. Removal of occluded surface vanadia should restore some of the initial activity in spent equilibrium FCC.

In the early '70s, the Engelhard Corporation successfully commercialized a new family of fluid cracking catalysts (FCC) characterized by superior density and attrition resistance, obtained by the hydrothermal treatment of calcined kaolin microspheres in diluted NaOH solutions *(1-3)*.

Kaolin is a clay mineral consisting of layers of Si-atoms in tetrahedral coordination joined to layers of Al-atoms in octahedral coordination *(4)*. Since kaolin is a non-swelling clay, it is possible to produce slurries containing 50-60 wt% solids

0097–6156/94/0571–0271$08.54/0

that can be spray-dried into hard microspheres about 60 μm to 150 μm in diameter. This hydrous aluminosilicate with composition $Al_2O_3.2SiO_2.2H_2O$ on heating in the 550-600°C range undergoes a dehydroxylation reaction to form metakaolin. When the temperature is raised above 925°C, an aluminosilicate spinel ($Si_3Al_4O_{12}$) is produced that converts to mullite ($Si_2Al_6O_{13}$) near 1050°C (4). The calcined (at 980°C) kaolin microspheres, when heated in a NaOH solution at 95°C for 10h-16h, form near their surface a layer of zeolite with the faujasite structure. The bulk crystallinity (with respect to a sample of NaY such as UOP LZY-52) typically is in the 20-40% range while the microspheres surface area increased from about 40m²/g to 220m²/g (5). The zeolitized microspheres, after multiple washing with hot de-ionized water and exchange with 1M NH_4NO_3 solutions to reduce the % Na_2O level below 0.5%, have cracking activity at microactivity test conditions (MAT) comparable to that of similarly steam aged (760°C for 5h with 100% steam) fresh commercial FCCs prepared by incorporating a zeolite with the faujasite structure and kaolin into an aluminosilicate matrix. (5).

The similarity between the cracking properties of a commercial FCC and the zeolitized calcined kaolin microspheres seems to indicate that the catalytic cracking of gas oil fractions is probably controlled by the physicochemical properties of its surface and that in an 80 μm diameter microsphere, cracking occurs mainly within the microsphere top 10-15 μm. Thus the architecture near the microsphere's surface could control FCC properties and deserve careful examination.

The literature contain detailed descriptions of the role that zeolite levels, zeolite type, and matrix composition have on the FCC activity and selectivity properties (1). By contrast, the effects of the surface structure on catalyst performance has received little attention. It is the purpose of this paper to examine in detail the surface topology of several FCC microspheres. Atomic Force Microscopy will be used for the first time to identify and report the nature (and architecture) of the various pore structures on the FCC surface that are available to hydrocarbon sorption, diffusion, and therefore, gas oil cracking.

Experimental

The cracking catalyst used in the present study (GRZ-1) was obtained from Davison. After drying in air at 400°C, a solution of vanadyl naphthenate in toluene was used to metal-load the dried microspheres according to an established procedure (6). The naphthenate solution was obtained from Pfaltz and Bauer, Inc., and contained 2.9 wt% V. The decomposition of the naphthenate was accomplished by heating in air at 540°/10h. Steam-aging was then performed by passing steam at 760°C over the catalysts for 5h.

The FCC microspheres were sprinkled over a steel disk covered with a film of epoxy resin. After the glue dried, the AFM tip was placed onto the microspheres. The AFM used for these experiments (7) was a contact mode microscope based on the optical lever cantilever detection design of Amer and Mayer(8) and Alexander, et al. (10). The AFM works like a record player. An xyz piezoelectric translator raster scans a sample below a stylus attached to a cantilever. The motion of the cantilever, as the stylus moves over the topography of the surface, is measured by reflecting a laser beam off the end of the cantilever and measuring the deflection of the reflected laser light with a two segment photodiode. A digital electronic feedback loop is used to keep the deflection of the cantilever and hence the force of the stylus on the surface constant. This is accomplished by moving the sample up and down in the z direction of the xyz translator as the sample is scanned in the x and y directions. The images contain either 256 x 256 data points and nearly all images were acquired within a few seconds. This Si_3N_4 cantilevers (with integral tips) used for imaging are 120μm in length and possess a spring constant of approximately 0.6 N/m. The force applied for these images ranged from 10-100 nN. Approximately 600 images have been acquired by examining a variety of microsphere surfaces.

Results and Discussion

GRZ-1 is a commercially available fluid cracking catalyst (FCC) obtained from Davison. The catalyst contains an estimated 35% CREY (calcined, rare-earth exchanged zeolite Y), 50% kaolin and 15% binder. The calcined (400°C/2h) microspheres BET surface area of $220m^2/g$ decreased to $160m^2/g$ upon steaming. After loading 2%V or 4%V, the steam-aged FCC surface area further decreased to $18m^2/g$ and $12m^2/g$ respectively.

SEM images (Figure 1) show that the FCC consists of spheroids about 80-90 μm in diameter fairly regular in size and shape. This surface appears to contain cracks and indentations oriented in a random manner (Figure 1A). After loading with 4%V calcination and steaming, the surface becomes coarser and granular, see Figure 1B. By contrast, AFM images (Figure 2) indicate that at greater magnification it is possible to observe on the FCC surface the existence of large pits, kinks, plates and plate aggregates, valleys, ridges, and crevices, all of which are believed to contribute to the catalyst's surface porosity and reactivity.

Large-scale images (Figures 2A-2B) show the presence of a large pit 4.2 μm in diameter circumscribed by folded and bent plates (or platelets aggregates) 2.2 μm wide. Granular and smaller in size components of the FCC surface are evident in Figure 2C–2D and Figure 4A. The gross images in Figures 2C and 2D show the inhomogeneity of the microsphere surface and that the catalyst components agglomerate in a random manner. In Figure 2E, a wedge seems to lift a plate 2.3 nm thick over a neighboring plate thus generating a large crevice on the surface. Well-defined pits (or pores) are evident in Figures 2F-2H. In Figure 2F the pore walls are formed by the face-to-face stacking of platelet aggregates. In Figure 2G, the wall of the pore is instead defined by the vertical agglomeration of plates; at the bottom, cracks about 7.5 nm wide are present. Figure 2G shows another example of a pit about 3 μm in diameter and 2 μm deep having walls formed by terraces separated by steps about 0.25 to 0.3 μm in height (Figure 2I). These steps could act as docking sites where sorbed hydrocarbons are cracked. Images such as those shown in Figure 2I have been attributed to the presence of kaolin platelet aggregates that have avoided delamination during catalyst preparation. The morphology of kaolin crystals used in FCC preparation is shown in Figure 3.

Details of the topography of the microspheres surface are given in Figure 4. Components irregular in size and shape (Figure 4A) agglomerate in such a way to

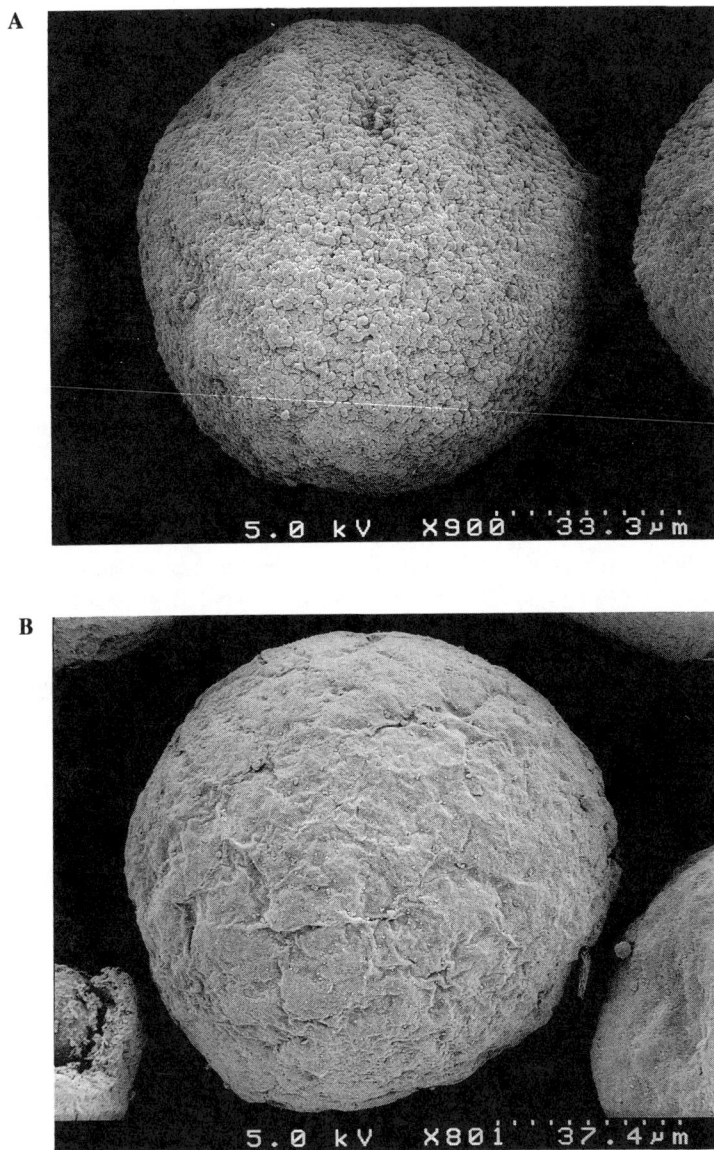

Figure 1. SEM micrographs of GRZ-1 microspheres (A) without and (B) with 4%V and steam-aging.

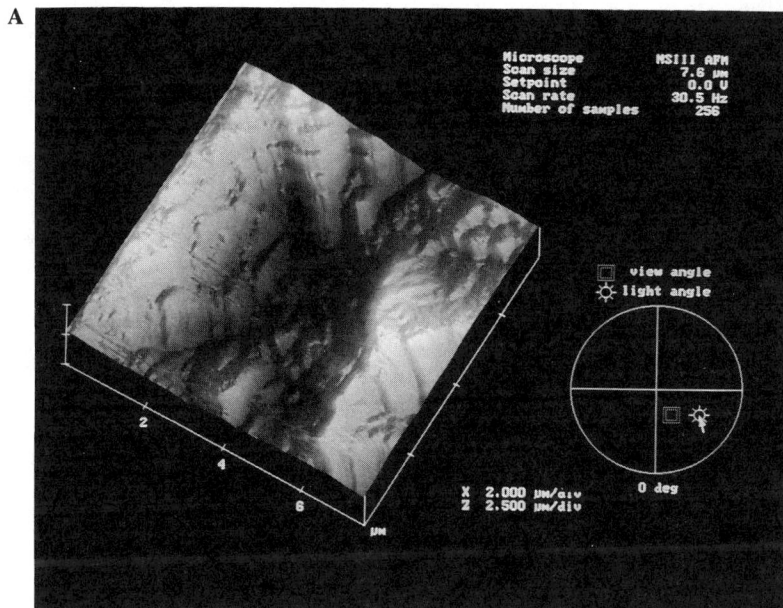

Figure 2. A-B. Two large-scale images of calcined GRZ-1 microspheres obtained with a Nanoscope III operating in a tap mode. The lowest parts in the image are black (0 µm) and the highest are white and correspond to about 6 µm. All other images were obtained with a Nanoscope II. Surface roughness is shown in C and D; crevaces in E. The scale bar (not shown) indicating the Z-axis height, show that the depth of the various pits is : F) 2 µm, G) 795 nm, and H) 2.0 µm. Details of the walls for the pore shown in H is given in the contour image shown in I.

Continued on next page

B

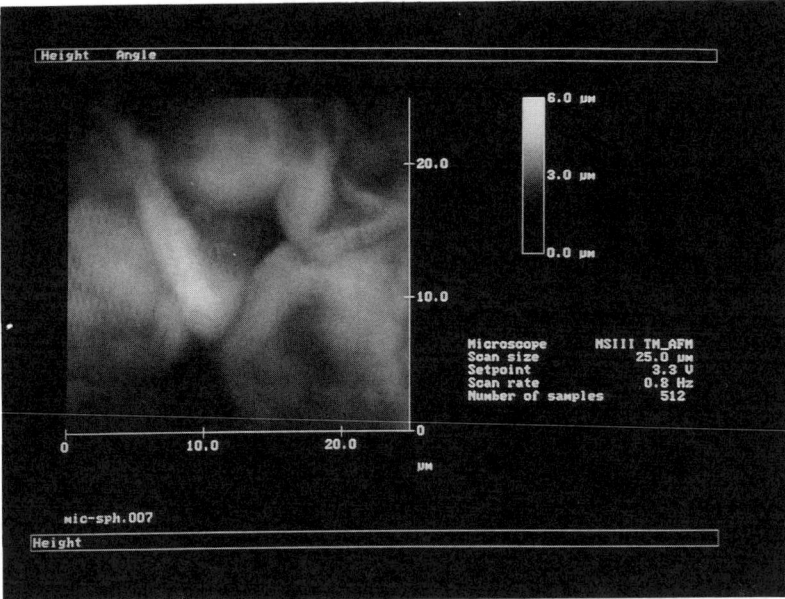

C

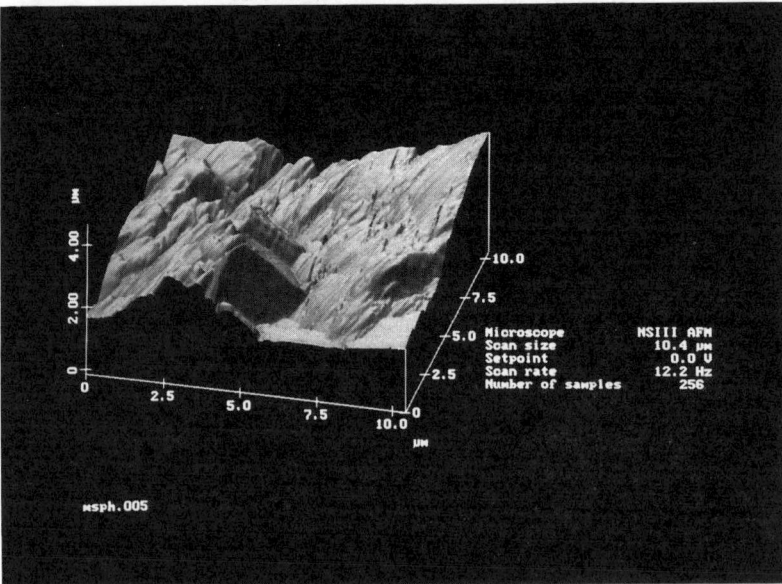

Figure 2. Continued.

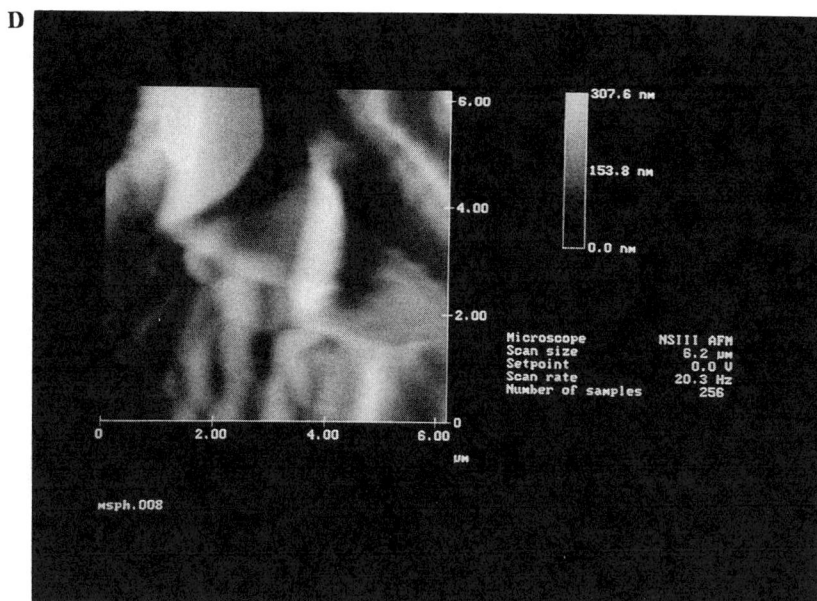

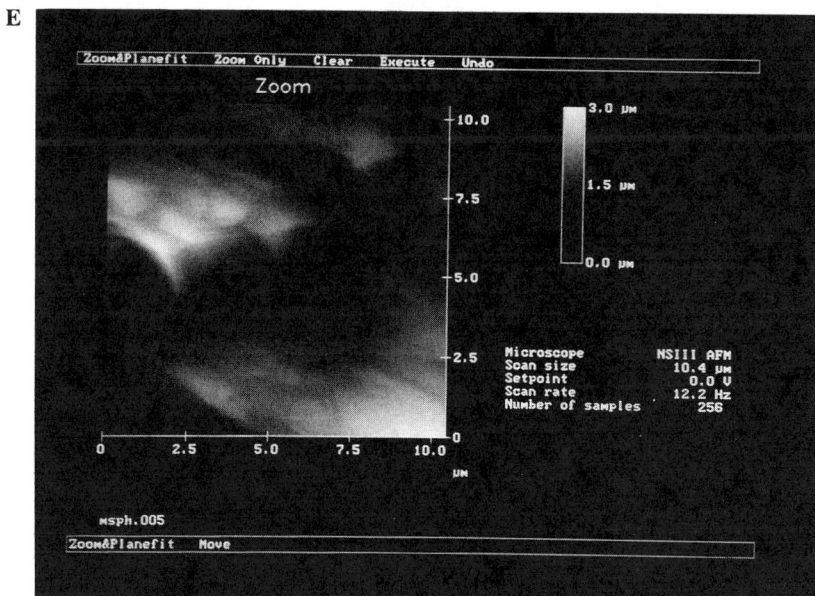

Figure 2. Continued.

Continued on next page

F

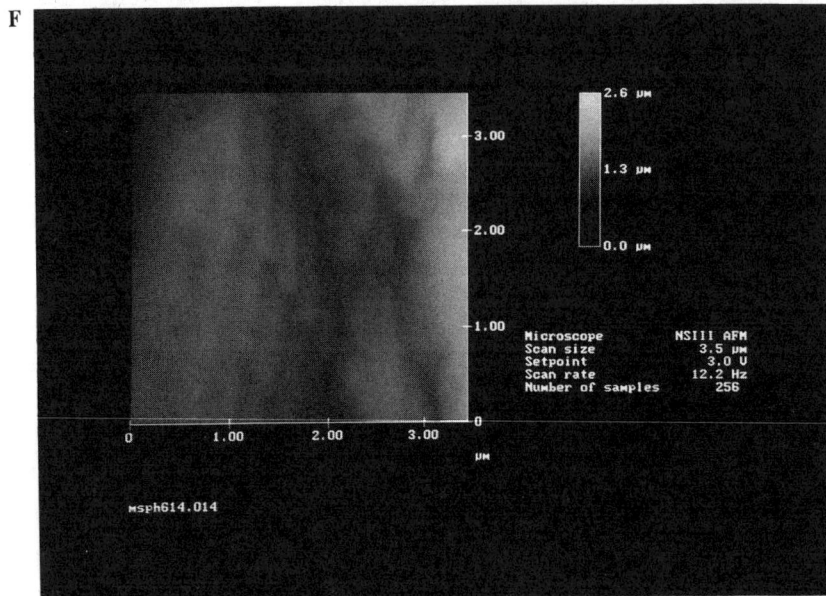

G

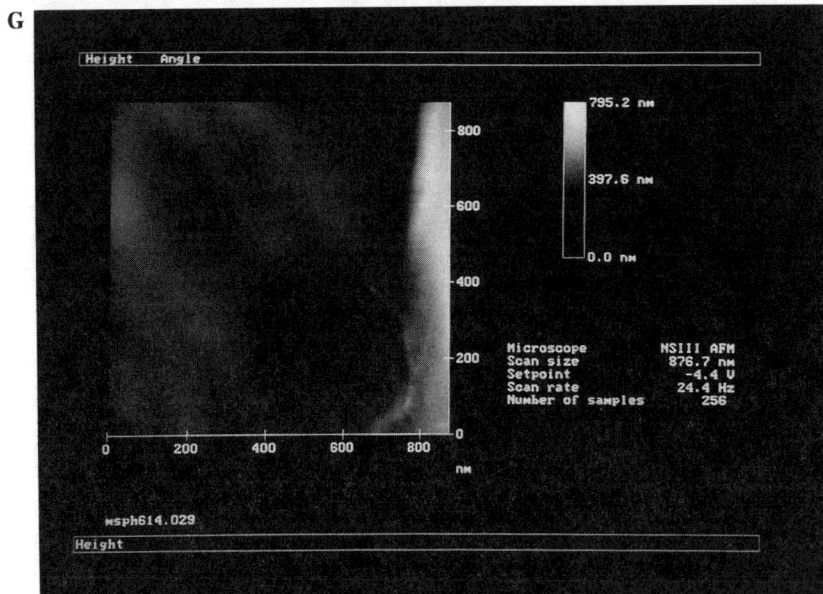

Figure 2. Continued.

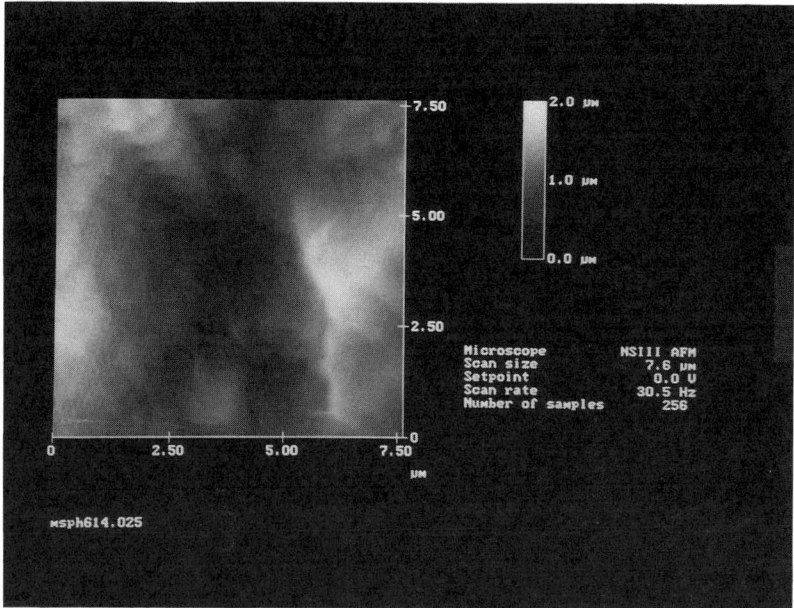

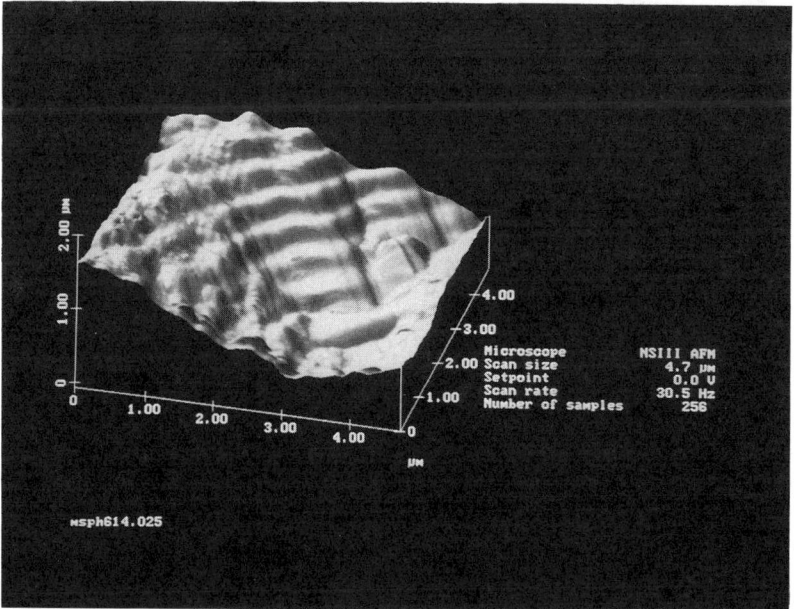

Figure 2. Continued.

Figure 3. SEM micrographs of a sample of Georgia kaolin of the grade used in FCC preparation.

form pits, cracks, and other openings of various dimensions (Figures 4B–4F). Plates and plates agglomerates are major components of the FCC surface and missing plates can form slit-like openings 70-80 nm in width (Figures 4B-4C). However, sectional analysis of the image in Figure 3A reveals that even adjoining plates are somehow wedged apart (probably by gel particles) thus offering elongated slits 5 to 9 nm in width, Figures 4D-4E. Openings of this type are very common and can also result inside the large pits or crevices shown in Figure 2. In fact, at the bottom of the pit in Figure 2G or 2H, slits 6–8 nm in width can be observed. These large pits could act as feeder-pores by occluding hydrocarbons and allowing their diffusion to the FCC interior where they are cracked.

It is believed that gel (the FCC binder) is evident on top and around the FCC components shown in Figure 4F. The 100-150 nm wide plates in Figure 4F are connected end-to-end by a thin layer about 20 nm thick of what could be gel. Atomic scale details of these plates consist of bright spots in a square arrangement that closely resembles the image of the oxygens in the AlO_6 layer in a sample of Georgia kaolin *(11)* (Figure 4G). The atomic scale image in Figure 4H cannot be attributed to kaolin and it is believed to represent the (111) surface of faujasite crystals. Steaming does not significantly change the local (1 μm x 1 μm) topography of the catalyst surface. The details and pits shown in the contour image in Figure 4I are typical also of the calcined FCC.

Following the oxidative decomposition (at 540°C/10h) of the naphthenate, 95% of V is still in the +4 state *(11)*. However, after steam-aging (at 760°C/5h, 100% steam) V is mostly present in the +5 state in a V_2O_5-like phase *(11-13)*. Since V_2O_5 melts at 658°C *(12)*, at the regeneration condition that exists in a commercial cracking unit, as well as during steam-aging prior to MAT evaluation, V_2O_5 has been observed to migrate on the FCC surface *(13)* and even from particle-to-particle *(14)*.

The effects of Vanadium on the surface topography are shown in Figure 5. Large pits are again seen also in the V-loaded, steam-aged FCC, Figures 5A-5B. However, areas 10μm x 10μm in size are visible in which V is believed to have filled valleys, cracks, and other surface openings (Figures 5B-5C). The top view image in Figure 5D contains white streaks about 100 nm wide believed to represent vanadia filling the interstitial space between platelets (side view of image in Figure 5E). The images in Figures 5F-5G show what is believed to represent V_2O_5-filled slits. In Figures 5F-5G (and in other images not shown) of V-contaminated FCC, V_2O_5 seems to accumulate near the V-filled surface openings (slits). Pores blockage by vanadia together with the irreversible destruction of zeolite crystallinity by V-compounds *(16)* are believed responsible for the drastic losses in surface area that occur when GRZ-1 microspheres are loaded with more than 1.0% V *(16)*.

In studying V-contaminated pillared rectorite catalysts *(10)*, images like those shown in Figures 5H-5I have also been observed. Since energy dispersive spectroscopy (EDS) results *(15)* have indicated possible V_2O_5-island formation, it is believed that the images in Figure 5H represents a 3.0 μm x 3.0 μm patch of V_2O_5 on the FCC surface. V islands formation is more easily seen in samples containing 4%V, (Figure 5I).

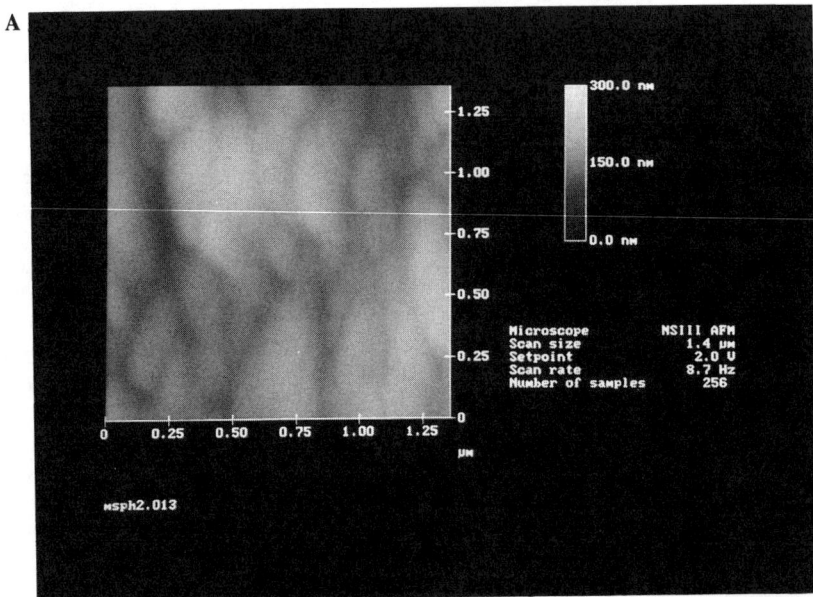

Figure 4. A) An image showing the various particle shape and sizes that form the surface of calcined GRZ-1. Openings or pores are formed by missing plates (B) and by voids between plates (C). Narrow slits such as those shown in (D) and (E) are the most common pore openings observed on the FCC surface. Stacks of adjoining plates appearing on the surface are shown in (F); arrows in (F) indicate the presence of gel on and between kaolin particles. Atomic scale detail of the plates in (F) are shown in (G). Atomic scale details shown in (H) have been attributed to the (111) surface of faujasite crystals. Contour profile of GRZ-1 after steam aging is shown in (I).

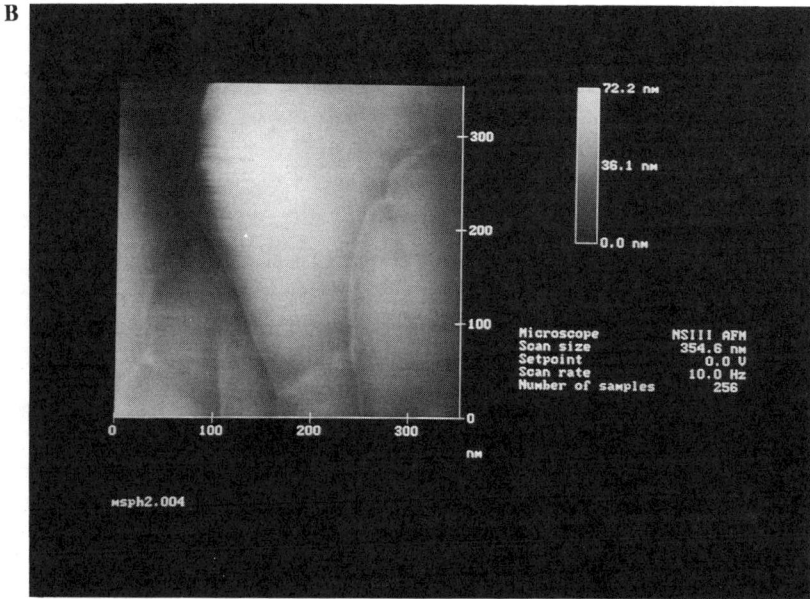

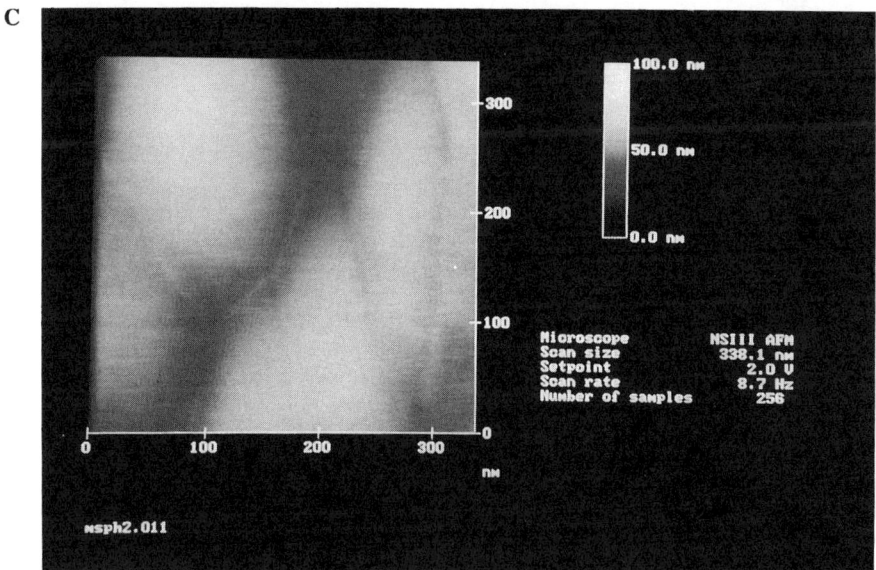

Figure 4. Continued.

Continued on next page

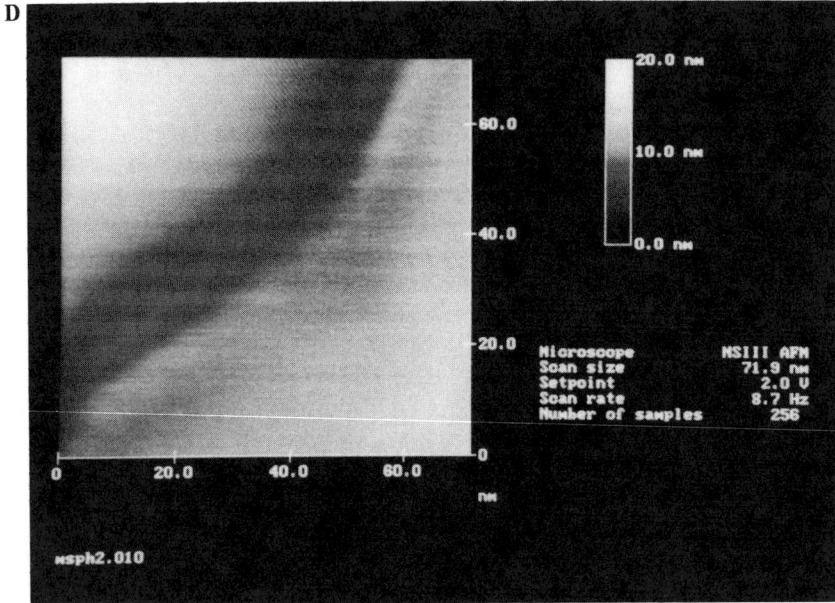

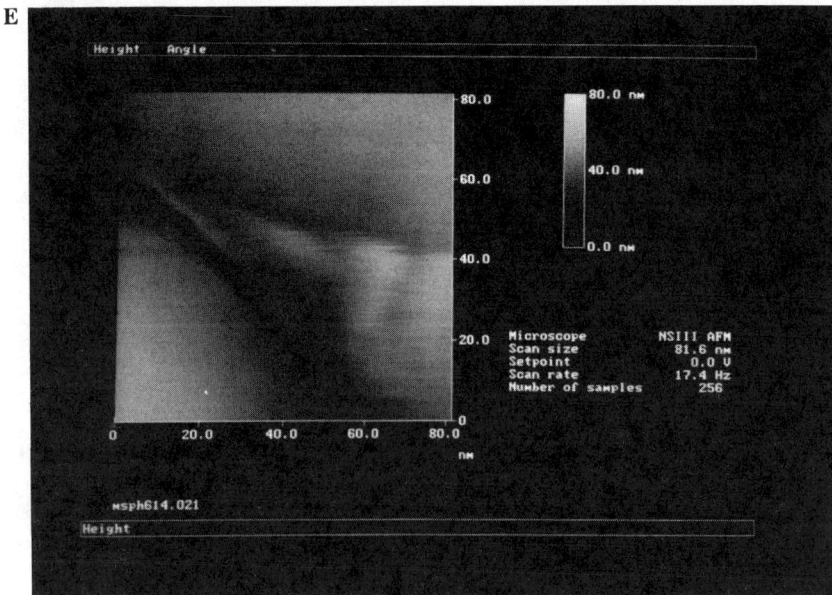

Figure 4. Continued.

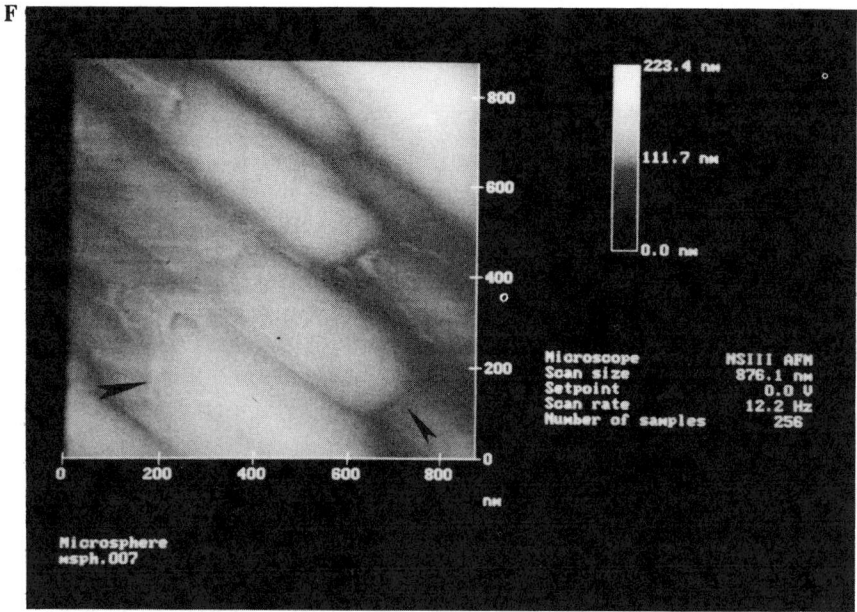

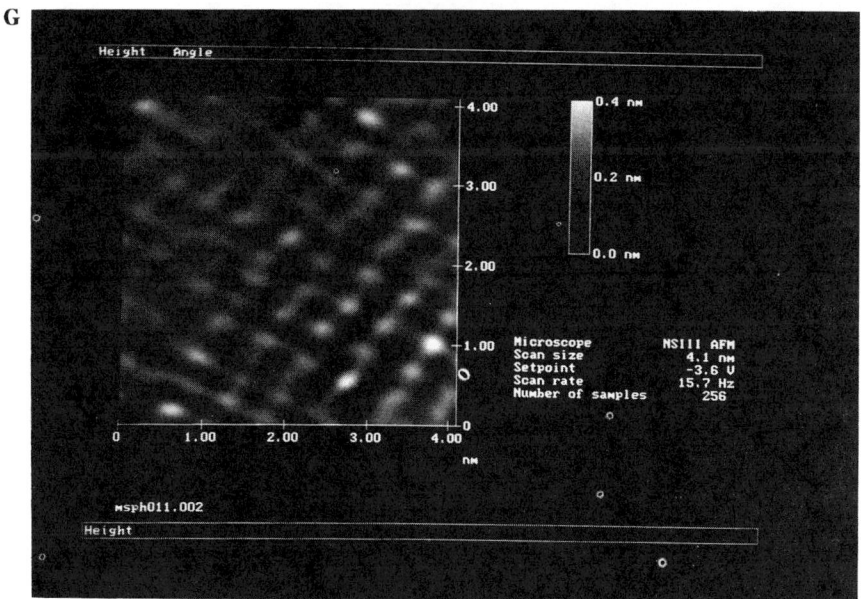

Figure 4. Continued.

Continued on next page

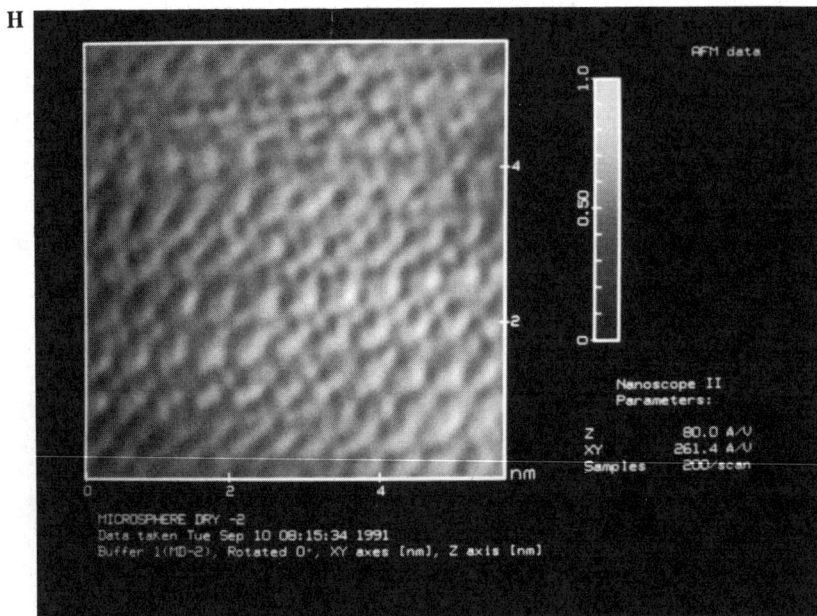

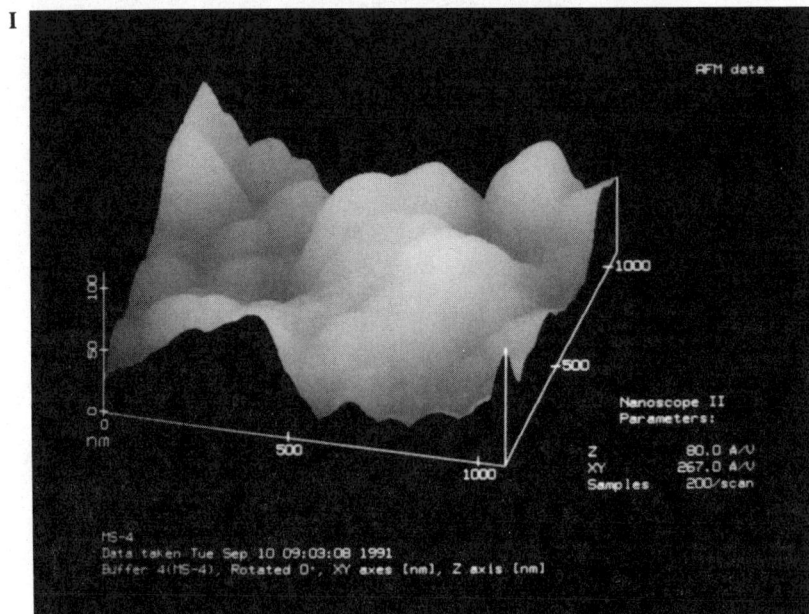

Figure 4. Continued.

A
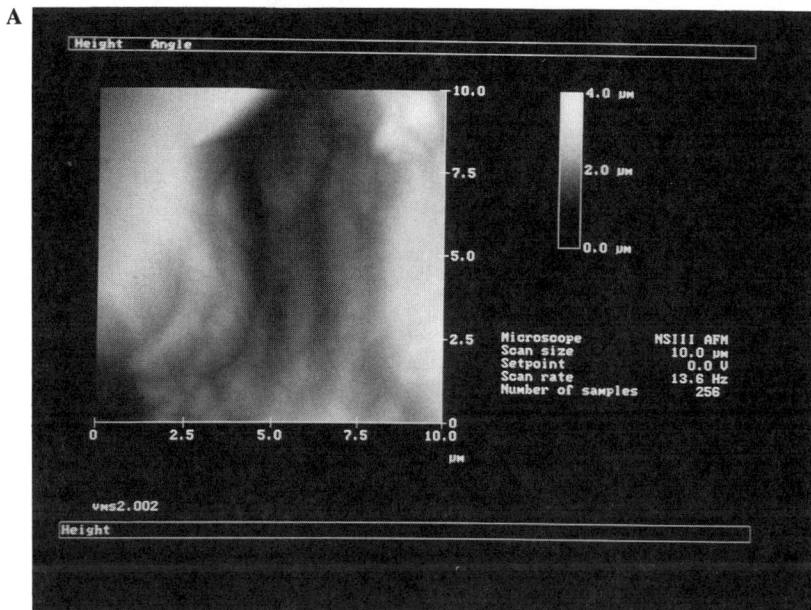

Figure 5. Large scale images of V-loaded steam-aged GRZ-1 microspheres: A) top view, 2%V; contour images of samples containing: B) 2%V; and C) 4%V. The top view (D) and contour image (E) show what is believed to represent the interstitial filling of platelets by vanadia. F) vanadia-filled slits after steaming. G) Ridges of vanadia near the surface of GRZ-1 steam-aged in the presence of 4%V. Vanadia islands can be observed in steamed microspheres containing: H) 2%V or I) 4%V. In both images, islands are about 500 nm high.

Continued on next page

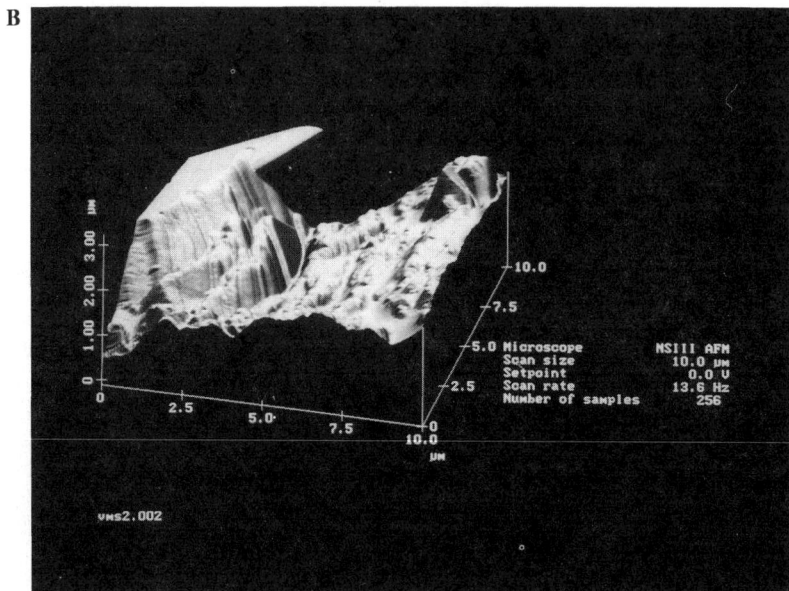

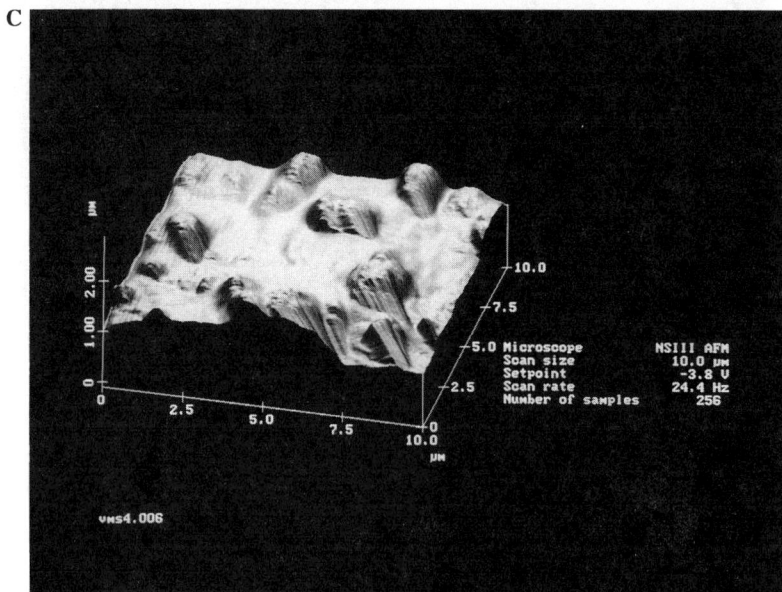

Figure 5. Continued.

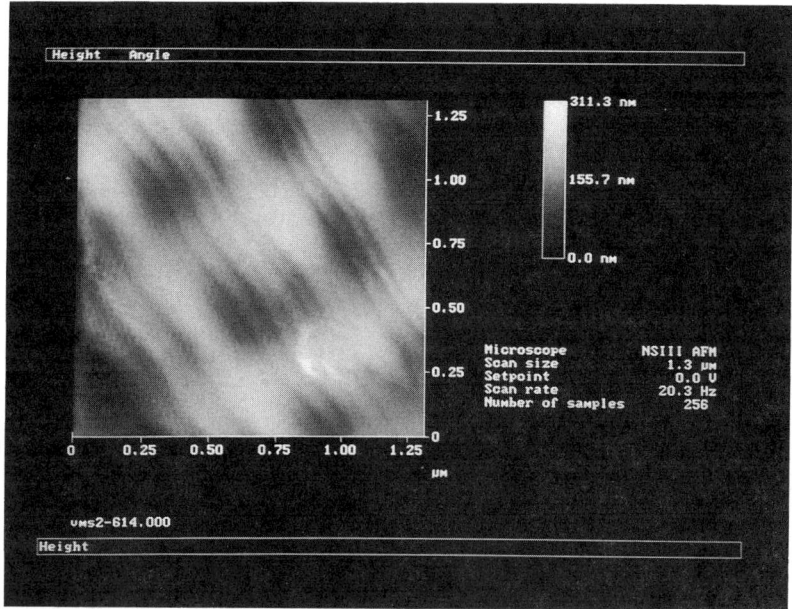

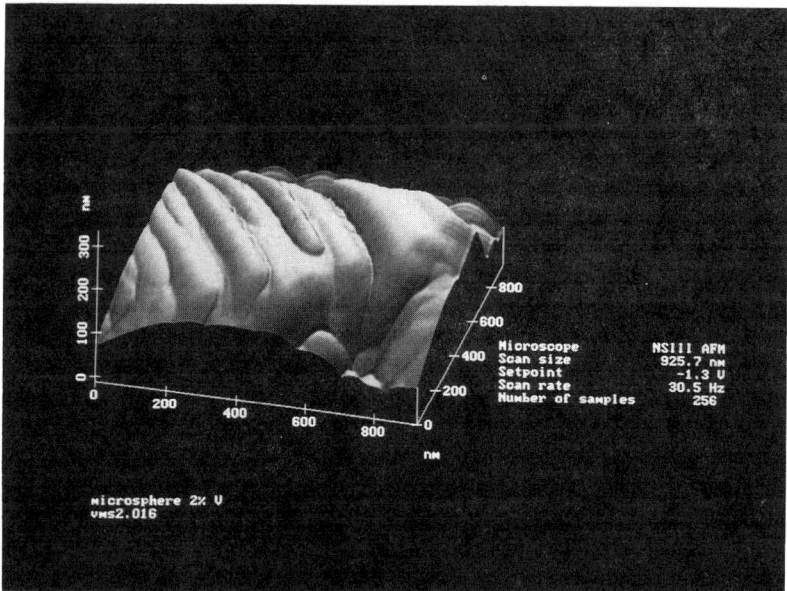

Figure 5. Continued.

Continued on next page

F

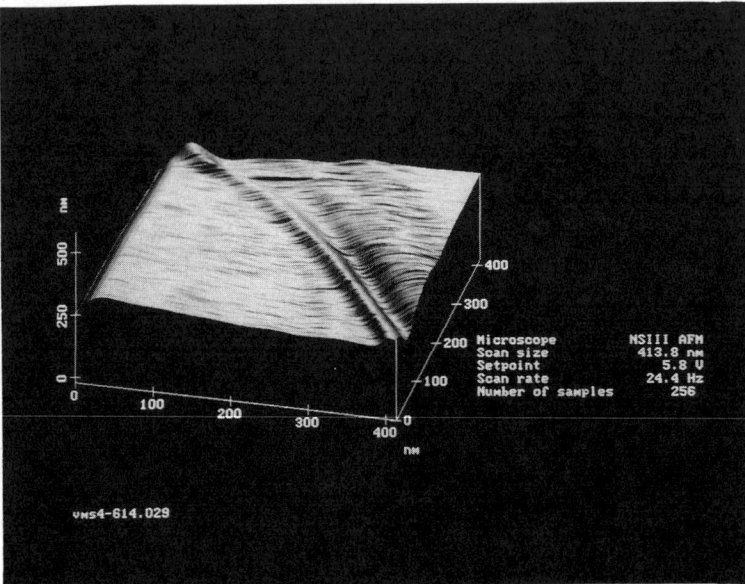

G

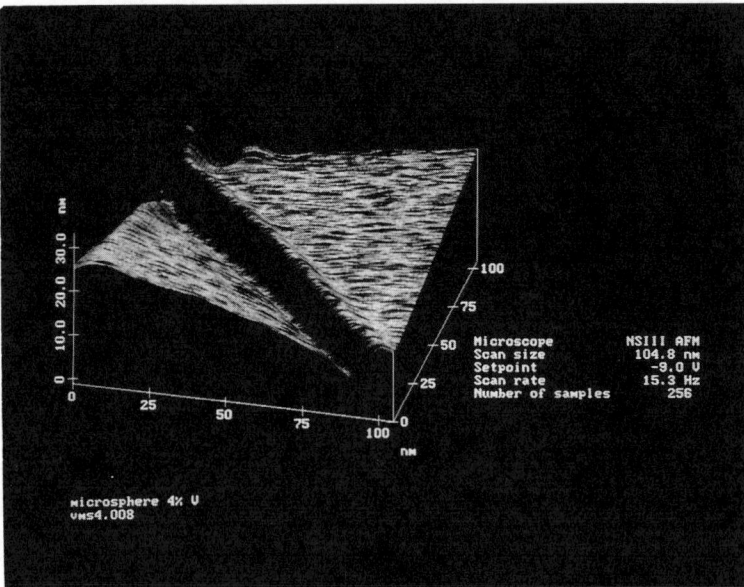

Figure 5. Continued.

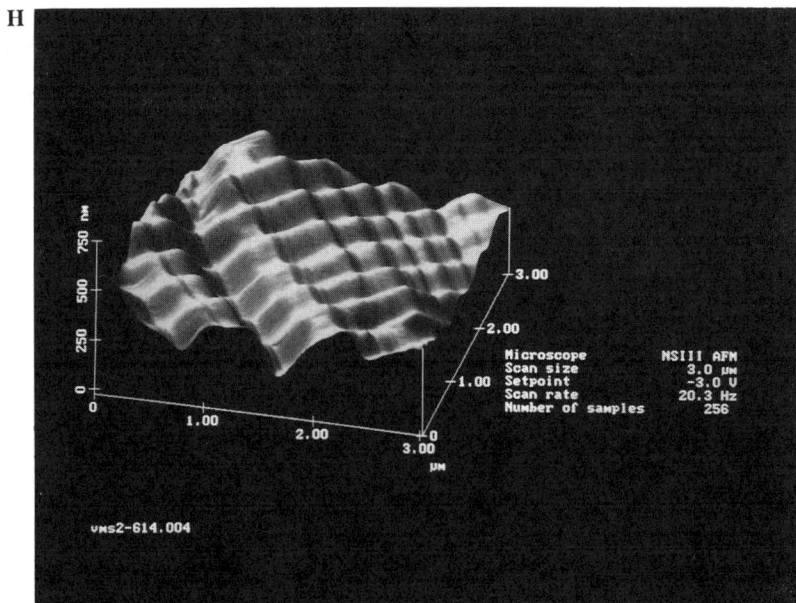

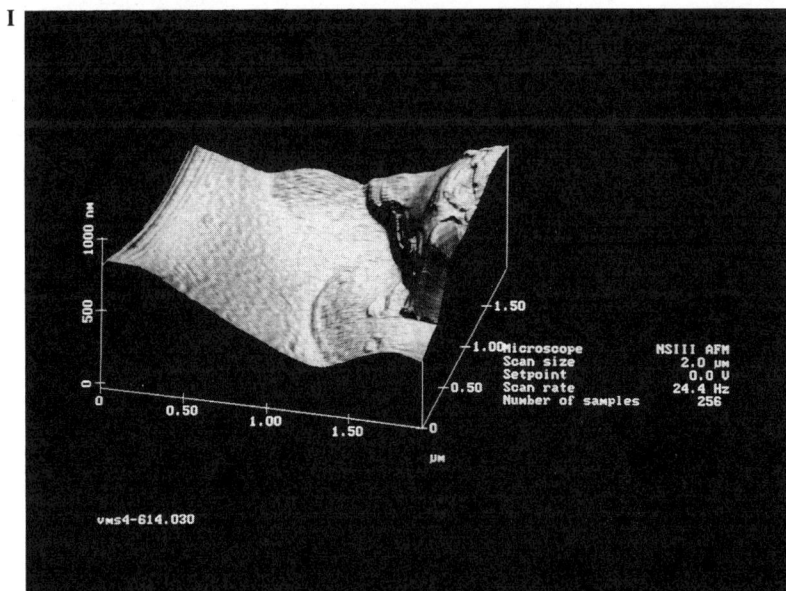

Figure 5. Continued.

Summary and Conclusions

The AFM can provide details of the surface topography of a FCC with unprecedent resolution. Atomic scale imaging has allowed the identification of kaolin and probably of faujasite crystals, two major components of the catalyst. The third component (i.e. the binder) appears as surface impurity and as a 20 nm thick layer between clay platelets as in Figure 4F. Identification of the catalysts components is, in general, tentative since the AFM cannot provide information concerning the chemical composition of the catalyst surface.

The catalyst components agglomerate in a random manner originating considerable surface roughness (Figures 2A-2E) and openings (or pores) with variable length (L) and width (W) irregular in size and shape. Slits with L/W>>>1 and width in the 6-9 nm range appear most frequently; valleys, kinks, and cracks originating from gel particles between the catalysts components are also present (Figures 4B-4E).

Large pits 3.0 μm in size are frequently observed. The architecture and structural features of these pores are defined by the mode of plates aggregation (Figures 2 and 4). The walls of large pits such as those found in Figure 2I offer terraces separated by steps about 0.2–0.3 μm high that could act as docking sites for hydrocarbon adsorption and cracking.

Large pores (such as those shown in Figures 2F, 2H, 2I, and 5A) could facilitate surface retention of gas oil during cracking and act as feeder pores for hydrocarbon dispersion to the FCC interior where they are cracked. In GRZ-1, gas oil components with diameter in excess of 9 nm will probably be retained in these surface pits as occluded hydrocarbons to be removed during steam-stripping or burned during FCC regeneration.

By comparing the SEM images in Figure 3 to the AFM images in Figures 2F, 2H, and 2I, it seems that pores with micrometer dimensions are formed whenever residual (short-range) aggregate of kaolin platelets are imbedded on the catalyst surface. These observations may help explain (in part) the high coke make of GRZ-1 and reaffirm the importance of properly delaminate the clay platelets in the clay-zeolite-binder mixture prior to spray drying. It is believed that in coke-selective microspheres, the clay forms a house-of-cards-structure that minimize surface macropores formation.

When contaminated with 2%-4% V, during steaming at 760°C, part of the catalyst surface appear to lose some of its surface roughness (Figure 5B-5C). The smaller pores (slits) appear blocked as in Figures 5F–5G and stacking of what is believed to be V_2O_5 layers has also been observed (Figures 5H–5I). Pore blockage and crystallinity losses are probably the two main causes of the drastic reduction in surface area and cracking activity suffered by GRZ-1 (and by others FCC) in the presence of 2%-4%V. (15) It is believed that the gain in cracking activity exhibited by spent equilibrium FCC after reactivation with the DEMET process (16) is probably due to the removal of vanadia impurities from the catalysts microporous structure.

Since the microscope cantilever does not damage or alter the surface, the AFM can provide the actual images of working catalysts. The AFM major limitation appears to be its inability to provide chemical composition data of the surface.

Acknowledgment

We would like to thank Digital Instruments for their generous contribution of equipment for this report and James Tsai for image acquisition and processing.

Literature Cited

1. Magee, J. S., and Blazed, J. J., ACS Monograph No. 171, J.A. Rabo Ed., Chapter 11 (1976).
2. Haden, W. L. and Dzierzanowski, F. J., U.S. Patent 3,657,154 (1972).
3. Haden, W. L. et. al. in U.S. Patent 3,663,176 (1972)
4. Grim, R. E., *Clay Minerology*, McGraw-Hill, New York , NY 1968..
5. Occelli, M. L., Unpublished results.
6. Mitchell, B. R., *Ind. Eng. Chem. Prod. Res. Dev.* **1980**, 19, pp. 209.
7. Digital Instruments, Inc., 6870 Cortona Dr., Goleta, CA 93117.
8. Meyer, G. and Amer. N. M. *Appl. Phys. Lett.*, **1988**, 53, pp 1095.
9. Alexander, et al., *J. Appl. Phys.*, 65, **1989**, pp. 164-167.
10. Occelli, M. L., Gould, S. A. C., and B. Drake, In preparation.
11. Anderson, M. W., Occelli, M. L., and Suib, S. L., *J. Catal,* 112,2, p.375.
12. Clark, R. J. in *The Chemistry of Vanadium and Titanium*, Elsevier, 1968.
13. Jaras, S., *Applied Catalysis*, Elsevier, Amsterdam, 1982, p. 207.
14. Occelli, M. L. in *Fluid Catalytic Cracking: Role in Modern Refining*, M. L. Occelli, Ed., ACS Symposium Series, ACS, Washington, DC, 1989, Vol. 375, p. 162.
15. Occelli, M. L., Eckert, H., Dominguez, J. M., *J. Catal.*, 141, 510, 1993.
16. Occelli, M. L., Psaras, D., and Suib, S. L., *J. Catal.*, 96, 363, 1985.
17. Elvin, F. J., Otterstedt, J. E., and Sterte, J. in *Fluid Catalytic Cracking: Role in Modern Refining,*, M. L. Occelli Ed., ACS Symposium Series, ACS, Washington, DC, 1989, Vol. No. 375, p.279.

RECEIVED April 19, 1994

Chapter 21

Kinetics of *n*-Decane Cracking at Very Short Times on Stream

A. Corma[1], A. Martínez[1], P. J. Miguel[2], and A. V. Orchillés[2]

[1]Instituto de Tecnología Química, Universidad Politécnica de Valencia—Consejo Superior de Investigaciones Científicas, Camino de Vera s/n, 46071 Valencia, Spain
[2]Departamento de Ingeniería Química , Universitat de Valencia, Doctor Moliner 50, 46100 Burjassot, Spain

The kinetics of n–decane cracking on USY and Beta zeolites has been studied by measuring instantaneous conversions from very short times on stream (TOS). Data at short TOS were seen to be crucial to obtain reliable kinetic parameters, especially when catalyst deactivation is fast. Different kinetic and decay models have been used to fit the experimental results. For both catalysts, the best fitting was obtained using a simple first–order kinetic model coupled with a dependent decay expression which considers cracking products as responsible for catalyst deactivation. On these bases, it has been shown that Beta zeolite is more active and decays slower than USY for cracking of n–decane.

Recently *(1–3)* the cracking of paraffins has been explained assuming two different cracking mechanisms. One of them accounts for the classical carbenium ion mechanism *(4)* which involves a bimolecular hydride transfer followed by β–scission. The other one, the monomolecular protolytic cracking, involves the attack of a C–C or C–H bond by a H^+ from the catalyst to form a carbonium ion that cracks to give a paraffin or hydrogen in the gas phase and an adsorbed carbenium ion. In the case of the large–pore Y zeolite, despite the fact that the mono and bimolecular cracking coexist, a first–order kinetic model has been applied *(5)*.

Riekert and Zhou *(6)* have established the contribution of the two cracking mechanisms by considering the equilibrium adsorption for olefins and assuming the hydride transfer and the formation of the pentacoordinated carbonium ion as the rate limiting steps in bimolecular and monomolecular cracking, respectively. They conclude that when olefin adsorption is not important the protolytic cracking dominates the overall process and the reaction can be described by a first–order kinetic expression.

0097–6156/94/0571–0294$08.00/0

Others *(7)* have fitted their kinetic results to a Langmuir–Hinselwood expression, considering only the monomolecular mechanism with the protolytic cracking as the controlling step and assuming competitive adsorption of reactant and cracking products. The authors justified the use of this kinetic expression by the good fitting of the experimental results and because they did not observe an induction period during cracking of paraffins.

In this work, we have studied the cracking of n–decane on USY and Beta zeolites while measuring instantaneous conversions at very short times on stream by means of a multisampling, computer–controlled valve located at the end of the reactor. With these data, a kinetic and decay expression has been developed and compared with previously reported kinetic equations for paraffin cracking.

The models used to fit the experimental results are the following:

A) A simple first–order kinetic model:

$$\frac{\partial X}{\partial (W/F_{AO})} = k_t \frac{(1-X)}{(1+\varepsilon X)} C_{AO} \phi \tag{1}$$

where X represents the degree of conversion for the paraffin feed, W is the weight of catalyst, F_{AO} is the molar flow rate of the paraffin, k_t is the global apparent kinetic rate constant, ε is the volume expansion coefficient, C_{AO} is the concentration of the reactant at the inlet of the reactor and ϕ is the catalyst activity at each time on stream. This model does not take into account the adsorption of reactants and products.

B) A Langmuir–Hinselwood model as proposed by Abbot and Wojciechowski *(7)*:

$$\frac{\partial X}{\partial (W/F_{AO})} = \frac{A (1-X)/(1+\varepsilon X)}{1+B (1-X)/(1+\varepsilon X)} C_{AO} \phi \tag{2}$$

where A and B are kinetic parameters which include kinetic rate and adsorption constants. The model considers competitive adsorption between the reactant and the cracking products.

C) Two catalyst decay models to relate catalyst activity with time on stream:

One of them directly derived from the time on stream theory *(8)*, which considers that catalyst deactivation is due to reactants or to reactants and products to the same extension, was used in ref. *(7)*:

$$\phi = (1 + G t)^{-N} \tag{3}$$

where G and N are decay parameters and t is the time on stream.

The other one, we propose here, is a dependent   decay model which considers the products responsible for catalyst decay:

$$\frac{\partial \phi}{\partial t} = - K_{md} \, \phi^m \, X \qquad (4)$$

where K_{md} and m are the decay parameters.

A model like this is consistent with the observed fact that a direct relationship exists between the amount of coke, which is a secondary product in paraffin cracking, and catalyst decay. Equation 4 can also be derived from the model proposed by Froment, et al. *(9)*.

Experimental

Materials. The USY and Beta zeolite catalysts were prepared following the procedures described in ref. *(10)*. The final unit cell size of USY catalyst was 24.46 Å, with a Na_2O content of 0.12 wt%. The Si/Al ratio for the synthesized Beta catalyst was 14.

The n–decane had a purity higher than 99% and was not further purified. N_2 (99.999% purity) was used as a carrier gas.

Reaction Procedure. The experiments were performed in a fixed–bed continuous glass reactor at 500ºC. A mixture of N_2 and n–decane (molar ratio 9) was fed to the reactor.

The system was connected to a multisampling, computer–controlled, heated valve. A hydrocarbon detector was located at the outlet of valve to detect in a very precise way the moment when hydrocarbon filled the first loop. This moment was considered the zero reaction time, and automatically the sample was kept in the first loop. The rest of the loops were automatically filled, as programmed, at 6, 18, 39, 90, 180, 330 and 600 seconds time on stream (TOS).

Results and Discussion

With the experimental set used, it is possible to determine the instantaneous conversion at the exit of the reactor and to follow its evolution with TOS. The results obtained for the USY zeolite at different W/F_{AO} are shown in Figure 1. In this figure, it is possible to see that in a very few seconds the catalyst has lost most of its activity. Therefore, it becomes clear that the results obtained at times on stream lower than 20 seconds are crucial if reliable kinetic rate constants in absence of decay have to be obtained.

Since in our case we can measure conversions at zero TOS at different contact times, it should be possible to fit these data to the different cracking kinetic models without needing to introduce the decay expressions. Thus, by integration of equations 1 and 2 (when $\phi=1$) the following expressions are obtained:

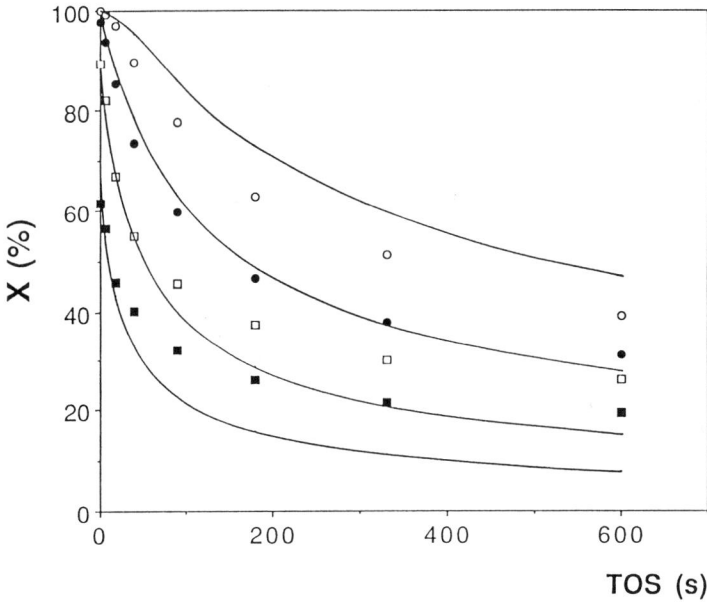

Figure 1. Influence of time on stream over instantaneous conversion in n-decane cracking on USY zeolite. Experimental values: W/F_{AO} (Kg.s.mol^{-1}) (○) 13.008, (●) 6.524, (□) 3.262, (■) 1.624. Calculated (——) equations 1,3.

$$- \varepsilon \ X|_{TOS=0} - (1+\varepsilon) \ Ln(1-X|_{TOS=0}) = k_t \ C_{AO} \ \cfrac{W}{F_{AO}} \qquad (5)$$

$$\cfrac{- (1+\varepsilon) \ Ln(1-X|_{TOS=0})}{X|_{TOS=0}} = A \ C_{AO} \ \cfrac{W}{F_{AO} \ X|_{TOS=0}} + (\varepsilon - B) \qquad (6)$$

Plots from Figure 2 show that when X at TOS=0 are considered, very good straight lines are obtained. We then must conclude that, from a kinetic point of view, both a pseudofirst-order and a Langmuir-Hinselwood model fit the results and, therefore, the contribution of the adsorption term may be marginal at the experimental conditions used here. From the parameters of the straight lines obtained at TOS=0, it is possible to calculate k_t, A, and B. The values obtained are given in Table I. From this table it can be seen that while k_t and A give very close values, the value of B is small. These results suggest again that the adsorption term in equation 2 has little significance, at least under our experimental conditions.

Table I. Parameters of equations 5 and 6 for n-decane cracking at 500ºC

Catalyst	$k_t(m^3s^{-1}kg^{-1})$	$A(m^3s^{-1}kg^{-1})$	B	ε^*
USY	0.396	0.404	0.139	0.144
BETA	0.438	0.451	0.125	0.171

*Calculated from experimental product distribution.

The Fitting of the Results Affected by Catalyst Decay. Figure 1 shows that when the experimental instantaneous conversions obtained at different TOS are fitted to a pseudofirst-order kinetics (Equation 1) coupled with the independent decay model (Equation 3) the fitting is not adequated. However, as soon as a catalyst decay dependent function is added (Equation 4), the fitting strongly improves and a very good fit of the experimental results is obtained for both USY and Beta samples (Table II and Figure 3).

**Table II. Parameters of pseudofirst-order kinetic equation
and independent or dependent decay model**

Decay model	USY		BETA	
	Independent	Dependent	Independent	Dependent
$k_t(m^3s^{-1}kg^{-1})$	0.395	0.375	0.417	0.413
G or $K_{md}(s^{-1})$	0.097	0.110	0.334	0.129
N or m	0.641	2.137	0.151	6.549
Σ (residuals)^2	0.123	0.007	0.026	0.009

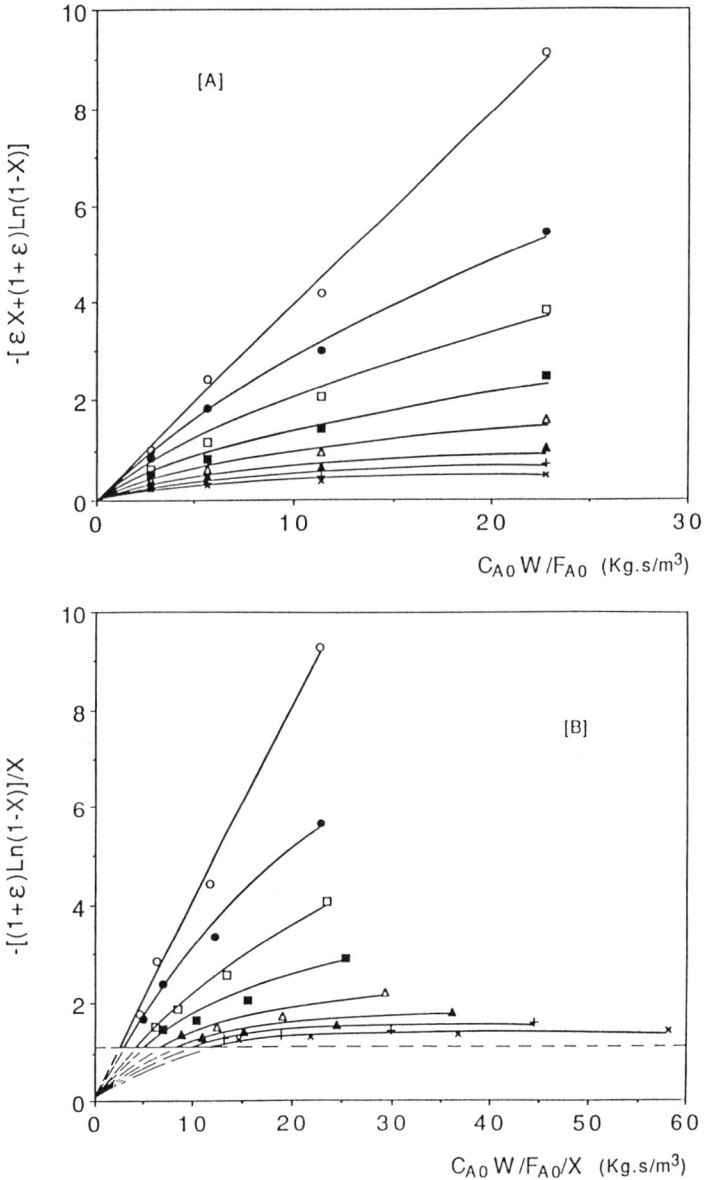

Figure 2. Kinetic plots for first order (A) and Langmuir–Hinselwood (B) models in according with equations 5 and 6 for n–decane cracking on USY zeolite. Times on stream (seconds). (o) 0, (●) 6, (□) 18, (■) 39, (△) 90, (▲) 180, (+) 330, (X) 600.

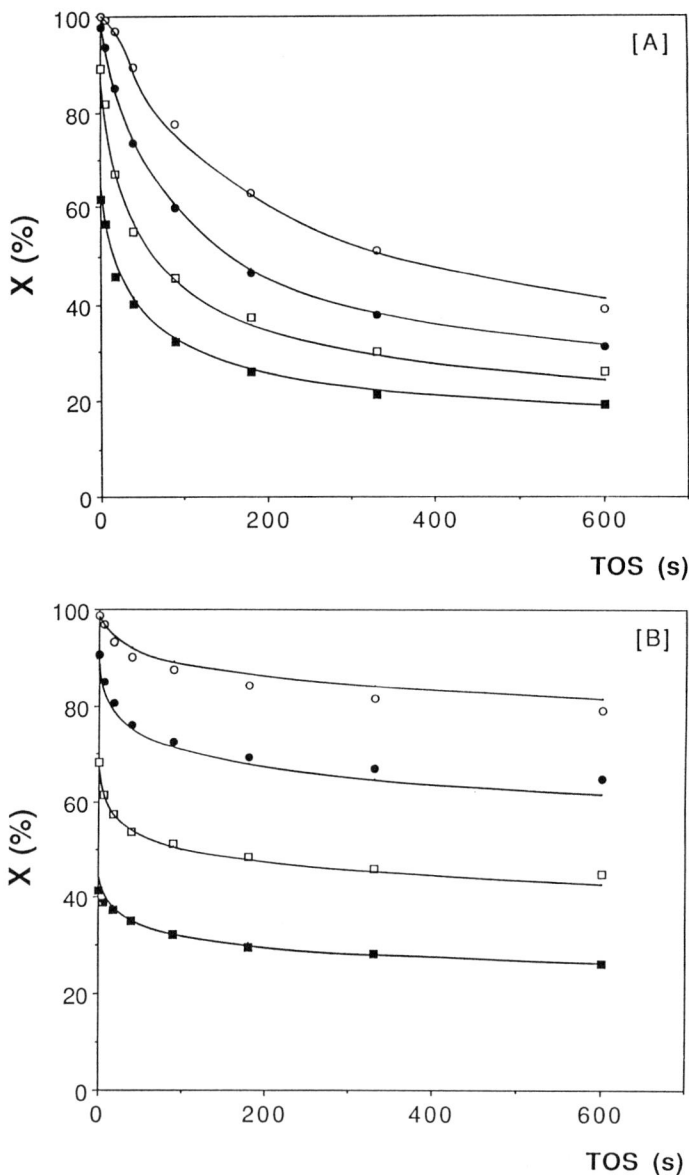

Figure 3. Influence of time on stream over instantaneous conversion in n–decane cracking. (A).– USY zeolite. Experimental values: W/F_{AO} (Kg.s.mol^{-1}) (o) 13.008, (●) 6.524, (□) 3.262, (■) 1.624. Calculated (——) equations 1,4. (B).– Beta zeolite. Experimental values: W/F_{AO} (Kg.s.mol^{-1}) (o) 6.511, (●) 3.262, (□) 1.624, (■) 0.845. Calculated (——) equations 1,4.

When a Langmuir–Hinselwood kinetic model, coupled with an independent catalyst decay function (Equation 3), is used, a good fitting is also observed (Figure 4 and Table III). However, the A and B parameters presented in Table III, and calculated when considering conversion data at different TOS, are different from those presented in Table I. On the contrary, the k_t parameters calculated by using the conversion data at zero TOS or dependent decay model are very similar (Tables I and II). From all this, and also from a conceptual point of view, it looks more logical to use a conversion dependent decay function since, as it is generally accepted, the secondary reactions of olefins are responsible for coke formation and consequently for catalyst decay.

Table III. Parameters of Langmuir–Hinselwood kinetic equation coupled with independent decay model for n–decane cracking

	USY	BETA
A $(m^3s^{-1}kg^{-1})$	0.207	0.327
B	−0.945	−0.370
G (s^{-1})	0.085	0.335
N	0.832	0.156
Σ (residuals)^2	0.030	0.015

To further discuss this, the theoretical behavior of the catalyst expected with independent deactivation can be considered for the two kinetic rate expressions. In both cases, straight lines with positive slopes should be generated at different TOS. This does not correspond to the observed reality. Indeed, results from Figure 2 for n–decane cracking on USY catalyst clearly show that the behavior is not linear for TOS>0 regardless of the kinetic model used. This is a confirmation that the independent decay model is not the most adequate to describe the catalyst deactivation during cracking of paraffins and that the good fitting obtained when combining equations 2 and 3 can be a mathematical artifact.

On the bases of the kinetic and decay parameters, it has been found that the Beta zeolite is more active and decays slower than the USY, for cracking of n–decane.

Conclusions

In conclusion, it has been presented that to obtain reliable results and to be able to compare zeolite cracking behavior when catalyst deactivation is fast, an experimental set should be used which allows measuring of instantaneous conversions, yields, and selectivities at very short times on stream. This will become even more critical when the new "ultrashort" contact time risers come into stream. Using this experimental procedure, the kinetics of n–decane cracking and catalyst decay have been established. A very good fitting of the experimental data

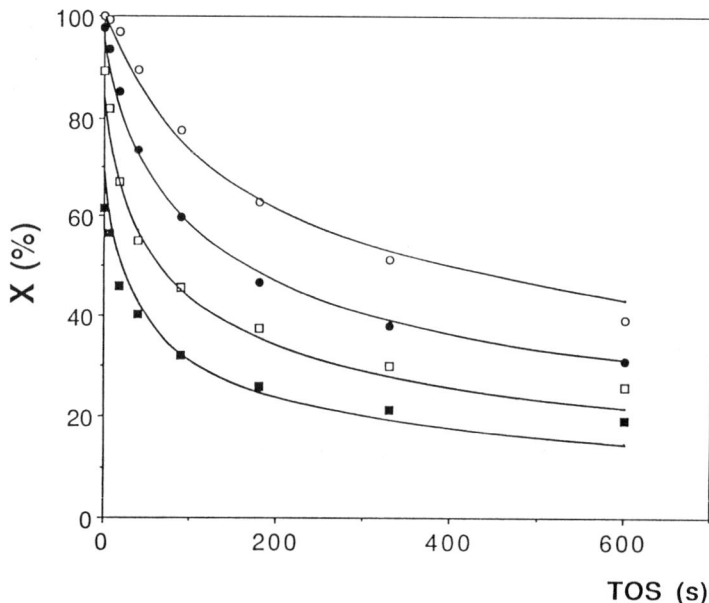

Figure 4. Influence of time on stream over instantaneous conversion in n–decane cracking on USY zeolite. Experimental values: W/F_{AO} (Kg.s.mol^{-1}) ($\circ$) 13.008, ($\bullet$) 6.524, ($\square$) 3.262, ($\blacksquare$) 1.624. Calculated (——) equations 2,3.

for both USY and Beta zeolites was obtained using a pseudo–first order kinetic model and a dependent decay expression.

Acknowledgements. Financial support by CICYT (MAT 91–1152) is gratefully acknowledged.

Literature Cited

(1)   Haag, W.O.; Dessau, R.M.; *Int. Congr. Catal. (Proc.)*, **1985**, 2, II 305.
(2)   Krannila, H.; Haag, W.O.; Gates, B.C.; *J. Catal.*, **1992**, 135, 115.
(3)   Corma, A.; Planelles, J.; Sánchez, J.; Tomás, F.; *J. Catal.*, **1985**, 93, 30.
(4)   Grensfelder, B.S.; Voge, H.H.; Good, G.M.; *Ind. Eng. Chem.*, **1949**, 41, 2564.
(5)   Wielers, A.F.H.; Vaarkamp, M.; Post, M.F.M.; *J. Catal.*, **1991**, 127, 51.
(6)   Riekert, L.; Zhou, J.; *J. Catal.*, **1992**, 137, 437.
(7)   Abbot, J.; Wojciechowski, B.W.; *J. Catal.*, **1987**, 104, 80.
(8)   Wojciechowski, B.W.; *Catal. Rev. Sci. Eng.*, **1974**, 9(1), 79.
(9)   Froment, G.F.; Bischoff, K.B.; *Chem. Eng. Sci.*, **1961**, 16, 189.
(10)   Corma, A.; Fornés, V.; Montón, J.B.; Orchillés, A.V.; *J. Catal.*, **1987**, 107, 288.

RECEIVED June 17, 1994

Chapter 22

High Activity, High Stability Fluid Cracking Catalyst SO_x Reduction Agent via Novel Processing Principles

L. M. Magnabosco[1] and E. J. Demmel

INTERCAT, P.O. Box 412, Sea Girt, NJ 08750

Novel processing principles have been developed that permit commercial-scale production of high activity/stability FCC SOX reduction agents. Prior to the development of this technology, high activity/stability SOX sorption agents could only be produced in laboratory batch quantities.

The novel processing principles involve intimate mixing of appropriate precursor materials followed by spray drying. During subsequent calcining, all unwanted constituents are driven off in gaseous form. Thus, the anions in the original, intimate mixture that is fed to the spray dryer have to be chosen such that all of the undesirable constituents end up as gases during the calcination step. These general principles and procedures can be employed in making a wide variety of catalysts and other performance materials including super-conductive materials.

X-ray diffraction data of these materials indicate that they are virtually free of undesirable and performance-impairing constituents. As an important consequence, the presence of the active, sulfur transfer agent phase Is maximized which makes these materials both highly active and stable.

Both pilot plant and commercial FCC SOX reduction data have confirmed the high activity/stability of the materials produced by these novel processing principles.

It has now been more than ten years since the first successful commercial test with a sulfur transfer agent, i.e. HRD-276, was announced (1). Subsequent testing of this material in other FCC units confirmed these early, favorable results (2). Continued research resulted in considerable efficiency improvements that were reported (3).

---

[1]Current address: 22001 Midcrest, Lake Forest, CA 92630

The commercial performance results of these SO_x transfer agents are summarized in Figure 1 where SO_x emission reductions are plotted as a function of agent concentration in catalyst inventory. As can be seen from this graph, HRD-277 is about five times as efficient as HRD-276, and HRD-280 (later named DESOX by Katalistiks) is about twice as active as HRD-277.

While the performance results of these materials, particularly HRD-280, are quite impressive, it should be pointed out here that all of these materials were produced by a so-called gel process (4,5) that does not lend itself to continuous, commercial-scale production. This is principally due to high water requirements for washing a filter cake that exhibits an extremely low filtration rate because of the high viscosity of the gelatinous precipitate. It is for these two reasons that alternate methods have been explored for commercial production of these types of materials.

FCC SOX TRANSFER FUNDAMENTALS

As the name implies, SOX transfer agents reduce SO_x emissions from Fluid Catalytic Cracking (FCC) units by capturing SO_x in the regenerator as metal sulfate and releasing the sulfur in the reactor as H_2S. The released H_2S is then converted to elemental sulfur in downstream processing along with H_2S produced in the cracking reactor. The incremental H_2S production is small and usually can be handled by existing sulfur recovery facilities. A schematic flow diagram of a typical FCC unit together with the postulated chemistry of SO_x removal by an SO_x agent, as reported in the literature (3,7), are shown in Figure 2.

In the FCC process, a mixture of fresh and recycled feed is mixed with hot regenerated catalyst in the transfer line or riser reactor inlet. The feed is catalytically cracked to distillate and ligher products, and a sulfur-containing coke is formed on the catalyst causing its deactivation. The sulfur in this coke originates from sulfur compounds in the petroleum feedstock. The coked catalyst is separated from the hydrocarbon products and unreacted feed, steam stripped of entrained hydrocarbons, and then contacted with air in the regenerator. Combustion of the sulfur-containing coke to restore catalyst activity results in a flue gas containing SO_2 and SO_3 in addition to CO_2, N_2, O_2 and minor amounts of NO and CO.

SOX TRANSFER AGENT PERFORMANCE REQUIREMENTS

A feasible SO_x transfer agent has to meet a well-defined number of critical performance criteria. These criteria have been listed in Table I. With hindsight it might be argued that common sense dictates that these criteria be observed. However, many R&D efforts failed because at least one of these criteria was being ignored.

Feasibility of sulfate formation requires that the resultant metal sulfate exhibits an SO_2 partial pressure significantly below that of the desired SO_x control level. It should be noted here that strictly speaking one should refer to the SO_3 partial pressure, since the solid metal sulfate is in equilibrium with its corresponding SO_3 partial pressure.

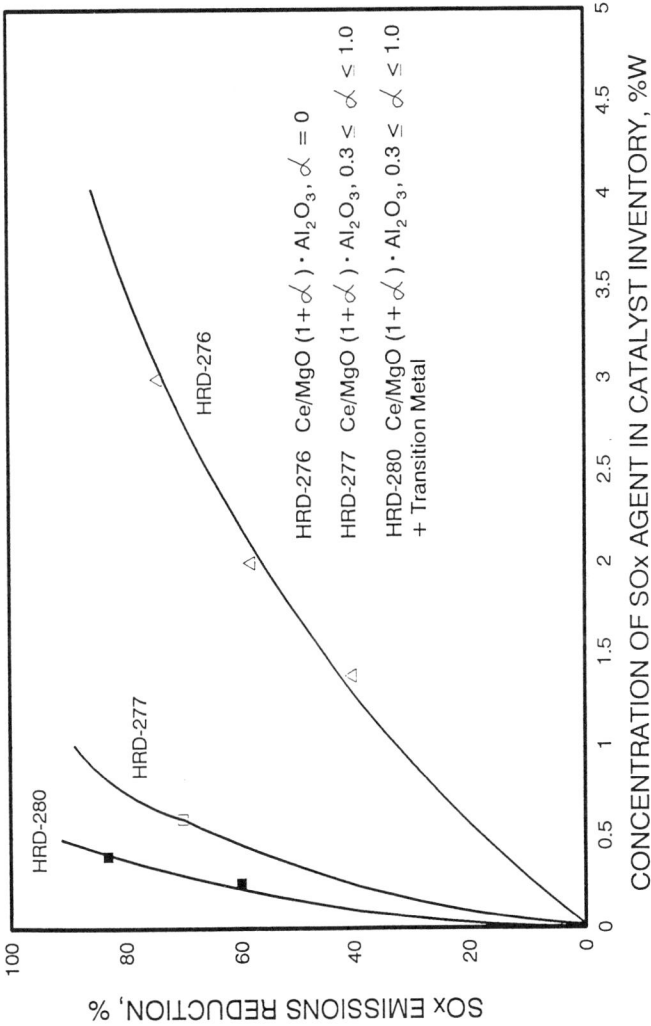

Figure 1. Activities of SO$_x$ removal agents observed in commercial trials

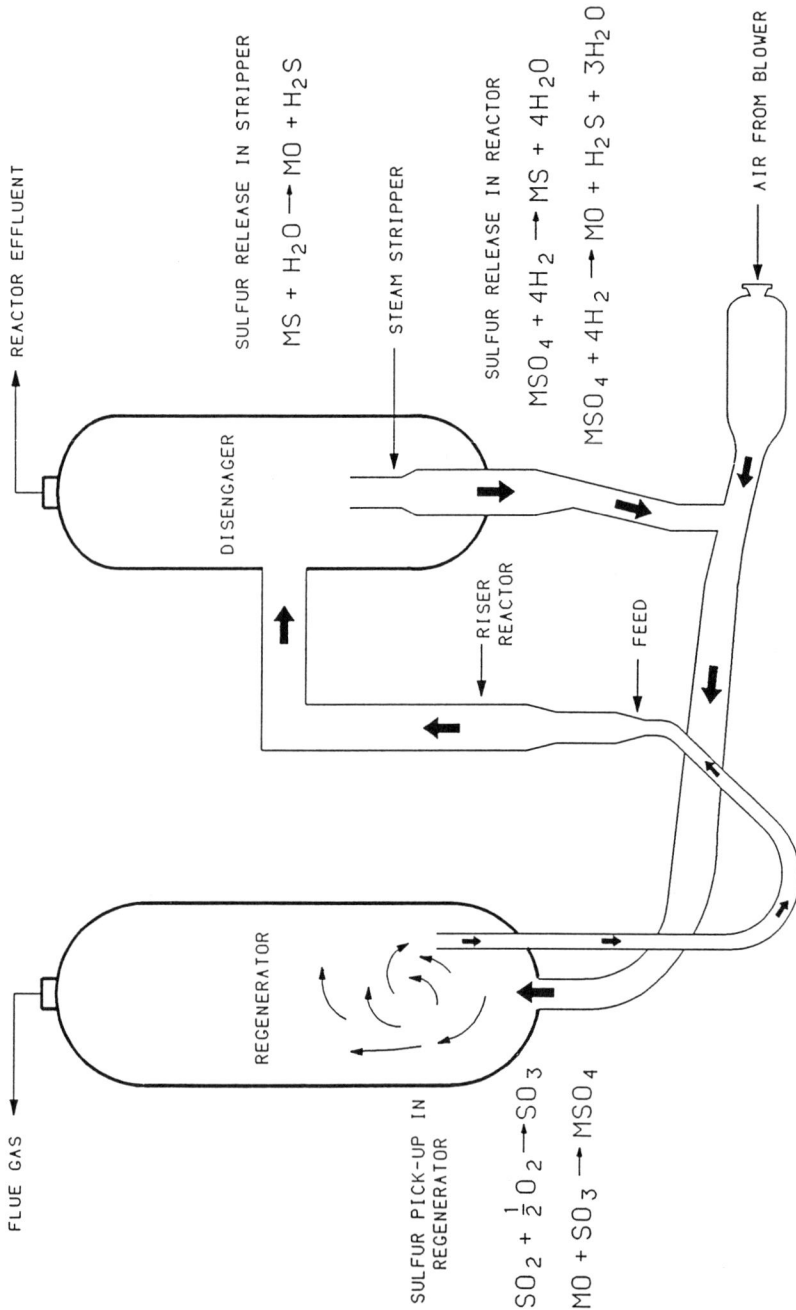

Figure 2. Schematic diagram of a typical FCC unit and postulated SO_x reduction chemistry

REACTOR EFFLUENT

SULFUR RELEASE IN STRIPPER

$$MS + H_2O \longrightarrow MO + H_2S$$

STEAM STRIPPER

SULFUR RELEASE IN REACTOR

$$MSO_4 + 4H_2 \longrightarrow MS + 4H_2O$$

$$MSO_4 + 4H_2 \longrightarrow MO + H_2S + 3H_2O$$

AIR FROM BLOWER

DISENGAGER

RISER REACTOR

FEED

REGENERATOR

FLUE GAS

SULFUR PICK-UP IN REGENERATOR

$$SO_2 + \tfrac{1}{2}O_2 \longrightarrow SO_3$$

$$MO + SO_3 \longrightarrow MSO_4$$

TABLE I. BASIC SOx TRANSFER AGENT PERFORMANCE REQUIREMENTS

---

1.   Thermodynamic Feasibility of Reaction in Regenerator

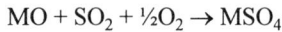

$$MO + SO_2 + \tfrac{1}{2}O_2 \rightarrow MSO_4$$

2.   Thermodynamic Feasiblity of Reaction in Reactor and/or Stripper

$$MSO_4 + 4H_2 \rightarrow MO + H_2S + 3H_2O$$
$$MS + H_2O \rightarrow MO + H_2S$$

3.   Kinetic Feasibility of Reaction in Regenerator.

4.   Kinetic Feasibility of Reaction in Reactor.

5.   Negligible Deactivation, i.e. High Stability.

6.   Sufficient Activity to Keep Agent Concentration Low.

7.   No Effects on FCC Yields and/or Product Qualities.

However, SO_3 under prevailing regenerator conditions is in equilibrium with SO_2, and hence for simplicity, reference to SO_2 only has been made.

The release reaction of sulfate to H_2S in the reactor has to be feasible.   Otherwise, SO_x will be captured in the regenerator, however, it can never be released in the reactor.

Criteria 3 and 4 are also important.   If for a given transfer agent only one of the two reactions, i.e. either the capture of SO_2 in the regenerator or the release of sulfate to H_2S in the reactor is too slow, the transfer agent cannot be used.   Similarly, a transfer agent has to exhibit a high degree of stability, Criterion 5, otherwise agent consumption and concentration of deactivated agent in catalyst inventory become too high.
Criterion 6 stipulates a high activity so as to keep agenct concentration low.   Criterion 7 requires that there be no effect, at least not measurable effect, on yields and/or product qualities.   Note that Criteria 6 and 7 are interrelated in that a high activity sorption agent results in a low agent level in inventory which in turn helps to meet Criterion 7.

For development of a commercial SO_x transfer agent, a number of additional performance criteria have to be met.   These are listed in Table II.   For easy reference, these criteria have been numbered 8-10 to emphasize continuity from the seven criteria listed in Table I.   As pointed out in the introduction, feasible commercial production, Criterion 8, is of crucial importance.   As a matter of fact, Criterion 8 and 10, the cost criterion, are related.   Obviously, if the cost becomes prohibitive, alternate SO_x control technologies may become more attractive.

TABLE II. COMMERCIAL SOx TRANSFER AGENT PERFORMANCE REQUIREMENTS

---

8.   Commercial Production Feasibility.

9.   Attrition Resistant, Proper Particle Size Distribution, Proper Density $\rightarrow$ Assurance Agent Stays in Unit.

10.   Cost.Physical characteristics are combined and listed as Criterion 9. These material properties have to be within certain limits to assure that the agent stays in the FCC unit.

Development of a suitable SO_x transfer agent presented quite a task, since all of the performance criteria listed in Tables I and II had to be met. Moreover, there are numerous interactions between several of the criteria that also have an influence on agent performance. For instance, Criterion 3, kinetic feasibility of release reaction in the reactor impacts on the kinetics of SO_x capture in the regenerator, Criterion 4, and vice versa. And both of these criteria influence stability of the transfer agent. Thus, it is quite clear that a simple SO_x pick-up test could not suffice for the successful development of the HRD series of SO_x transfer agents. The R&D tool relied on most heavily was a circulating pilot plant, as reported earlier (6). In this unit, the commercial-scale FCC process is simulated realistically while cracking a typical FCC hydrocarbon feedstock. Furthermore, SO_x capture and SO_x release interact with each other. This interaction then results in establishment of a dynamic equilibrium, i.e. a steady-state level of sulfur on catalyst.

MAGNESIUM ALUMINUM SPINELS

As indicated in Figure 1, the commercially successful SO_x transfer agents are magnesium aluminum spinels that contain cerium. HRD-280 contains both cerium and a transition metal.

If properly prepared, magnesium oxide and aluminum oxide spinel solid solutions can be synthesized over a wide range of compositions, as indicated in Figure 3, where at the magnesia-rich side the composition is about 50% w MgO/50% w Al_2O_3, whereas at the alumina-rich end, the composition is about 3.4% MgO/96.6% Al_2O_3. The data in Figure 3 have been taken from the literature (5,7). In this graph, the (440) XRD diffraction peak has been plotted against composition as defined by:

$$y = 0.01469 \, x + 65.24 \qquad\qquad (1)$$

y = (440) XRD diffraction peak
x = composition

where x is defined as follows for magnesia-rich spinels:

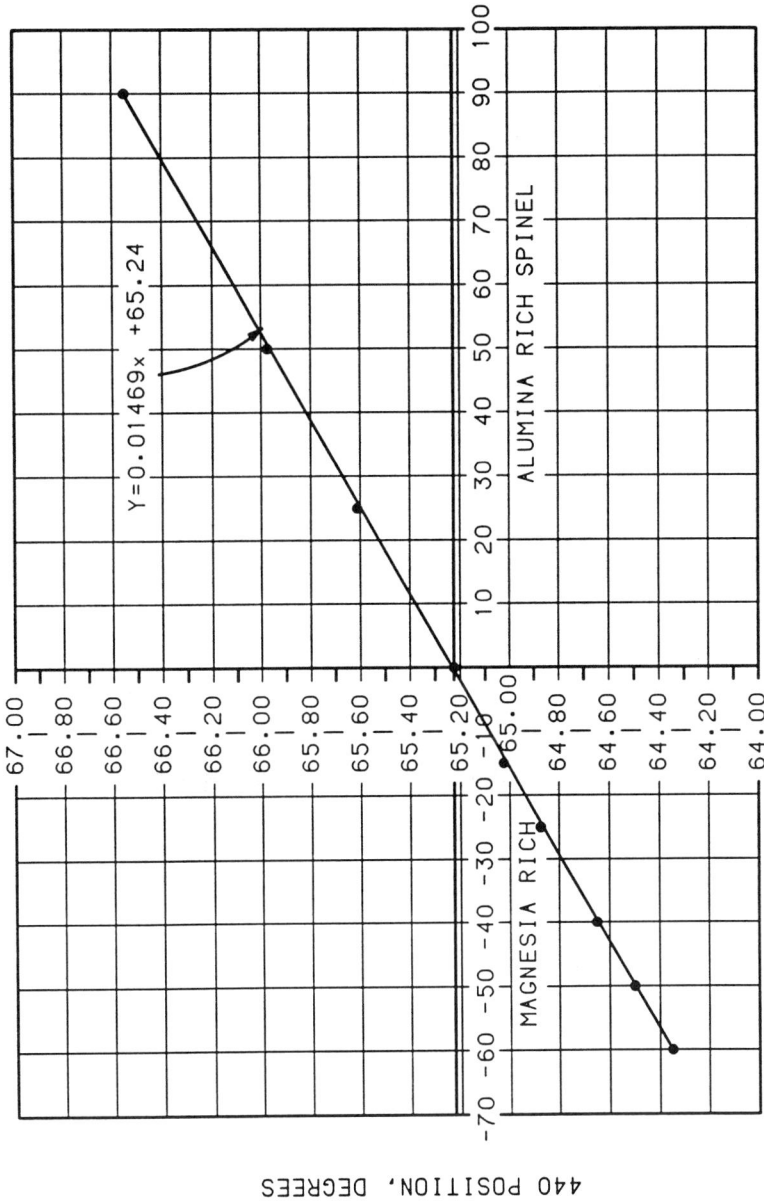

Figure 3. X-Ray diffraction 440 position vs. composition for solid solutions of Al$_2$O$_3$ or MgO in spinel

$$MgO(1 + \alpha) \cdot Al_2O_3$$
$$\alpha > O$$

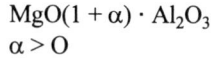

$$x = -\frac{\alpha \, 100}{1+\alpha} \qquad (2)$$

For alumina-rich spinels,

$$MgO \cdot Al_2O_3 \, (1 + \alpha)$$

$$\alpha > O$$

$$x = \frac{\alpha \, 100}{1+\alpha} \qquad (3)$$

and $x \equiv 0$ for stoichiometric spinel.

The relationship in Figure 3 together with overall compositional information regarding concentration of MgO and Al_2O_3, respectively, can be used to determine:

1.   Amount and type of spinel, i.e.
         $x = 0 \;\rightarrow\;$ stoichiometric spinel
         $x < 0 \;\rightarrow\;$ magnesia-rich spinel
         $x > 0 \quad\rightarrow\;$ alumina-rich spinel

   and

2.   Amount of "free" magnesia or "free" alumina.

Note that the calculation procedure depends on a mass balance for both MgO and Al_2O_3 and is explained in detail in the Appendix.

Production of materials devoid of free MgO is quite difficult even on a laboratory scale as pointed out by several researchers (4)(5)(7)(8). In references (5) and (7), it was pointed out that materials containing no free magnesium oxides were preferred over mixtures containing both spinels and free magnesium oxide. This is due to the fact that magnesium oxide sulfates quite readily in the regenerator of an FCC unit. However, reduction of the magnesium sulfate formed on a magnesium oxide phase cannot be reduced at typical FCC reactor temperatures of 850-1000°F. In contrast, magnesium sulfate formed on a magnesium oxide aluminum oxide spinel solid solution can easily be reduced under these conditions. Experimental evidence by Habashi et al. (9) indicates that magnesium sulfate cannot be reduced with hydrogen at temperatures below 1350°F. This is surprising since thermodynamic calculations clearly show that this reaction should be feasible at 800-1000°F. Thus, it must be concluded that the magnesium sulfate reduction with hydrogen is controlled by kinetics. This is confirmed by the observation that sulfate formed on a magnesium aluminum spinel (even magnesia rich) can easily be reduced with hydrogen at 800-1000°F.

The fact that magnesium sulfate formed on magnesium oxide cannot be reduced at FCC reactor conditions is the reason that spinel solid solutions of magnesium oxide/aluminum oxide are preferred over mixtures consisting of magnesium aluminum spinels with free magnesium oxide.

MAGNESIUM ALUMINUM SPINEL PREPARATION

Magnesium aluminum spinels have been made by the following methods:

1.   Powder Method.   Finely ground powders of alumina and magnesia are intimately mixed and then pressed together followed by high temperature calcination at temperatures of at least 1200°C (10).

2.   Gel Method.   An intimate mixture of magnesium hydroxide and aluminum hydroxide is formed from, for example, magnesium nitrate and sodium aluminate by mixing aqueous solutions of these compounds and then adding sodium hydroxide until a pH of about 9.5 is reached.   The precipitate is then filtered and washed with large amounts of water to remove undesirable sodium ions.   The gelatinous and high viscosity precipitate is then dried and subsequently calcined at 1350°F (4,5).

3.   Slurry Peptizing Method.   This procedure is a combination of methods 1 and 2, in that aqueous slurries of high-reactivity alumina and high-surface area magnesia are intimately mixed together with a monoprotonic acid that is added for peptization of the alumina.   After drying, the material is again calcined at 1350°F (8).

Method 1 is obviously not suitable for commercial production of SO$_x$ transfer agents, since Criteria 8 and 9, in Table II, cannot be met.   It would be particularly difficult to meet Criterion 9, i.e. produce materials with proper particle size distribution and acceptable attrition resistance.   In addition, spinels cannot be produced that are devoid of free magnesia.   Thus, there is always a phase of free magnesium oxide in addition to magnesium aluminum spinel, principally stoichiometric spinel.

Use of Method 2 permits to produce suitable spinels, including the more desirable magnesium-rich spinels.   As pointed out in the introduction, commercial use of this procedure is not possible due to slow filtration rates of the gelatinous precipitate and the high wash water requirement to wash out undesirable sodium ions.   Even in laboratory-scale preparations, it has been difficult to produce magnesia-rich spinels devoid of free magnesia (5).

Method 3 is well-suited for commercial production.   However, this procedure is unable to produce desirable magnesia-rich spinels devoid of free maganesia (8). Depending on the details of the procedure, including proportion of magnesia and alumina used, free magnesia varied between 8 and 17% of total material produced (8). An XRD plot for a typical material that has been produced with this method is shown in Figure 4. As shown in this plot, the presence of free magnesia is indicated by the "shoulder region" at a 2-theta value of about 62° and the spinel peak is indicated at a 2-theta value of 64.65°. A 2-theta value of 64.65° indicates this particular spinel to be 40% magnesia rich as per Figure 3.

An XRD pattern of a material produced with this method that is even less preferred is shown in Figure 5. In this plot, free magnesia is indicated by a distinct peak at a 2-theta value of about 62.5° and the spinel peak appears at 65.2°. The fact that two distinct peaks appear on this plot points out that this material has more free magnesia, and is therefore less suitable as SO_x transfer agent than that for which the XRD pattern has been displayed in Figure 4.

NOVEL PROCESSING SCHEME

This procedure is the quintessential extension of the Slurry-Peptizing Method to its ultimate physical limitation. It consists of intimate mixing of magnesium oxide and alumina in water whereby both of these reactants consist of particles less than 5 nanometers (nm). In the extreme, one or both of the reactants may be present in the form of ionic species, i.e. magnesium may be present In the form of magnesium hydroxy acetate, magnesium hydroxy nitrate or similar, and aluminum may be present in the form of an aluminum oxy/hydroxy sol with a particle size of less than 2 nm. This intimate mixture is spray dried at severe conditions so as to rapidly evaporate the water and thereby preserve the status of intimate mixing. In the subsequent calcination step at 1350°F, remaining water and all unwanted constituents (acetic acid, nitric acid and their respective decomposition products) are driven off and the desired spinel is formed. It should be noted here that this method was developed in recognizing that spinel precursor material could include any material comprising magnesium and aluminum provided that:

1.   Reactants can be intimately mixed on an "atomic" or "molecular" level.

2.   Water can be evaporated fast enough to preclude segregation of magnesium and/or aluminum reactants.

3.   Anions for both reactants, i.e. magnesium and aluminum have to react to gaseous products during calcination so that the desired spinel can be formed (7).

An XRD for a magnesia-rich spinel that has been prepared according to this processing scheme is shown in Figure 6. As indicated, the (440) peak for this spinel is at a 2-theta value of about 64.65° which translates to a 40% magnesia-rich spinel based on the relationship displayed in Figure 3. More importantly, the XRD (440) is devoid of a shoulder that would indicate presence of undesirable free magnesium oxide. As a consequence, the presence of the active, sulfur transfer agent phase is maximized which makes these materials both highly active and stable after impregnation with $Ce(NO_3)_3$ and a transition metal to impart oxydation function.

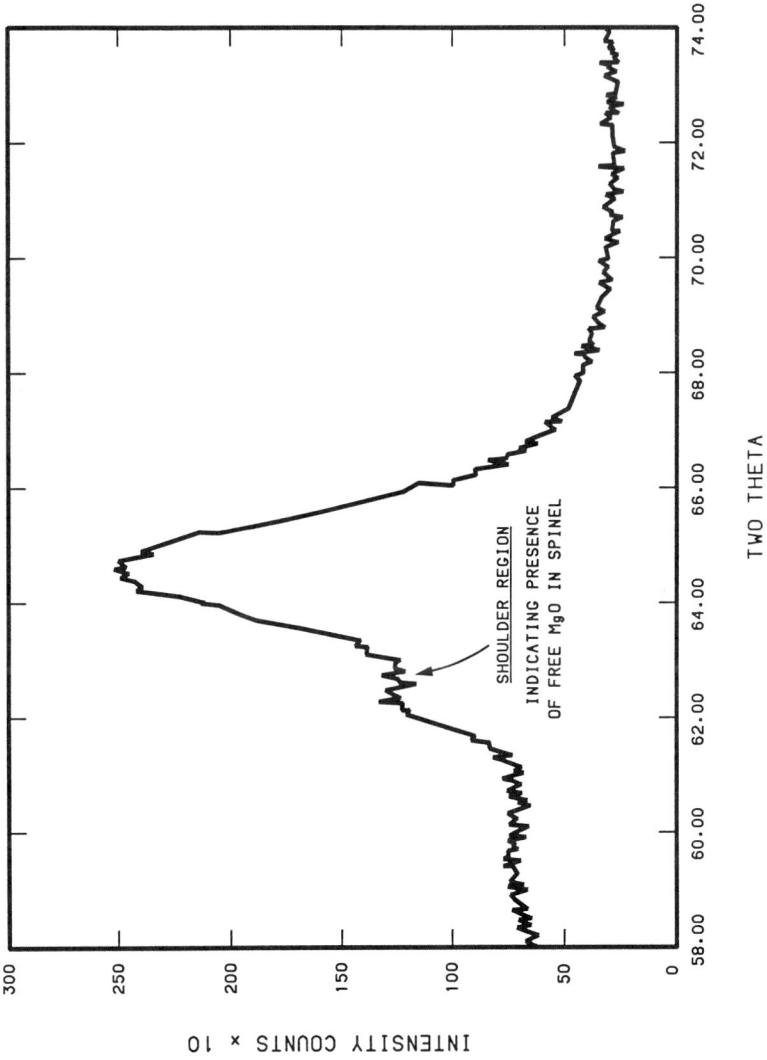

Figure 4. XRD for magnesium aluminum spinel produced via slurry-peptizing method

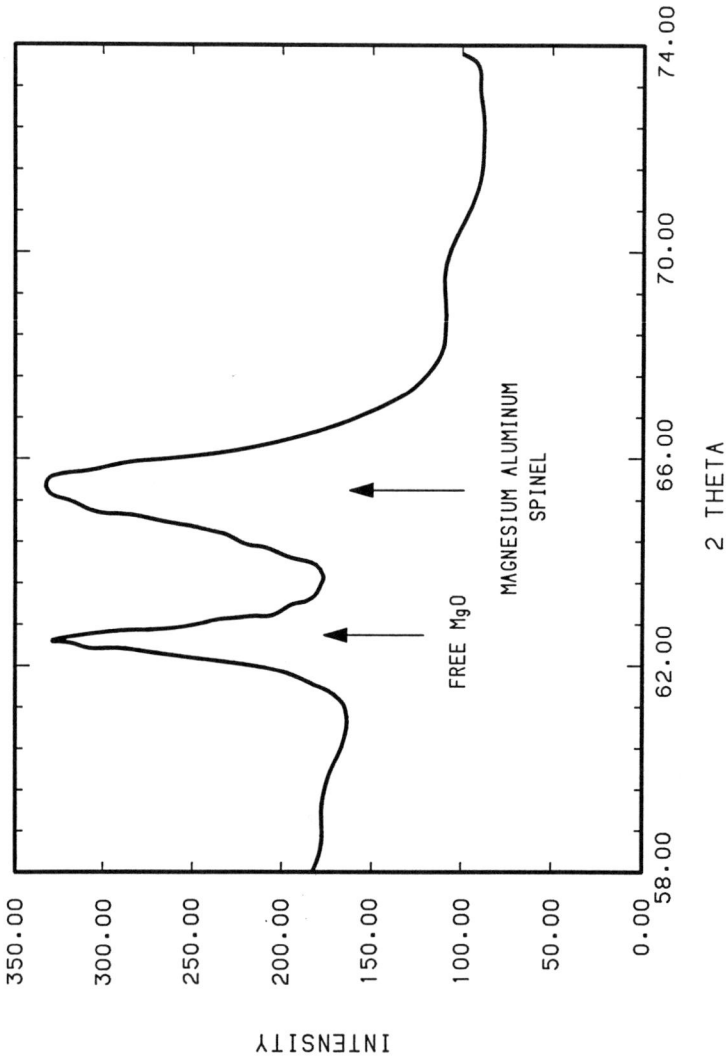

Figure 5. XRD for magnesium aluminum spinel produced via slurry peptizing method

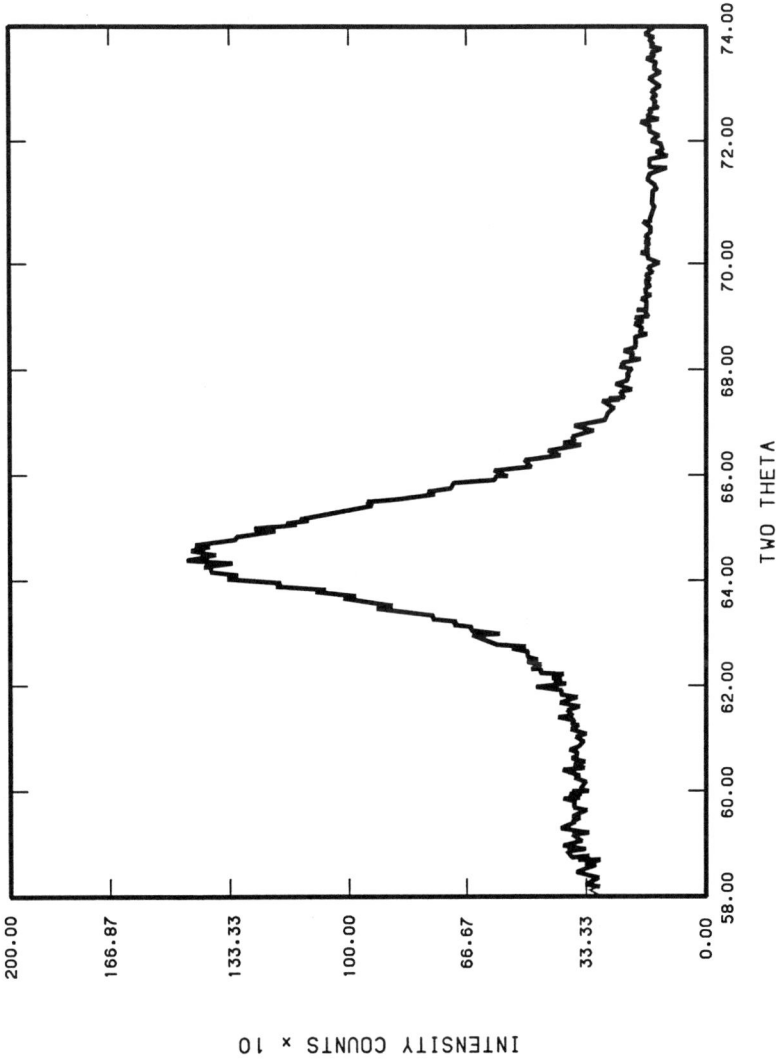

Figure 6. XRD for magnesium aluminum spinel produced via novel processing scheme

It should be pointed out here that the general principles and procedures can also be employed in making a wide variety of catalysts and other performance materials, including super-conductive materials.

TRANSFER AGENT PERFORMANCE COMPARISON

Materials prepared according to the "Slurry Peptizing Method" (8) and the "Novel Processing Scheme" (7) were provided with cerium (to a level of about 10%) and a transition metal to impart the necessary oxydation function. The material made according to our processing scheme is named NOSOX and is used successfully on a commercial scale. A performance comparison of a variety of materials is presented below.

A performance comparison of SOX transfer agents is presented in Figure 7, where SO_x emission, $KgSO_x$/Mbbl, are plotted against time on stream. As indicated on this graph, performance decreases in the sequence Additive A to D. Agent A is a magnesium-rich spinel with 15% free magnesium and has been made according to the "Slurry Peptizing Method". Agent B is an alumina impregnated with cerium. Agent C is also an alumina impregnated with cerium and a transition metal. Agent D is an alumina impregnated with lanthanum. These data show convincingly that a transfer agent based on magnesium aluminum spinel is superior to other competitive materials, even though free magnesia is present.

A pilot plant comparison between NOSOX and Agent A is presented in Figure 8. This graph shows quite clearly that NOSOX is both more active and more stable than Agent A which has "free MgO". Note that NOSOX has no "free MgO". Its magnesia-rich spinel peak is similar to that shown in Figure 6.

Another pilot plant comparison between NOSOX and two commercial agents, Agent A and Agent E is displayed in Figure 9. Base emissions in this test were 150 kg. NOSOX again exhibits superior activity. Average emissions for NOSOX amounted to about 17 kg/Mbbl whereas for both Agent A and Agent E, average emissions amounted to about 25 kg/Mbbl.

Results of a commercial trial with NOSOX are presented in Figure 10. After establishing an emissions base line of 500 ppmV SO_x, about 210 lbs. of NOSOX were added per day for eight days for a total NOSOX addition of about 1700 lbs. The total daily addition was split in three to four additions of 40 to 80 lbs. each. Emissions dropped to zero after eight days and then slowly increased towards the base line level of 500 ppm. The base line emissions level was reached seven days after NOSOX addition was stopped. There was no effect on yields and/or product qualities during the test.

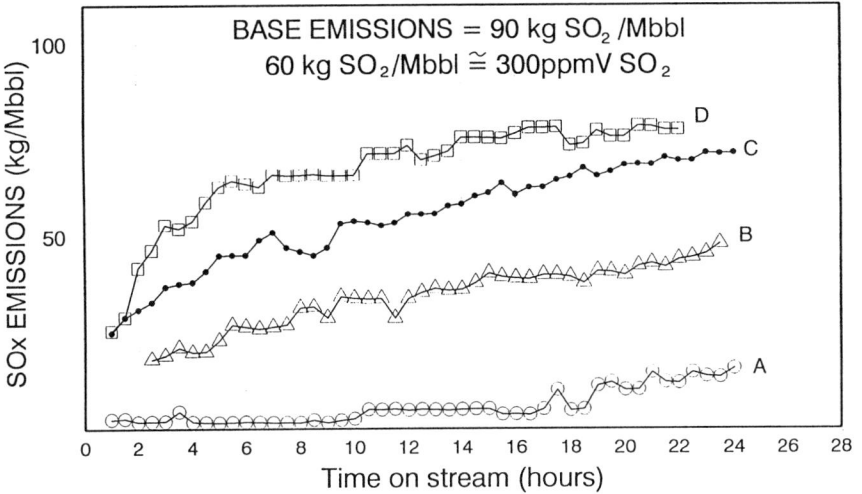

Figure 7. Comparison of FCC SO$_x$ additives
Circulating pilot plant data
1 wt% fresh additive in inventory

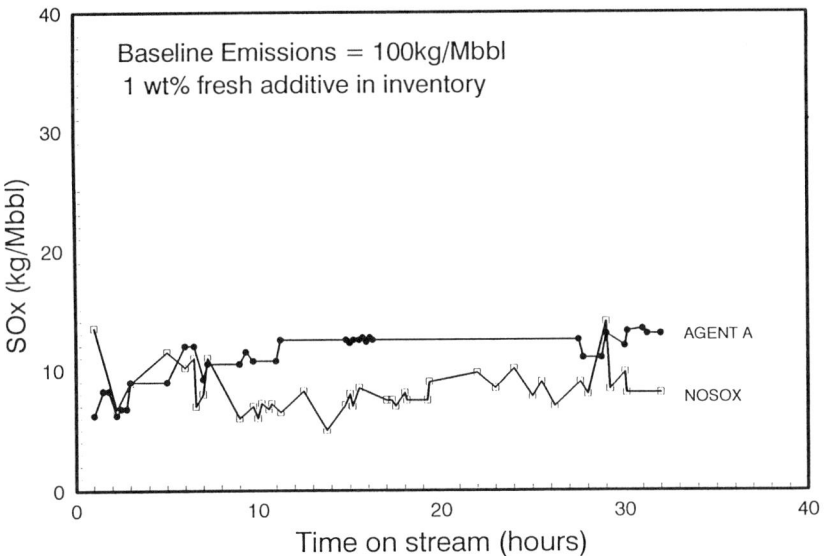

Figure 8. Pilot plant evaluation of FCC SO$_x$ additives

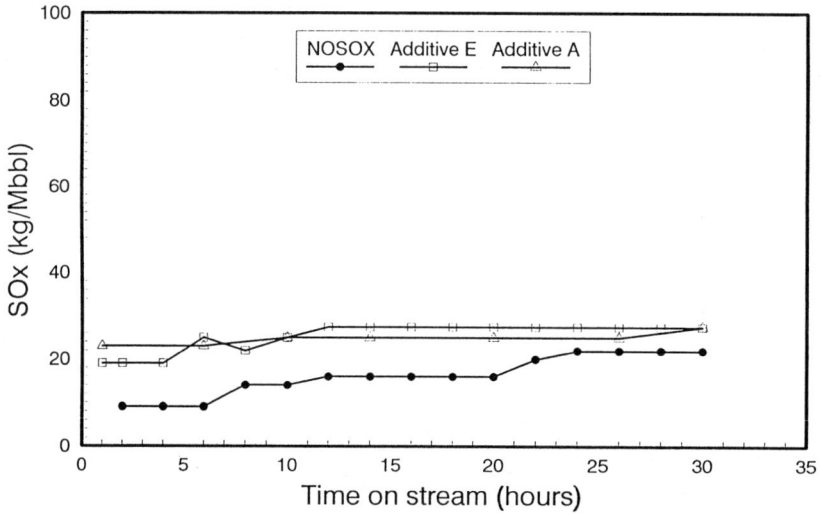

Figure 9. Pilot plant performance of SO_x additives
FCC SO_x vs Time-on-Stream

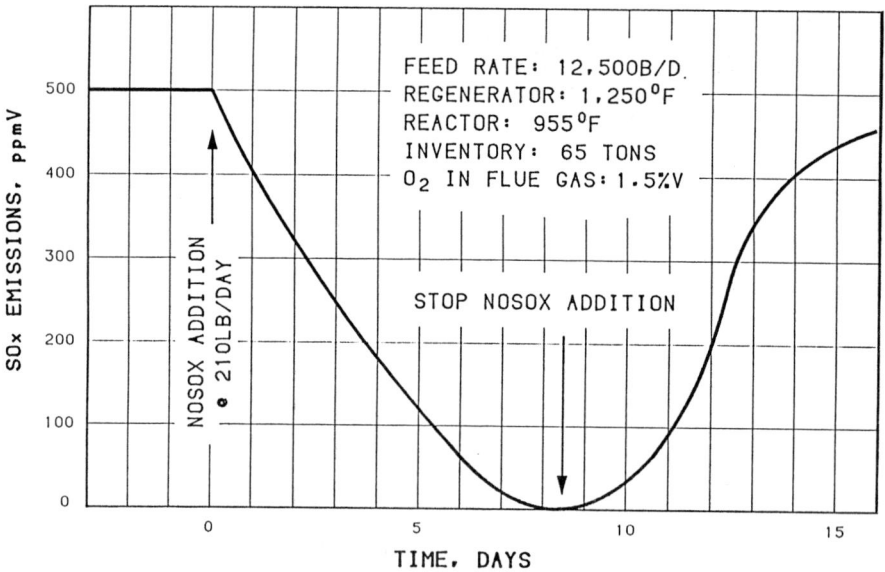

Figure 10. Commercial NOSOX Trial

Proper assessment of additive performance requires knowledge of a number of performance parameters including:
1) activity
2) stability
3) initial loss of agent
4) attrition
5) type of catalyst inventory.

An analysis of these commercial data and comparison with earlier commercial trial results (1-3) indicate NOSOX performance to be virtually identical with that of HRD-280 which is depicted in Figure 1.

Thus, it has been demonstrated that it is possible to produce SO$_x$ transfer agents, i.e. NOSOX, on a commercial scale such that these materials exhibit a process performance equal to the materials produced via the gel process.

ACKNOWLEDGEMENT
The concepts and ideas that led to the relationship displayed in Figure 3 were developed by Mr. M.F.L. Johnson.

LITERATURE CITED
1. Magnabosco, L.M. and Powell, J.W., "Commercial Evaluation of SOx Catalyst," Katalistiks Fourth Annual Fluid Cat Cracking Symposium, Amsterdam, The Netherlands, May 18-19, 1983.
2. Powell, J.W., Chao, J., Edison, R.R. and Magnabosco, L.M., "FCC Sox Reduction Via ARCO Catalyst Technology," 1984 NPRA Annual Meeting, Paper AM-84-56, San Antonio, Texas, March 25-27-1984.
3. Tamborski, G.A., Magnabosco, L.M., Powell, J.W. and Yoo, J.S., "Catalyst Technology Improvements Makes SOx Emission Control Affordable", Katalistiks Sixth Annual Fluid Cat Cracking Symposium, Munich, Germany, May 22-23 (1985).
4. Yoo, J.S., Jaecker, J.A., U.S. Patent 4,472,267 (1984).
5. Siefert, K.S., Yoo, J.S., Burk, E.H., U.S. Patent 4,471,070 (1984).
6. Wagner, M.C., Humes, W.H., Magnabosco, L.M., "Fully Automated Catalytic Test Unit", Plant/Operation Progress, 3, 4, 222-226 (1984).
7. Magnabosco, L.M., Demmel, E.J., U.S. Patent 5,108,979 (1992).
8. Bhattacharyya, A., Cormier, W.K., Wolterman, G.M., U.S. Patent 4,728,635 (1988).
9. Habashi, F., Shaheer, A.M., Vo Van K., Can.J.of Chem., (54) 23, 3646-3650 (1976).
10. Alper, A.M., McNally, R.N., Ribbe, P.H., Doman, E.C., J.Am.Ceram.Soc., 1962, 45(6), 263.

APPENDIX

DETERMINATION OF SPINEL COMPOSITION

A = Normalized Amount of MgO, % w

B = Normalized Amount of Al_2O_3, % w

A + B = 100

$$\alpha' = \frac{A}{MMO} \quad , \quad \beta' = \frac{B}{MAl}$$

MMO = Mole weight MgO

MAl = Mole weight Al_2O_3

1. If $\alpha' = \beta'$ and (440) peak position, 2-theta is equal to 65.24°, the sample consists of stoichiometric spinel.

   formula : $MgO \cdot Al_2O_3$

2. If $\alpha' > \beta'$ and (440) peak position, 2-theta, y < 65.24.

$$x = \frac{(y - 65.24)}{0.01469}$$

x = Amount of MgO in magnesia-rich spinel, %w

$$\alpha = \frac{|x|}{100 - |x|}$$

$$\gamma = \frac{\alpha'}{\beta'} \quad , \quad \gamma \geq \alpha$$

$$\delta = \frac{(\gamma - \alpha) \cdot MMO \cdot 100}{\gamma \cdot MMO + MAl}$$

δ = Amount of free MgO in magnesia-rich spinel, % w

Thus, the formula for magnesia-rich spinel is:

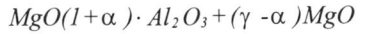

$$MgO(1+\alpha) \cdot Al_2O_3 + (\gamma - \alpha)MgO$$

3.   If $\alpha' < \beta'$ and (440) peak position, 2-theta, y > 65.24

$$x = \frac{(y - 65.24)}{0.01469}$$

x = Amount of Al_2O_3 in alumina-rich spinel

$$\alpha = \frac{x}{100 - x}$$

$$\gamma = \frac{\beta'}{\alpha'}, \ \gamma \geq \alpha$$

$$\delta = \frac{(\gamma - \alpha) \cdot MAl \cdot 100}{MMO + \gamma MAl}$$

δ = Amount of free Al_2O_3 in alumina-rich spinel, %w

Thus, the formula for the alumina-rich spinel is:

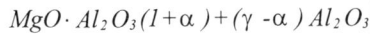

$$MgO \cdot Al_2O_3(1+\alpha) + (\gamma - \alpha) Al_2O_3$$

RECEIVED June 17, 1994

Chapter 23

Resid Cracking Catalysts

Shiying Tu, Zubi Chen, and Zhongbi Fan

Research Institute of Petroleum Processing, China Petrochemical Corporation (SINOPEC), 18 Xue Yuan Road, 100083 Beijing, China

This paper introduces the R&D activities at RIPP in FCC catalyst field. In recent years, efforts have been paid to heavier and more inferior feedstocks. Varieties of dealuminated ultrastable Y type such as DASY, SRNY were developed and commercialized in mid-1980s. RSADY, a newly developed ultrastable Y, was in commercial production at the end of 1990. Matrices of different formulations and functions have also been studied and developed. Highlights of commercial FCC performance in China are included.

Catalytic cracking is the most important process in the petroleum refining industry in China. RIPP has paid great efforts to the FCC CATALYST RESEARCH & DEVELOPMENT PROGRAM which was set up as a major field of the institute since its founding in 1956. RIPP has also been continuously contributing to the R&D of FCC process technology, with its catalyst capability and close cooperation between the refineries and design institutes. All these together help promote China's progress in FCC technology.

The development and production of fluid catalytic cracking (FCC) catalysts in China has about 30 years of history. RIPP pioneered the R&D work and was involved in most of the technological innovations of FCC catalyst advancements in the past years which, in a sense, reflect RIPP's continuous efforts in this field.

Main Challenges in the 1990s

FCC is a major conversion process in petroleum refining. It is relatively inexpensive to construct and operate and is quite flexible. Since most of the Chinese crudes are low in sulfur and paraffinic in nature, FCC process development was emphasized.

The challenges encountered in the 1990s dealt mainly with the feedstocks for FCC. The residual oil ($530°C^+$) of Chinese crudes averaged about 40% of the crudes. Major

0097–6156/94/0571–0322$08.00/0

concerns about processing these heavy ends are becoming increasingly crucial. As shown in Table I, to cope with the increased capacity of FCC, the straight-run VGO was not enough for full capacity operation. Resids and other secondary processed product slates such as CGO and DAO were blended as feedstocks to the unit. The total FCC feedstocks processed in 1991 compared with those in 1988 increased by 708×10^4 MT/Y, in which VGO increased merely 84×10^4 MT/Y, resid and inferior CGO plus DAO reached 832×10^4 MT/Y. It is expected that more heavier and inferior stocks are to be blended in FCC feedstocks in the coming years.

Table I. The Changing FCC Feedstocks

10^4 MT/Y	1988	1990	1991
Total Processed	2,506	2,870	3,214
Breakdown:			
VGO	2,298	2,256	2,382
ATR	—	245	299
VR	—	150	237
CGO	—	134	174
DAO	—	85	122
ATR+VR+CGO+DAO	208	614	832
% (on total)	8.3	21.4	25.9

As the feedstocks became heavier, problems emerged for both process and catalyst. Table II shows the main properties of some Chinese resids. Table III shows the properties of some CGOs and DAOs.

Table II. Characteristics of Some Chinese ATRs in Crude

Crude Source	Daqing	Shengli	Liaohe	Zhongyuan
Boiling Range, °C	>350	>350	>350	>370
% in Crude, wt %	71.5	68	68.9	55.5
Density(20°C), g/ml	0.8959	0.9463	0.9436	0.9062
°API	25.8	17.5	17.9	24
Elemental Analysis, wt%				
C	86.32	86.36	87.39	85.37
H	13.27	11.77	11.94	12.02
S	0.15	1.2	0.23	0.88
N	0.2	0.6	0.44	0.31
Pour Point, °C	44	40	30	47
Carbon Residue, wt%	4.3	9.6	8	7.5
V, ppm	<0.1	1.5	—	4.5
Ni, ppm	4.3	36	47	6

From Table II and Table III, the following main problems are obvious:
- (1). Carbon residue increased;
- (2). High boiling fractions increased;
- (3). Heavy metals, especially Ni, increased;
- (4). Basic nitrogen content increased;
- (5). More refractory hydrocarbon content increased (CGO and DAO in particular).

Regarding solutions to these problems, what process can do mainly concerns the removal of excess heat and better atomization of feed to alleviate the coke make. Catalyst, on the other hand, is the key factor of concern.

Table III. VGO, CGO, DAO and VR Characteristics of Selected Crudes

Crude Source	Shengli			Liaohe		Luning		
	VGO	DAO1	DAO2	VGO	CGO	VGO	CGO	VR
Density(20°C), g/ml	0.8935	0.9239	0.9375	0.9249	0.9057	0.8758	0.8755	0.9647
Carbon Residue, wt%	0.14	3.20	6.26	0.20	0.20	0.05	0.12	14.7
Basic Nitrogen, ppm	—	—	—	860	2400	347	1536	—
Elemental Analysis,wt%								
C	86.74	86.58	86.82	87.07	86.70	85.63	86.11	86.05
H	12.61	12.60	11.33	12.11	12.16	12.93	12.45	11.30
S	0.14	0.44	0.52	0.20	0.31	0.68	0.92	1.31
N	0.48	0.84	0.95	0.22	0.49	0.12	0.38	0.59
Heavy Metal, ppm								
V	<0.01	<0.1	<0.1	—	—	—	—	3.8
Ni	0.21	7.3	16.4	—	—	—	—	33.6
Group Analysis, wt%								
Saturates	69.8	47.4	28.7	59.1	56.8	69.7	64.5	22.9
Aromatics	23.6	39.4	40.6	35.7	36.1	27.2	29.8	32.5
Resins & Asphaltenes	6.6	13.6	30.7	5.2	7.0	3.1	5.7	44.6

Regarding products, the goal is to meet the market demands. To maximize gasoline and LCO remains the primary objective. LCO as blending stocks for diesel oil, LPG as gas fuel, olefins as raw materials for petrochemicals, and high–octane gasoline, are all in great demand.

In view of this situation, FCC catalyst R&D work needs to catch up with the ever-changing feedstock and product demands, and catalysts developed should be more functional and flexible.

Considerations of Designing a Resid Cracking Catalyst

Basic requirements pertinent to resid cracking catalysts are:
- (1). High bottoms cracking capability;
- (2). High stability to endure severe hydrothermal aging in the FCCU;
- (3). Good selectivity, particularly coke selectivity;

(4). Tolerance of metal poisons, especially the high nickel contents in Chinese crudes.

To satisfy the diversified demands of product slates and the varying characteristics and operating constraints of a specific unit, tailoring and formulating specific catalysts for use is necessary.

The resid cracking catalysts explored and developed in recent years by RIPP and joint efforts of related catalyst manufacturers and refineries are mainly constituted by various kinds of dealuminated stable Y zeolites (e.g. DASY, SRNY and RSADY), rare-earth Y zeolites (REYs), rare-earth hydrogen Y zeolites (REHYs), pentasils and so on, combined with functional matrices by proprietary matrix technology. Proper configuration of acidity, pore system and essential physical and mechanical properties are thus provided for making a good catalyst.

In the following sections, five series of resid cracking catalysts— ZCM, LCH, LCS, RHZ and RMG—will be presented with results of commercial performance tests. Catalyst properties are listed in Table IV. CAT-A and CAT-B are catalysts from world market.

Table IV.   Fresh Catalyst Properties

Catalyst	ZCM7	LCH	LCS7	RHZ200	RMG	CAT-A	CAT-B
Chemical: wt%							
Al_2O_3	42.2	48.2	33	22	42.9	33.9	31.7
Na_2O	0.27	0.07	0.29	0.29	0.13	0.35	0.27
RE_2O_3	0.8	0.6	2.1	1.5	—	0.07	1.1
Fe_2O_3	0.25	0.55	0.2	0.2	0.49	0.63	—
SO_4^{2-}	—	—	1.1	0.7	0.85	—	—
Physical:							
SA, m^2/g	225	218	256	325	206	256	246
$PV(H_2O)$,ml/g	0.32	0.36	0.5	0.47	0.3	0.38	0.33
ABD, g/ml	0.72	0.74	0.56	0.55	0.79	0.73	0.83
A.I., wt%/hr	0.9~1.5	1.8~2.5	3	2.5	0.7	1.2	3.3
MAT activity, wt%							
800°C-4hr-100%H_2O	66	70	76	74	74	61	64
U.C.S., Å	24.49	24.48	24.65	24.65	—	24.52	—

ZCM Series

This catalyst series was developed in 1986 and commercialized in 1987. It comprises a proprietary dealuminated stable Y (DASY) zeolite and a formulated matrix. Laboratory evaluation, using fixed fluidized bed and MAT unit to compare ZCM-7 with CAT-A and other catalysts (Fig.1 and 2), shows that at the same coke yield ZCM-7 has higher dynamic activity and better selectivity. The compared catalysts CAT-A and CAT-B are USY-type catalysts. ZCM-7, CAT-A and CAT-B have all been used in FCCUs in China. Performance data on the same unit were taken and tabulated in Table V, which

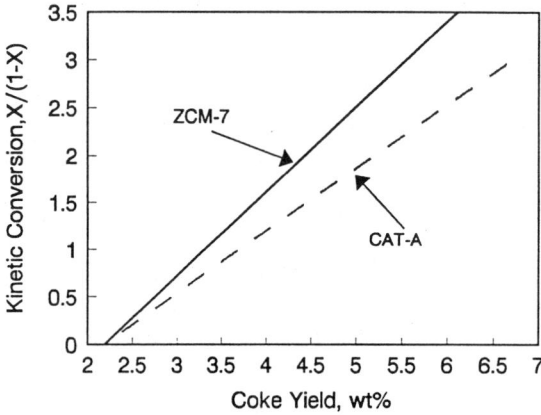

Fig 1. Dynamic Activity

(MAT Data)

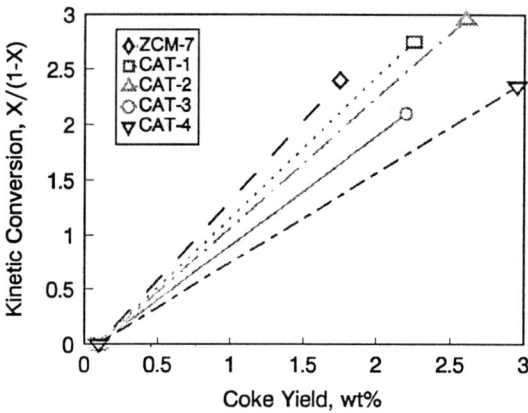

* CAT-1, 2, 3, 4 are catalysts from foriegn manufactures

Fig 2. Dynamic Activity Comparison

(Fixed Fluidized Bed Unit)

shows that ZCM-7 has a better gasoline selectivity, so octane-barrel is higher, and has less dry gas and LPG, in spite of 2% more 500°C$^+$ fractions blended in feed.

Several resid FCCUs in China have been using the ZCM-7 catalyst and have proved its good gasoline selectivity and low coke yield. Feedstocks for these RFCCUs have Conradson carbon from 3 to 6 wt%, and fractions boiling above 500°C averaged 30 v%. Ni on equilibrium catalyst often lies in the range of 5000~8000 ppm. Gasoline yields are in range of 45~47 wt% (on FF), with RONC 92~94, MONC 80~83, which depends on feedstock properties.

Table V. Performance Test Results (Commercial RFCC unit)

Catalyst	ZCM-7	CAT-A	CAT-B
In inventory, %	75	>90	75
Feedstock Properties:			
D(20°C), g/ml	0.9057	0.9079	0.9054
Con. C, wt%	3.32	3.64	2.3
UOP K	12	12.1	12
500°C$^+$, v%	22.5	20.5	20
Group Analysis, wt%:			
Saturates	56.96	55.77	66.7
Aromatics	24.51	28.27	16.3
Resins	18	15.14	16.1
Asphaltenes	0.53	0.32	0.9
Operating Conditions:			
Riser Outlet temp., °C	515	518	518
Regenerator temp., °C	698	698	717
Product yields, wt%			
Dry gas	3.9	4.7	4
LPG	10.5	12.9	12.5
GLN	52.2	51.4	48.9
LCO	16.5	14.2	18.7
Bottoms	9.9	9.7	9.4
Coke	7	7.1	6.5
Conversion , wt%	73.6	76.1	71.9
Gasoline Octane			
RONC	93.4	—	93.2
MONC	80.5	80.2	80.4

LCH Series Catalysts

LCH catalyst, a recently developed series containing a new type of high Si/Al ratio stable Y zeolite was designated as RSADY with special formulated matrix. Table VI shows the pilot results of RCH (LCH at its R&D stage) compared with CAT-A. It is seen that both fixed fluidized bed and riser pilot test results show that RCH converts more heavy ends and produces more gasoline with less dry gas and coke.

The first commercial trial performance test was conducted in a S&W–designed RFCC unit from December 12, 1991 until February 28, 1992. Data of feedstocks and test results are listed in Table VII. It shows that the LCH yields 3 wt% more gasoline at the expense of LCO and slurry, in spite of the inferior feedstock quality. It is obvious that LCH has the bottom cracking ability better than that of the compared catalyst, and has a better coke selectivity, as shown by less delta coke. Commercial performance test data show the same predications as pilot test results.

Table VI.   Pilot Test Results of RCH and CAT-A
(The same feedstock)

	Fixed-Fluidized Bed Unit		Riser Pilot Unit(1.5BBL/D)	
Catalyst	RCH	CAT-A	RCH	CAT-A
Reaction Temp., °C	510	510	524	524
C/O	7	7	7	7
WHSV, h^{-1}	11	11	—	—
Residence Time, sec.	—	—	2.24	2.16
Conversion , wt%	73.5	75.1	74.2	72.8
Product yields, wt%				
Dry gas	2.3	2.4	3.6	4.0
LPG	17.1	20.0	19.7	21.6
GLN	46.7	45.1	45.7	41.6
LCO	17.6	14.6	17.2	16.1
Slurry	8.9	10.3	8.6	11.1
Coke	7.4	7.6	5.2	5.6
Δ Coke	1.05	1.08	0.74	0.8

LCS and RHZ Series

Both series are for maximizing gasoline plus LCO. LCS7 is for normal VGO+VR. Table VIII shows the commercial performance test results. The base catalyst is of conventional REY type. The result shows that at 60% of LCS7 in inventory, the unit can operate with 5% more VR with the same LCO and gasoline yields.

**Table VII. LCH Catalyst: 1st Commercial Performance Test
(W Refinery, Dec. 12. 1992-Feb. 28. 1992)**

Catalyst	LCH	REF	
In inventory, %	90	~100	
Feedstock Properties			
D(20°C), g/ml	0.9166	0.9022	
Con. C, wt%	2.5	2.9	
C_7 insolubles, wt%	0.5	0.43	
UOP K	11.65	11.79	
Elemental Analysis:			
C, wt%	86.2	86.5	
H, wt%	12	12.6	
S,	0.9	0.6	
N, wt%	0.22	0.3	
Ni, ppm	4.8	6	
V, ppm	0.6	0.6	
Na, ppm	0.7	0.6	
500°C+ in feed, v%	40	33.5	
Reaction conditions:			
Riser exit temp., °C	518	520	
Regeneration temp., °C	693	716	
Cat/Oil	7.6	7.2	
Product yields, wt%:			Δ
Dry gas	4.6	4.4	+0.2
LPG	12.5	12	+0.5
GLN	50.9	47.8	+3.1
LCO	15.1	17.5	-2.4
Slurry	9.4	10.9	-1.5
Coke	7.5	7.4	+0.1
Conversion , wt%	75.5	71.6	+3.9
Δ Coke	0.99	1.03	
GLN MONC	80.6	80.3	
RONC	92.8(LCH50%)	—	

**Table VIII. LCS7 Catalyst: Commercial Performance Test
(Y Refinery, Riser FCCU, 80×10^4 MT/Y Capacity)
VGO+VR, Slurry Recycle**

Catalyst	LCS7	Base
In inventory, %	60	100
Feedstock Properties:		
D(20°C), g/ml	0.88	0.8736
Con. C, wt%	2.95	1.9
VR in Feed, wt%	30.1	25.8
Product yields, wt%:		
Dry gas	3.3	3.1
LPG	7.4	8.2
GLN	52.8	52.9
LCO	29.6	29.2
Coke	6.9	6.6
Conversion, wt%	71.4	70.8
GLN+LCO, wt%	82.4	82.1

Table IX shows that the feedstock was   blended with 6.7 wt% VR and the throughput increased by 5T/H, yet the yield of LPG+GLN increased by 3.9 wt% at the expense of LCO with minimal increase of coke yield. In Table X, in spite of the fact that the feedstock quality is rather inferior and coker gas oil was blended in, by using RHZ200, the conversion increased by ca. 5 wt%. The product yields shift from slurry to gasoline and LPG; meanwhile, RON and MON of the gasoline increase 1.1 and 0.9 units, respectively.

LCS7 and RHZ200 catalysts are now widely used in FCCUs in China.

RMG Series Catalysts

RMG catalysts, being specially formulated for the MGG process, which is a newly developed and commercialized catalytic conversion process by RIPP in 1992, aimed at maximizing the products of high–quality gasoline and olefinic gases (C_3 and C_4 olefins). They appear to be of high activity, high stability and excellent selectivity.

The first commercial trial of RMG catalyst performance was put on stream in August 1992 at a revamped FCCU of 40×10^4 MT/y capacity. To provide the basic data for engineering design and for operation. parameters of a commercial trial, different feedstocks with the same RMG catalyst were run in riser pilot plant . The riser pilot unit has a capacity of 1.5 bbl/day. As shown in Table XI, RMG catalyst, combined with MGG technology, can process feedstocks of VGO blended with some 20% VR. It is good for paraffinic–base feedstocks and also applicable for naphthenic–base feedstocks. Results show that RMG catalyst has some distinguishing features. High activity makes high conversion of 83~87 wt%. The LPG yields can reach 30~35 wt% with the olefinicity of about 70 wt%. In a once-through operation mode, gasoline yields at 40~45 wt%, GLN+LPG+LCO at 83~87 wt%. While in full-recycle operation mode, LPG+GLN maximized over 85 wt%. In both modes, C_3+C_4 olefins yield at 20~27 wt%. Octane number of product gasoline lies in 80~83 for MONC and 92~95 for RONC, depends on feedstock properties.

Table IX. RHZ200 Catalyst: Commercial Performance Test (1)
(T Refinery, Riser FCCU, Slurry Recycle)

Date	1992.12.18	1992.7.3
Catalyst	RHZ200	Base
In inventory, %	76	100
Feedstock Properties:		
VR in Feed, wt %	6.7	0
D(20°C), g/ml	0.8842	0.8795
Con. C, wt %	0.6	0.09
C, wt %	87.02	86.45
H, wt %	12.74	13.36
S, wt %	0.13	0.11
N, wt %	0.11	0.08
Ni, ppm	2.5	0.2
V, ppm	<0.1	<0.1
Boiling Range, °C	241~551(91%)	244~539(95%)
Group Analysis, wt %		
Saturates	68.7	74.9
Aromatics	23.7	19.2
Resin+Asphaltenes	7.6	5.9
Fresh feedrate, T/H	123.2	118
Recycle, T/H	52	46.7
Riser exit temp., °C	496	479
Regeneration temp., °C	683	680
Cat/Oil	4.01	3.58
Product yields, wt %:		
Dry gas	4.3	4.8
LPG	10.7	9.2
GLN	53	50.6
LCO	27.3	30.8
Coke	4.7	4.6
LPG+GLN+LCO, wt %	91	90.6
Conversion , wt %	72.7	69.2

Table X. RHZ200 Catalyst: Commercial Performance Test (2) (Q Refinery, Riser FCCU, VGO+CGO)

Catalyst	RHZ200	Base
In inventory, %	~80	100
Feedstock Properties:		
D(20°C), g/ml	0.9071	0.9032
Con. C, wt%	0.22	0.2
N, ppm	3100	~1000
Basic N, ppm	1031	~300
Feedrate, T/H	151.3	152.7
Reaction temp.,°C	500	495
Regeneration temp., °C	716	720
Cat/Oil	4.58	4.11
Product yields, wt%:		
Dry gas	3.9	3.6
LPG	10.1	8.8
GLN	45.4	42.1
LCO	30.7	30
Slurry	4.7	10.4
Coke	5.2	5.2
Conversion, wt%	64.6	59.7
GLN+LCO, wt%	76.1	72.1
GLN Octane		
RONC	92.6	91.5
MONC	80.6	79.7

Table XI. RMG Catalyst: Riser Pilot Plant Test (MGG Process)

	A	B		C	D	E
Description	DQ VGO	SB VGO+18%VR		XQ VGO	XQ VGO+20%VR	LH VGO
Operation Mode[a]	O-T	O-T	F-R	F-R	O-T	O-T
Feedstock Properties:						
D(20°C), g/ml	0.8546	0.8612		0.8572	0.8743	0.9230
Con. C, wt%	0	2.18		0.11	1.9	0.12
Basic N, ppm	152	600		457	—	690
Ni, ppm	—	3.3		0.1	5.2	3.6
UOP K	12.5	12.6		12.1	12	11.5
Product yields, wt%:						
Gas	37.6	39.7	40.3	34.4	33.2	25.7
H_2~C_2	2.5	4.5	4.7	3.5	3.3	3
C_3~C_4	35.1	35.2	35.5	30.9	29.9	22.7
C_5+GLN	46	41.9	52.3	57.3	43.3	42.8
LCO	11.3	7.3	0	0	15.2	13
HCO	1.2	5.6	0	0	2	11.5
Coke	3.8	5.5	7.4	8.3	6.3	7
Conversion, wt%	87.4	87.1	100.0	100	82.8	75.5
$C_3^=$, wt%	11.2	12.9	12.8	9.3	9.8	7.1
$C_4^=$, wt%	13.2	14.5	14.7	9.9	10.5	7.7
C_3~C_4+GLN+LCO,wt%	91.4	84.4	87.8	88.2	88.4	78.5
GLN RONC	92.7	92.3	93.4	92.2	95	95.2
MONC	81.1	80.2	80.5	81.5	83	82.8

[a] O-T: Once-through mode
F-R: Full-recycle mode

Summary

The trends of FCC in China, likely also worldwide, are processing heavier crudes, and are moving toward integration with petrochemicals. With the advent of mandated oxygenates content in reformulated gasoline, light olefins from FCC are drawing great attention to the refineries. RIPP currently has focused its efforts on the development of resid FCC catalysts and technologies. Varieties of catalysts were successfully developed. Resid FCC for high–octane barrels, resid FCC for maximum liquid yield, and resid FCC for maximum high–octane gasoline plus LPG (MGG), etc. were successively developed and commercialized. And MGG is attracting the attention of refineries both in China and abroad. FCC is indeed very flexible. It is true that the technology is far from mature. RIPP, with its accumulated experience in FCC process and catalyst R&D, shall continue to do its best in FCC advancement.

Acknowledgments

The information in this paper is based on research work at FCC Catalyst and Process Department of RIPP. The authors wish to express their gratitude to the staffs involved and to Mr. Wenbin Jiang in preparing this manuscript.

RECEIVED June 17, 1994

Chapter 24

Impact of Additive Usage in Distillate Fluid Catalytic Cracking Operation

S. Mandal, D. Bhattacharyya, V. B. Shende, A. K. Das, and S. Ghosh

Research and Development Center, Indian Oil Corporation Ltd., Faridabad 121007, Haryana, India

Additives have a major impact on the performance of distillate FCC operation. Such improvement in performance is even more pronounced than that of the gasoline-mode FCC units. In this paper, the effect of various FCC additives, e.g., ZSM-5, CO promoter, bottom cracking additive, SOx additive, have been studied mainly for low-severity distillate-mode operation and compared with the results of gasoline-mode units. The basic observations in the laboratory units for these additives were also corroborated with commercial plant data.

Fluid catalytic cracking (FCC) is the most widely used secondary conversion process for producing gasoline, olefins and middle-distillate from heavy petroleum stocks. Depending on the desired product, FCC units can be designed and operated in either gasoline or distillate-mode.

In gasoline-mode operation, the per-pass conversion is maintained at a high level (> 70%), whereas in distillate-mode, the operation is adjusted to avoid the overcracking of middle-distillate by reducing the reaction severity. These units are generally equipped with partial combustion regenerators and are typically operated at low reaction temperature, contact time and catalyst-to-oil ratio to keep per-pass conversion around 40 wt%. Air supply to the regenerator is restricted in the partial combustion mode, resulting in high coke on regenerated catalyst (CRC). The differences in the two modes of FCC operation are summarized in Table I.

Commercial FCC units always operate with many constraints which are even more prominent in distillate-mode operation. Additives are used to enhance the operating flexibility by reducing some of the major unit limitations.

To optimize the performance of these low-severity units in India, many additives have been tried in commercial scale after detailed laboratory studies.

0097–6156/94/0571–0335$08.00/0

This paper summarizes the laboratory results and plant performance with these additives in distillate-mode FCC operation.

TABLE I. Comparison of Gasoline and Distillate Mode FCC Operation

Parameters	Distillate Mode	Gasoline Mode
Yields, wt%(Total feed basis)		
Dry gas	1.50	4.00
LPG	7.00	18.00
Gasoline(C_5-150°C)	22.00	41.00
Heavy Naphtha(150-216°C)	10.00	13.00
LCO(216-370°C)	26.00	15.00
TCO(150-370°C)	36.00	28.00
Bottom(370$^+$°C)	30.00	4.00
Coke	3.50	5.00
216$^-$ Conversion	44.00	81.00
Operating Conditions		
Combined/Fresh Feed	1.37	1.04
Riser Top Temp. °C	492	527
Regn. Dense Temp. °C	642	728
CRC, wt%	0.50	0.05

ZSM-5 Additive

The effect of ZSM-5 on the product yields have been studied by several researchers (*1-3*) and its performance in commercial scale has also been reported (*4*). Most of these findings are based on high-severity operation.

Das et al. (*5*) have recently reported the performance of ZSM-5 in low-severity FCC units. The experiments were carried out with equilibrium catalyst and combined feed at simulated test conditions. The feed injection time was maintained at 30 secs and reactor temperature was 510° C. The yield data for different conversions were obtained by changing the rate of feed injection without altering the injection time or the reaction temperature. The base catalyst was an equilibrium ReUSY type. The additive was obtained from M/S Intercat, USA, with brand name Zcat$^+$,which contains about 25 % ZSM-5 zeolite in an inert matrix. The ZSM-5 sample was steamed at 760 °C for 5 hrs using 100% steam before testing.

In Figure 1, the laboratory results for yields of total cycle oil (TCO, 150-370°C) and gasoline (LCN,C_5-150°C) are plotted for base catalyst with and without ZSM-5. It is seen that at the same severity level, the gasoline yield increases at the cost of TCO yield. Details of these results for other products are reported elsewhere (*5*).

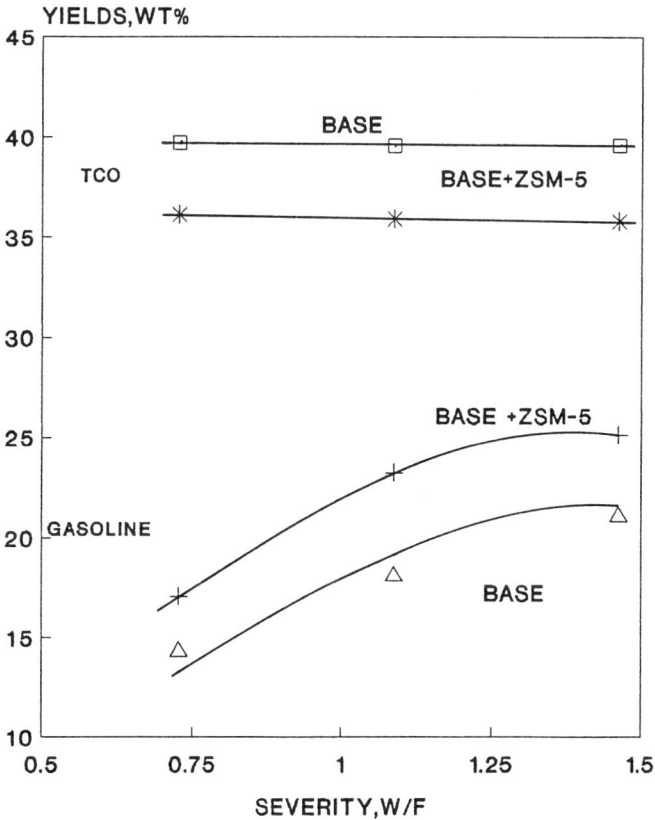

Figure 1. Gasoline and TCO yields with (+,*) and without (△, □) ZSM-5 additive in distillate-mode FCC operation based on simulated MAT results.

Based on the laboratory findings, commercial trials were conducted with ZSM-5 additive in many Indian FCC units. These results, along with the laboratory prediction, are summarized in Table II. From Table II, it is observed that the laboratory findings are well matching with commercial observations, except the gas yield.

Table II. Comparison of Laboratory Prediction vs. Actual Plant Data on ZSM-5 Additive Performance in Distillate Mode (Additive Concentration in Catalyst Inventory: 1wt%)

Products	Delta Yields(% wt on Fresh Feed Basis)			
	Lab. Prediction	Plant A	Plant B	Plant C
Dry Gas	- 0.95	+0.30	0.00	+0.19
LPG	+3.28	+2.40	+2.90	+1.60
GASOLINE	+4.85	+4.80	+3.30	+3.50
TCO	- 4.13	- 3.20	-6.70	-4.00
CLO(430°C$^+$)	- 3.05	-4.50	+0.20	-1.30
COKE	00.00	00.00	+0.20	0.00

Further, the performance of ZSM-5 additive in low-severity was compared with that of high-severity operation. Based on data for high-severity operation *(1-3)*, it is noticed that ZSM-5 primarily cracks gasoline to LPG. This results in significant reduction of gasoline yield. In contrast, at low-severity, ZSM-5, in fact, increases gasoline yield at the cost of TCO. The overall differences of ZSM-5 performance in these two modes are summarized in Table III.

TABLE III. Comparison of ZSM-5 Performance in Two Modes of FCC Operation

Parameters	Gasoline Mode	Distillate Mode
LPG yield	Significant increase	Significant increase
LCN(C$_5$-150°C)	Significant drop	Significant increase
LCO(216-370°C)	Remained constant	Significant cracking to gasoline & LPG
216°C-Conversion at equal severity	Constant	Increased
Gasoline octane	Significantly improved	Marginally improved
TCO pour point	Constant	Reduced
Bottom Conversion	Constant	Improved

Increase in gasoline yield, with simultaneous gain in octane number and reduction in TCO pour point, makes the use of ZSM-5 in low-severity units more attractive and beneficial.

The different behavior of ZSM-5 could be explained by looking at the concentrations of the three lumps of hydrocarbons in the reactor having boiling ranges of C_5-150°C, 150-216°C and 216-370°C. At low-conversion, the concentration of the third lump is more while at higher conversion that of the first two lumps is higher. Again, the crackability of the lightest lump is the least and that of the heaviest lump is the most. Therefore, acid sites of ZSM-5 cracks the paraffins of the third lump to the maximum extent in low-conversion process. In any case, for low-severity operation, ZSM-5 could significantly increase octane barrel by simultaneously increasing the gasoline yield and octane number, whereas in gasoline-mode the overall octane barrel is actually reduced.

CO Promoter

For a regenerator operating under total combustion mode, CO promoter additive is mainly used to achieve complete combustion within limited regenerator catalyst inventory. On the other hand, a partial combustion regenerator is designed to produce less heat by restricting air supply and hence the conversion of CO to CO_2. Therefore, under normal circumstances, the use of CO promoter is not envisaged in partial combustion operations.

However, there are situations where one can consider the use of such additives in a partial combustion mode. For example:
1) due to poor air distribution, the efficiency of oxygen utilization in the regenerator dense-bed is low, leading to excessive burning of CO in the dilute phase ("after burning"),
2) in cases where the regenerator dense-bed temperature remains unusually low and any attempt to increase the temperature by supplying more air leads to afterburning without increasing the dense-bed temperature. Low regenerator temperature is not desirable since the CRC in such cases go as high as 1 wt%.

In both these situations, CO promoter additive comes in handy, particularly in the second case where the available cushion in the dense-bed temperature could be utilized to reduce the CRC.

In one of our FCC units, the second condition prevailed. The regenerator temperature was typically at about 620°C and CRC was at the level of 0.5-0.6 wt%. Any attempt to increase the air rate only increased the cyclone temperature to its limiting value without increasing the dense-bed temperature or reducing the CRC. After careful analysis of the problem, use of CO promoter was conceived. Laboratory studies were carried out to establish the possible benefits of using this additive (*6*). In the first instance, experiments were conducted to see the effect of CO promoter addition on the flue gas composition and regenerator temperature. The FCC pilot plant, which was used for this study, usually operates in isothermal mode which means that both the regenerator and reactor temperatures are controlled by independent heaters and controllers. The sources of heat in the regenerator are from burning of coke and from heat supplied externally. The additional heat produced by excess CO oxidation using CO promoter could be estimated from the temperature rise in the regenerator, if the external heat supply is kept at a constant level.

To establish the base case, the dense-phase temperature was maintained at 646 °C, while no CO promoter was present with the base catalyst. The regenerator conditions in the pilot plant were similar to the commercial partial combustion FCC operation. The flue gas sample was collected and analyzed. Before adding CO promoter, the controllers of the regenerator section were disconnected and the heat supply were maintained at the fixed level, allowing the excursion of the regenerator temperature. The CO promoter additive was obtained from M/S W.R. Grace & Co, USA, with brand name CP3A. About 2 gm of the additive /kg of catalyst was added into the regenerator after establishing the base case, and the change in temperature of the regenerator was monitored. An increase of 9°C temperature was observed after CO promoter was added and the unit was allowed to attain steady state. The sample of flue gas was collected after stabilization and analyzed. The flue gas composition before and after addition of CO promoter is shown in Table IV.

Table IV. Comparison of Flue Gas Composition with and without CO Promoter (Pilot Plant Data; Conversion level: 40-45 wt%)

Volume%	Without CO Promoter	With CO Promoter
O_2	0.58	0.54
CO	1.63	0.32
CO_2	6.64	7.41
Regen. Temp.°C	646	655

Results indicate that a small amount of CO promoter additive could increase the regenerator dense-bed temperature significantly. This is expected to reduce the CRC, hence increasing effective catalyst activity at the riser bottom. And the increase in CO_2/CO ratio at the exit of the dense-bed will reduce the $\triangle$ T between the dense and dilute bed. In short, CO promoter can shift the heat generation zone from dilute phase back to the dense-phase even in a low-severity regenerator.

The laboratory results, along with the plant operating conditions, were used for simulation studies with the help of our in-house simulation model (FCCMOD). The effect of CO promoter on regenerator dense-bed temperature was established from heat of combustion of carbon and CO. Since CRC and cat/ oil ratio are automatically changed with regenerator temperature in adiabatic FCC operation, the model could predict the overall effect of CO promoter on

FCC performance by considering the effectiveness of the additive to promote the CO conversion to CO_2. The predictions based on laboratory results and plant data with CO promoter additive in a low-severity unit are shown in Table V. It is observed that the predictions closely match the plant performance, except the feed throughput. With the addition of CO promoter, the level of afterburning in the regenerator was reduced and consequently, cyclone temperature came down. This ultimately allowed the refinery to increase the regenerator air rate and thereby enhance the feed throughput level.

TABLE V. Effect of CO Promoter in Partial Combustion FCC Operation

Parameters	Delta	
	Laboratory Predictions	Plant Results
Regen. dense temp. °C	+22.00	+18.00
Air NM³/hr	-775.00	-775.00
CRC, wt %	-0.25	-0.13
Feed rate, t/d	0.00	+80.00
Gasoline, wt %	+2.30	+3.00
Coke, wt %	-0.60	-0.50

It is interesting to note that at low-severity operation, there is an improvement in yield pattern with CO promoter. The effect of CO promoter in different modes of FCC operation would be clearly understood if regenerator-reactor interactions were followed.

Figure 2 shows the prediction of our in-house FCC model demonstrating the effect of changing regenerator severity on FCC performance. Since CO combustion produces more heat resulting in higher dense temperature, the cat/oil ratio automatically drops, leading to lower conversion. However, in partial combustion operations, the effect of reduced CRC predominates over the effect of reduced cat/oil until CRC reaches about 0.3 wt%. Similar results were also reported by Krishna et al.(7). Further reduction of CRC leads to detrimental effect on unit conversion and yield pattern, particularly distillate yield. This has also been established in our earlier publication (8) while studying the effect of CRC on distillate FCC units. It was reported that in case of distillate-mode FCC operation, there lies an optimum coke level around 0.2-0.3 wt% on catalyst at which the distillate yield is maximum. Therefore, it is clear that CO promoter is more effective for partial combustion regenerators as long as the base case CRC is above the typical optimum value of around 0.3 wt%.

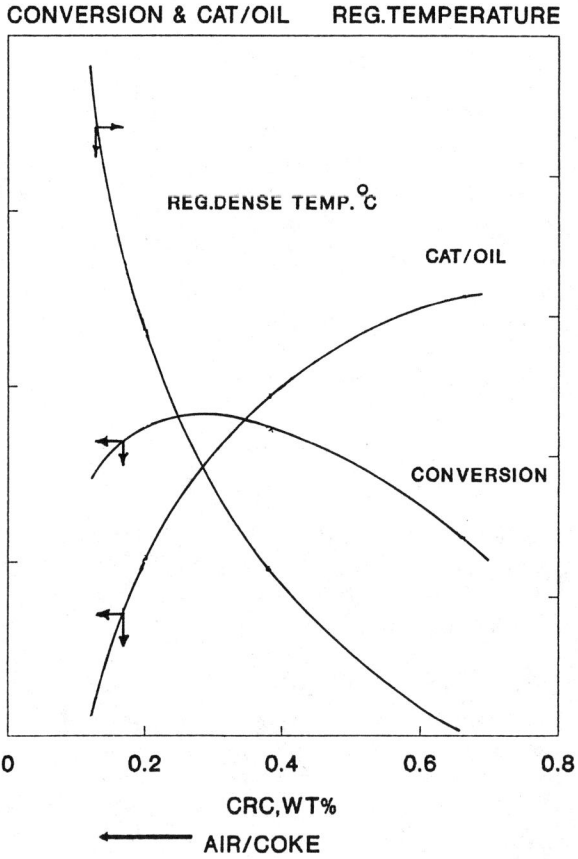

Figure 2. Effect of regeneration conditions on FCC performance.

Bottom Cracking Additive (BCA)

For FCC units where the middle-distillate is to be maximized, the question of bottom reduction by increasing the operating severity does not arise. This is because the conversion level of such units are maintained for maximum distillate yield, which drops at higher conversion. Again, the use of high-matrix catalyst may not serve the purpose as it will increase the coke and dry gas yields. Furthermore, frequent change in feed quality requires change of the base catalyst property which means replacement of the entire catalyst inventory by a suitable alternative catalyst. To overcome this problem, use of separate solid particle additive was conceived. Unlike the catalyst with incorporated matrix, these additive particles are expected to improve the bottom conversion without increasing coke make. The validity and economic viability of the separate additive particles have been clearly established on the laboratory(*9*) as well as the commercial scale(*10*).

BCA additive selectively cracks the heavier fractions without increasing the undesired product yields. The BCA was obtained from M/S Intercat, USA, with brand name BCA-105. Performance of the BCA has been evaluated in laboratory micro-reactors simulating distillate-mode FCC conditions. The product selectivities with BCA additive were established from the simulated data taken at different reaction severity levels and concentrations of BCA. The effect of BCA additive on bottom upgradation and coke selectivities are shown in Figure 3. It is interesting to note that there is an optimum level of BCA (about 5 wt%) at which the bottom (370 °C$^+$) and coke yields are minimum and that of TCO is maximum. Commercial performance of BCA has been predicted based on laboratory results using in-house FCC process simulator. Incremental shifts from baseline in product yields and process conditions are summarized in Table VI.

It is seen from Table VI that BCA additive could effectively improve bottom cracking to the desired LCO at reduced coke yield. As a result, regenerator temperature comes down by about 10 °C which could be utilized to optimize the FCC performance where regenerator temperature limits operation. Thus, as shown in column 2 of Table VI, further benefit in terms of conversion and yield could be derived by enhancing recycle rate and feed preheat temperature. The advantage of coke reduction with BCA additive is quite interesting, particularly in the context that matrix addition in the catalyst increases the coke and dry gas. Nevertheless, in coke-limited FCC operation, this aspect of BCA will be quite useful where bottom upgradation is realized with reduced regenerator temperature.

In the recent past, Ellison et al.(*10*) have reported commercial trial results of BCA additive in both distillate and gasoline-mode FCC operations. The commercial results confirm the R&D prediction and that in both the modes, BCA additive is effective to upgrade bottom.

The effect of BCA at higher operating severity highlights that BCA produces gasoline at the cost of bottom. In low-severity operation, as shown in Table VI, we find that the bottom is upgraded to LCO and, in fact, there is

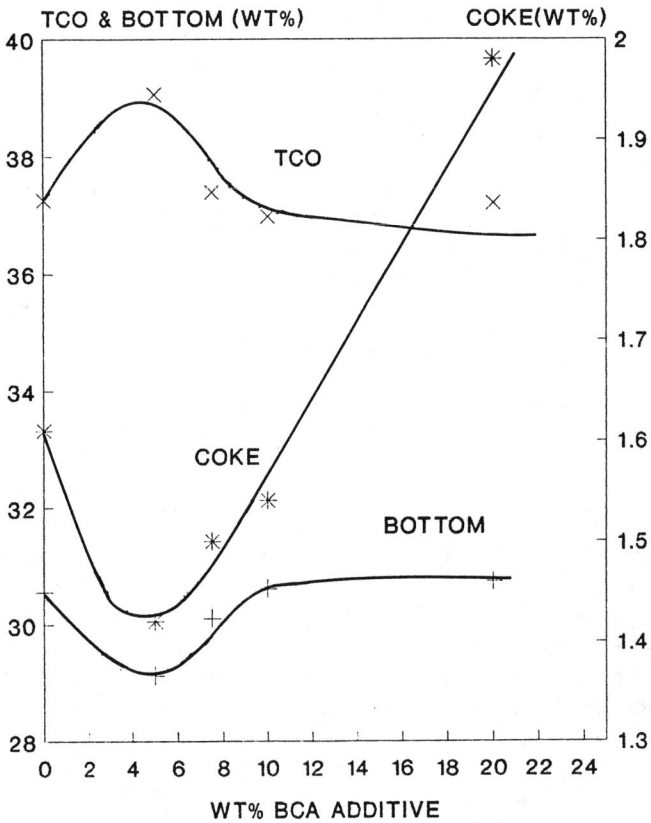

Figure 3. Effect of BCA on TCO(x), coke(*) and Bottom(+) yields at equal conversion (42 wt%) based on simulated MAT results.

Table VI. Effect of BCA on Product Yields and Process Conditions in Distillate Mode FCC Operation

Basis: At equal plant coke level

Parameters	Incremental Shifts From Base Case	
	Case I Same T'put	Case II With Enhanced Recycle
BCA Concentration,wt%	5.00	5.00
Yields,wt%(Fresh Feed Basis)		
216°C Conversion	+0.78	+2.85
Dry Gas	+0.08	+0.03
LPG	+0.16	+0.64
Gasoline(C_5-150°C)	-0.13	+0.86
Heavy Naphtha(150-216°C)	+0.67	+1.12
LCO(216-370°C)	+1.48	+0.47
CLO(430 °C+)	-2.27	-3.33
Coke	0.00	+0.20
Process Conditions		
Fresh feed rate,M^3/hr	0.00	0.00
Recycle rate (% of fresh feed T'put)	0.00	+11.39
Cat/oil(on total feed)	+0.30	-0.07
Feed preheat temperature,°C	0.00	+10.00
Riser temp. °C	0.00	0.00
Reg. Dense Temp.°C	-11.00	-4.00

little reduction of light gasoline. Such difference in the behavior of BCA selectivities is mainly attributed to the difference in operating severity. In fact, for both the modes of operation, the BCA initially cracks the bottom to LCO which is further cracked to gasoline and LPG under high-severity condition. On the other hand, the subsequent cracking of LCO is restricted at low-severity. Thus, in gasoline-mode, BCA additive actually increases LPG and gasoline instead of LCO as in distillate-mode operation.

SOx Additive

Essentially, SOx additive converts SO_2 to SO_3 in the regenerator and releases H_2S in the riser along with product gas. In the first step, SO_2 is oxidized to SO_3 and then to metal sulfate in the regenerator. The sulfate-laden additive is reduced to H_2S and the regenerated additive is circulated back to the regenerator. The formation of SO_3 is favored by lower regenerator temperature and higher oxygen availability (*11*). In most of the partial combustion regenerators, comparatively low regenerator temperature favors this reaction while low availability of excess oxygen restricts the fixation of SO_2. For complete combustion units, the situation is reversed.

Laboratory studies in the pilot plant were carried out simulating the partial combustion regenerator to estimate the effectiveness of the SOx additive. The additive was obtained from M/S W.R. Grace & Co., USA, with brand name Additive R. The operating severities and the catalyst CRC levels were kept constant in the pilot plant. The product yields and SOx emission were obtained at different SOx additive concentrations. At 6 wt% additive level, about 60% reduction in SOx emission was observed. In addition, this additive also modified the yields favorably. In Figure 4, the yields of TCO and bottom are plotted against the additive concentration. It is seen that TCO yield increases at the cost of bottom with increase in additive level. Therefore, in distillate-mode, SOx additive not only helps to reduce SOx emission but also improves the most desirable product yield. For gasoline-mode units, most of the literature does not indicate any yield benefit with SOx additive.

Conclusion

Additives play a major role in improving FCC operation. For the units designed and operated to produce gasoline, use of such additives are well-established. However, for FCC unit operating in distillate-mode, application of these additives were not very common till recently.

In India, most of the FCC units are designed and operated to produce maximum middle-distillate. R&D efforts were directed to identify some of the additives to improve the operation of these low-severity FCC units. During the last few years, almost all the Indian units have started using ZSM-5 additive, not so much for increasing gasoline octane but to produce more LPG, with the added advantage of enhanced octane barrel. Recent plant trials with CO promoter additive show that under certain circumstances, CO promoter could

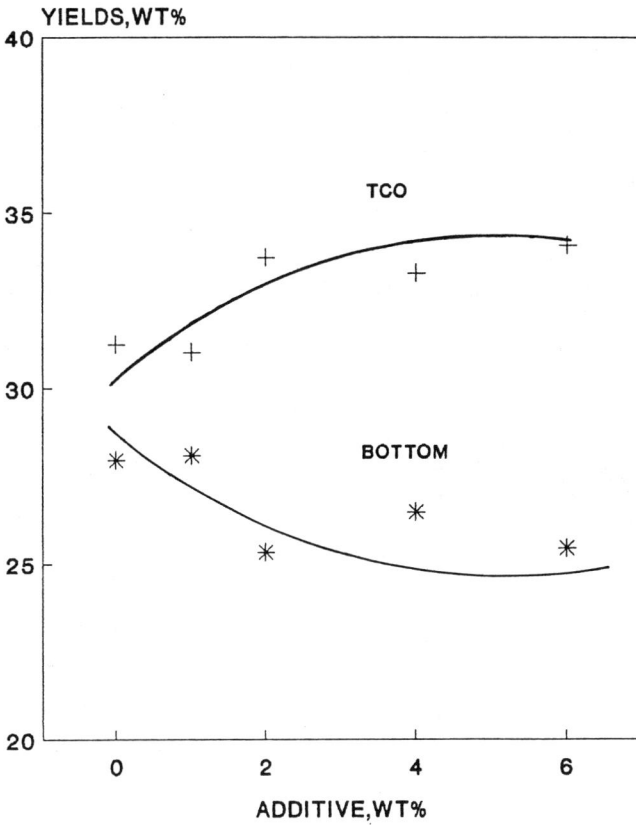

Figure 4. Effect of SOx additive on TCO(+) and Bottom (*) yields at equal conversion (42 wt%) based on FCC pilot plant results.

be used very effectively in partial-combustion regenerators, although such promoters are supposed to work only in complete combustion mode regenerators.

Laboratory studies indicate that both the BCA and SOx additives could be very useful for improving yield pattern and controlling SOx emission from FCCU. In fact, it appears that these additives could improve distillate FCC performance much more than what is expected in high-severity operation.

Acknowledgments

The authors are thankful to the management of Indian Oil Corporation, R & D Centre, for giving permission to publish this paper.

Literature Cited

1.   Biswas, J., Van Der Griend, J., KaaK, H.,and Maxwell, J., *Akzo Catalysts Symp.*, May 29-June 1, *1988*, Kurhaus, Scheveningen, The Netherlands.
2.   Rajagopalan, K., Young, G.W., *Preprints*, Div. of Petrol. Chem., ACS, *1987,32*(3), 627.
3.   Pappal, D.A., and Schipper, P.H., *Preprints*, Div. of Petro. Chem., ACS, *1989,34*(4), 727.
4.   Dwyer, F.G., Economide, N.L., Herbert St, S.J.A., and Gorra, F., *The Institute of Petroleum, Petroleum Review*, July, *1987*.
5.   Das, A.K., Kumar, Y.V., Lenin, V.R., and Ghosh, S., *Akzo Catalysts Symp.*, June *1991*, Scheveningen, The Natherlands.
6.   Mandal, S., Das, A.K., and Ghosh, S., *Hydrocarbon Tech. 1992,*Aug.15, Issue no.22.
7.   Krishna, A.S., Hsieh, C.R., English, A.R., Pecoraro, T.A., and Kuehler, C.W., *Hydrocarbon Proc., 1991,70*(11), 59.
8.   Mandal, S., Das,A.K., and Ghosh,G. *Ind.Eng.Chem. Res., 1993,32*(6),1018.
9.   Mitchell, M.M., Moore, H.F., and Goolsby, T.L., *AIChE*, Spring National Meeting, Orlando, Florida, March *1990*.
10. Ellison,T.W.,Demmel,E.J. Jr., Steves,C.A., and Johson,C.R.*Fuel Reformulation,1993,3*(3),18.
11. Atiken, E.J., Baron, K., McArthur. D.P., and Mester, Z.C., *Katalistiks' 6th Annual Fluid Cat Cracking Symp.*, May 22-23, *1985*, Munich, Germany.

RECEIVED July 18, 1994

Chapter 25

Fluid Cracking Catalyst Performance and Development

Now and in the Future

John S. Magee[1] and Warren S. Letzsch[2]

[1]Catalytic Science Associates, 2201 Old Court Road,
Baltimore, MD 21208
[2]Stone and Webster Engineering Corporation, 1430 Enclave Parkway,
Houston, TX 77077-2023

Fluid catalytic cracking (FCC) has maintained its premier position in the refinery over the years due to its ability to handle a wide range of feedstocks and upgrade them into more valuable lighter liquid products. A steady progression of FCC process and catalyst compositional changes have occurred since the introduction of zeolite containing FCC catalysts in 1962. While hundreds of different catalysts compositions are available commercially, only three major compositional characteristics account for the catalyst's activity and selectivity: the zeolite unit cell size, the presence (or absence) of a shape selective molecular sieve and the similar presence or absence of a catalytically active matrix. Process changes have primarily centered around both reactor and regenerator where catalyst/oil contact time has been shortened to improve gasoline and olefin yields and regenerator design changes have been geared to lessen catalyst deactivation by using two stage systems. However, current and evolving environmental regulations impose a number of new boundaries on FCC catalysts and operations. The present paper explores how the FCC process and FCC catalyst compositional changes can influence the production of environmentally friendly products into the 21st century.

Developments in the FCC area since 1963, both process and catalyst, have been nothing short of astounding, especially since the field was considered to be totally mature at that time. In place cracking capacity was expanded by 100% with the introduction of the new zeolitic FCC catalyst systems without major capital expenditure on the FCC units themselves. The ever present "push-pull" of catalyst versus process technological advancement soon led to equipment changes. Short catalyst/oil contact times were required to utilize the higher catalyst activity levels and high temperature regenerators were necessary to lower carbon-on-regenerated-catalyst to unblock the zeolite's activity and selectivity. Fortunately, the catalyst's

0097-6156/94/0571-0349$08.54/0

constantly improving hydrothermal stability allowed the catalysts to be used successfully at high severities.

We believe that we are at another pivotal point in the development of FCC technology. A high point of the "S-Curve" has been achieved and amazingly the catalyst technologists and process technologists each believe that they are more advanced on the "S-Curve" waiting for the other to catch up. That the process is a success can be demonstrated by the fact that today over 12 million barrels of gas oil per day are catalytically cracked using over 1400 tons/day of FCC catalyst (1).

Current discussions of fluid catalytic cracking unit (FCCU) design technology involve considerations of short contact time risers, multiple risers or feed injection points, improved feed atomization, new stripper designs, catalyst coolers, resid cracking units and high temperature regenerators. Catalyst discussions and innovations revolve around ways to control hydrogen transfer reactions, unit cell size control, small and large pore shape selective catalyst components, metals tolerance, zeolites selective to light olefin formation, and high silica/alumina ratio Y type zeolite formation and modification.

In the present chapter we will explore these currently evolving catalyst and process technologies to see how they will lead us to consumer oriented products which are compatible with currently evolving environmental regulations.

CATALYSTS

- Is there life after silica/alumina?
- Will we ever be able to control acidity? Do we need to?
- Is more structural control needed?
- Can we ideally make only light olefins, isomeric gasoline, n-cetane and coke?
- What can be done about hydrogen balance given the feeds that must be cracked to the "ideal" product distribution?

PROCESS

- Is it time for segregated risers and/or feeds to afford better reactor control of selectivity?
- Can catalyst/oil contact time ever be too short?
- How important is feed atomization in selectivity control?

ENVIRONMENT

- How will the current and evolving regulations influence what can and should be done above?

We may consider the above questions in light of Catalyst, Process and Environmental (C-P-E) changes which have gotten us to our present state in FCC and how C-P-E changes will move us to the early years of the 21st century.   Figure 1 illustrates the general aspects of present and future C-P-E issues which will be further described in the present paper.

The present and future C-P-E positions are strongly linked through the environmental portion of the technology and much of the driving force for future advances in C and P come from this linkage.   Much of the technology base for these advances are, of course, derived from Catalyst and Process advances made since the introduction of zeolite containing FCC catalysts in the early '60s.   However, the basic initial drive in C & P was for increased gasoline and throughput and both catalyst composition and unit configurations reflected those incentives (2).   A basic understanding of catalyst activity control and unit designs to optimize cat/oil contact time and coke handling was achieved.   Superimposed on this time frame was the need to process resid feeds caused by the OPEC imbargoes of the 70's, the need to increase FCC gasoline octane caused by the environmentally dictated elimination of tetraethyl lead, and the changes brought about in the reactor/regenerator heat balance inter-relationships with the discovery of CO oxidation catalysts designed for regenerator dense bed oxidation of CO to CO_2.   This discovery led to better carbon-on-regenerated catalyst (CRC) control and generally resulted in a more energy efficient operation (3).   Downstream CO boilers were converted to waste heat recovery or taken out of service.

CATALYST AND PROCESS CHOICES FOR THE FUTURE

These early C-P-E interactions have brought us to our current position which, at the least, can be described as a position of multiple choice - from a wide variety of catalyst formulations and unit configurations, to process a variety of feedstocks into a variety of environmentally "friendly" products while protecting the environment.

Beginning with the broad spectrum of catalyst choices available to meet current and future FCC needs, we can simplistically illustrate present and future directions in Figure 2 below.

What we see here are three basic catalysts types for the three main FCC conversion areas which are the primary areas of focus now and for the next decade.   A petrochemical application in the late 90's or early 21st century may also be anticipated but the extent of FCC support to this is not clear at this time (4).   For the three principal areas of interest there are today more than 700 catalyst grades available, most of which are literally "fine tuned" in activity and selectivity for specific FCC unit use (more than 350 FCC units are in operation world-wide).

In spite of the seeming diversity possible, there are only three main controls of activity/selectivity shown in Figure 2.   That is, zeolite unit cell size (U.C.S.), the presence (or absence) of a shape selective molecular sieve and the presence or

FCC TO 2000+

Catalyst   FCC   **RFG1 Simple Model**
Process   PRESENT   **(1995)**
Environment   POSITION   • 1% Benzene max.
                (1994)   • 2 wt% oxygen min.
                            • RVP 7.2 → 8.1
                            • Hydrogen Balance

                Gasoline Catalysts   **RFG2 Complex Model**
                                               (1997)
                Octane Catalysts   • Gasoline T90 Reduction
                                               • S Reduction
                Resid Catalysts   • Aromatic Reduction
                                               • Olefin Reduction
                Riser Units   • Hydrogen Balance

                Resid Units   **Resid Processing**
                (RFCC)   • Metals Passivation
                • 2 Stage Regen   • Catalyst Architecture
                • Cat Coolers   • Hydrogen Balance

                                               Petrochemicals
                Additives   • C_2 - C_4 Olefins
                • Octane   • Aromatics
                • SO_x
                • CO
                • Bottoms   **Unit Configurations**
                                               • Reactor/Regenerator
                Oxy Fuel   • Segregated Risers
                                                 and/or Feeds
                                               • Feed Atomization

                                               General Environmental
                                               Issues
                                               • Catalyst Handling (TCLP)
                                               • Toxic Emissions
                                               • HCO Use/Disposal

FIGURE 1

Catalyst-Process-Environmental Position: Now And In The Future

REFINING - PETROCHEMICAL CATALYSTS	REFORMULATED GASOLINE/OCTANE	RESID CONVERSION	HIGH GASOLINE YIELD
– Low U.C.S. ≤ 24.50 SiO₂/Al₂O₃ ≥9.0	– Low U.C.S. 24.50 - .58 SiO₂/Al₂O₃ 9.0 - 6.1	– Moderate to low U.C.S. 24.58 - .62 SiO₂/Al₂O₃ 6.1 - 5.5	– High U.C.S. 24.62 - .68 SiO₂/Al₂O₃ 5.5 - 4.65
– Shape selective small pore ≤ 10-ring	– Shape selective small pore molecular sieve	– Shape selective large pore molecular sieve or PILC	– Possibly a shape selective large pore molecular sieve or PILC
– Zero activity matrix to high activity depending on feedstock properties.	– Active matrix	– Low to moderate	– Active matrix

FIGURE 2. CATALYST ROADMAP

absence of an active matrix whose catalytic and physical properties may be changed during manufacturing (6). The catalyst may also contain (or be used in mixtures with) a second particle to aid in bottoms cracking, SO_x emission control, octane enhancement, CO oxidation and metals passivation. How changes in the three principal catalyst activity/selectivity controls impact both resid conversion catalysts and catalysts which give enhanced yields of products useful in reformulated gasoline (e.g., light olefins, gasoline range isomers-"RFG catalysts") are of importance to both the catalyst user and manufacturer. The hundreds of different catalyst grades available today attest to the fact that both parties recognize the importance of catalyst formulation changes as they impact product selectivity changes.

Unit Cell Size

By now it is clear that U.C.S. controls Y zeolite (Faujasite structure) acid site concentration and strength (7,8). Because of this, the ratio of cracking to hydrogen transfer can be controlled and product selectivities will be affected (9). This aspect of catalyst activity and selectivity control will be discussed in more detail later.

Matrix

It is also well understood that a catalytically active, relatively large pore diameter (compared to the zeolite pores) matrix is needed to "pre-crack" large molecules that would not otherwise be effectively cracked by the small pore Y-type zeolite, the primary catalyst activity promoter (9, 10). Estimations have been made which show that $\sim 50\%$ of all gaseous feedstock molecules are larger than the pore opening in type Y zeolite (7.4 Å) (11).

Shape Selective Component

Catalysts in use today contain the medium pore (12 ring, 7 Å opening) type Y and not infrequently the relatively smaller pore (10 ring, 5 Å opening) ZSM-5 generally used as an additive (12). It seems quite clear that both smaller (8 ring, 4 Å opening, e.g. erionite) and larger shape selective materials such as VPI-5 (18 ring, 12 Å opening) or pillared interlayer clays with spacings near 20 Å (13) will be needed to further refine the selectivity of future catalysts used in resid cracking and for RFG applications.

There is one variable which can profoundly effect all issues involved in Catalyst-Process-Environmental control of FCC products in the next decade. This parameter is the hydrogen balance around the refinery and more specifically the hydrogen content of the feed to the cat cracker (14). Ideally, virtually all of the problems foreseen could be solved if the cat cracker produced predominantly ethylene, propylene, isobutylene, isomeric saturated gasoline, n-cetane and enough coke to stay in heat balance and if the ratio of these products could be changed at will with catalysts and process changes. Unfortunately, a feedstock containing about 14.7 wt% hydrogen would be needed to supply the H/C ratio necessary to produce this "ideal" product distribution. However, any significant change to higher feed H/C ratios either by feed hydrotreatment or by using carbon rejection processes substantially changes both catalyst and process requirements. We will illustrate these hypothetical changes later in the present paper but for now we will consider C-P-E issues based on typical hydrogen balances generally available in most operations.

FUTURE OCTANE AND RFG CATALYST AND PROCESS CHANGES

Taking into account the C-P-E needs in the future as shown in Figures 1 and 2, we can characterize catalyst and process changes which are likely to occur as follows:

RFG 1 and 2 (Simple and complex model, respectively)

By 1995 the EPA simple model for RFG will be in effect (15). Benzene levels in gasoline will be controlled at 1 vol% maximum, oxygen content will be 2 wt% and RVP will be controlled between 7.2 to 8.1 lbs/in^2 (in specified areas in the U.S.A.). Oxygen supply to gasoline via ether or ethanol addition was started in November 1992 (OxyFuel) and early supply problems did not develop (16). Technology for producing ethers (methyl tertiary butyl ether, ethyl tertiary butyl ether, tertiary amyl methyl ether, tertiary amyl ethyl ether - MTBE, ETBE, TAME, TAEE, respectively) within the refinery is under constant, energetic development so that the refiner has less dependence on merchant suppliers (17, 18).

Benzene Reduction in FCC Gasoline

Some catalytic control appears viable since strong hydrogen transfer catalysts (high equilibrium U.C.S.) produce more benzene at constant conversion then low hydrogen transfer catalysts (low equilibrium U.C.S.). This latter type catalyst system is also the choice for making high yields of light olefins to prepare ethers (19). From the process side, shorter contact times decrease benzene yields. In contrast, higher reactor temperatures and increased conversion levels increase benzene yield (20). Since higher reactor temperatures are appropriate for increasing light olefins, a compromise of catalyst and process conditions may be called for. However, since a vast majority of the benzene in the gasoline pool comes from the catalytic reformer, benzene control from the FCCU will not be critical in the RFG I stage for most refiners (21).

Oxygen Content of Gasoline

Ether and ethanol usage in OxyFuel to regulate gasoline oxygen content is already established and the C-P-E related issues for RFG I, the simple model, are at least partially known. FCC catalysts will contain high silica/alumina Y zeolites (HSAY) which are well known to increase the cracking/hydrogen transfer ratio yielding increased quantities of C_3, C_4, and C_5 olefins, The latter two of which are used to prepare MTBE and TAME (and ETBE or TAEE) (7, 8, 22, 23). Propylene can also be used to prepare isopropyl alcohol (IPA) for gasoline oxygen content control, though IPA is not without its own set of problems, such as high blending RVP compared to MTBE and fungibility issues typical of all alcohols (24).

Data available to date shows that the amount of C_5 and lighter olefins formed is directed toward equilibrium values as the SiO_2/Al_2O_3 ratio of the HSAY is increased (25). In addition, catalysts believed to contain beta zeolite yield isobutylene approaching equilibrium ($iC_4=/$total $C_4= \approx 0.45$) (26). Use of ZSM-5 or other 10-ring zeolites is well known to enrich C_3 and C_4 olefin streams at the expense of low octane gasoline range molecules (11). The ratio of olefins produced with ZSM-5 are reported to be 4, 3 and 1 for propylene, butylenes and amylenes, respectively. While some optimization of the HSAY zeolite component of RFG catalysts has been done, other variables need to be examined along with the Si/Al ratio for the best performance in either RFG 1 or 2 applications.

An active matrix that would pre-crack large molecules into straight chain fragments which could then be easily and selectively cracked to light olefins would be a distinct benefit. However, matrix cracking control as presently practiced is frequently non-selective. The major matrix issues which have been addressed in the recent past relate primarily to its ability to crack heavy cycle oil into light cycle oil and gasoline thus minimizing the amount of HCO remaining at basically all costs. This desire is not at all likely to change because the demand for residual fuel is not likely to increase since it must compete with coal as a residual fuel (27). Asphalt applications are more profitable but have limited markets.

Thus, additional studies of matrix porosity and acidity function for controlling large molecule conversion to proper feed fragments for the 10-ring shape selective component and the primary HSAY promoter are needed. Porosity control from the zeolite surface outward (micro → meso → macro) will no doubt influence the feed fragments composition and the final product desorbed from the catalyst surface (28, 29). What has not been adequately controlled or recently studied are ways to control matrix acidity. That this can control matrix cracked molecular structure was demonstrated with silica/magnesia matrices vs. silica/alumina matrices in the 60's (30). Selectivity differences caused by the differences in acidity between a CREY promoted silica/magnesia and a CREY promoted silica/alumina catalyst are shown in Table 1 (31).

Compared with SiO_2/Al_2O_3, the lower acidity SiO_2/MgO matrix gave products related to less severe cracking: higher gasoline yield with lower C_3 and C_4 yields, higher gravity gasoline with corresponding lower research and motor octane, and a higher yield of improved quality light cycle oil. Silica/magnesia cracking catalyst did not enjoy widespread acceptance because of unsatisfactory hydrothermal stability.

CONTROL OF THE CRACKING/HYDROGEN TRANSFER RATIO (C/H-t)

Variations in type Y U.C.S., silica/alumina ratio and acid site distribution as they would impact the activity and selectivity of RFG, RFCC, together with conventional FCC gasoline oriented catalysts are summarized in Table 2 and Table 3.

If we consider RFG I and II catalysts to be an extension of the present octane catalysts, we can see from Table 2 that any improvements in selectivity will probably relate to HSAY structures with less than one aluminum site/unit cell. This would relate to some maximum in the ratio of cracking to hydrogen transfer (C/H-t) as shown in Table 3. Table 3 shows the C/H-t ratio as a function of U.C.S. assuming that the controlling feature is the number of aluminums with various combinations of next nearest neighbors (NNN). The NNN concept (O-NNN have only silica tetrahedra in the immediate two coordination spheres surrounding the aluminum tetrahedron) has been reviewed extensively (8). Regardless of the actual contribution of the various aluminum sites it is interesting to note the C/H-t ratio changes brought about by assuming different combinations of NNN. Thus, based on the total Al (IV) sites present at what would be an expected U.C.S. for an octane/RFG catalyst of today (24.25 Å), we see that the ratio of C/H-t could vary from infinity (100% cracking, 0% H-t) to 2.7, depending on the relative contribution of 0, 1, 2, 3 and 4-NNNs. More importantly, Table 3 data suggests methods of selectivity/activity control that are known but not well quantified. Since rare-earth exchanged octane catalysts (REHSAY) typically equilibrate at 24.30 U.C.S., a maximum in activity with minimal H-t(32), and gasoline catalysts (REY) typically equilibrate at 24.40-.55 U.C.S., the data in the vertical columns of Table 3 appear directionally appropriate. It seems to the authors that an important selectivity tool to develop would be to quantify the C/H-t ratio based on the breakdowns shown in Table 3. Variations in the ratio are likely to influence product distributions over the entire spectrum of type Y containing FCC catalysts. The importance of this can be inferred for RFG catalysts where the optimum U.C.S. may be below 24.25 and for resid catalysts where control of C/H-t in the zeolite is necessary to complement the resid catalysts matrix cracking activity (and possibly its hydrogen transfer ability).

Aside from variations in C/H-t ratios of the primary HSAY promoter or in the use of beta zeolite containing catalyst compositions, we may project that other 12-rings (e.g., Breck Structure-6, ECR 4/32, SAPO 37, L) may also be intensely investigated for use in RFG 1 and 2 applications. Also, the deliberate presence of both intergrowth and outergrowth of zeolite systems may advantageously affect product distributions in both octane and RFG 1 and 2 catalysts (33).

Small pore molecular sieve components (except for ZSM-5) have not been successful promoters for RFG light olefin production usually because of unacceptable steam stability. We may expect continuing efforts to solve this problem with perhaps some unique materials in commercial operation by the time of the Complex Model implementation in 1997.

TABLE 1:   MATRIX ACIDITY EFFECTS ON FCC YIELDS *

	<----- CREY ----->	
Matrix:	Silica/Magnesia	Silica/Alumina
Conv: Vol%	77.0	77.0
Tot. C_3's: Vol%	8.7	12.0
Tot. C_4's: Vol%	9.0	11.2
Gaso: Vol%	69.5	65.0
Gaso./Conv.:	0.90	0.84
LCO: Vol%	14.3	10.0
HCO: Vol%	8.7	13.0
Coke: Wt%	3.8	3.8
Gaso. Props.		
RON:	88.0	90.0
MON:	76.0	78.7
Gravity °API	61.1	56.7
LCO Properties		
Gravity: °API	20.8	16.0
Aniline pt: °F	90	78
HCO Properties		
Gravity: °API	7.0	11.2
Aniline pt: °F	118	148

* Constant severity, constant CREY content.

TABLE 2. UNIT CELL/SITE CHARACTERISTICS OF PRESENT AND FUTURE FCC CATALYSTS

	Octane-RFG	RFCC	FCC (Gasoline)
As prepd. U.C.S. Range	24.50-.58	24.58-.62	24.62-.68
As prepd. SiO_2/Al_2O_3	9.0-6.1	6.1-5.5	5.5-4.65
Equilibrium U.C.S. Range	24.25-.30	24.30-0.40	24.40-.55
Equil. # A1/Unit Cell [65]	1.1-6.4	6.4-17.1	17.1-33.2
Equil. A1 Site Distribution: [66]			
0-NNN	0.8-3.5	3.5-4.3	4.3-5.3
1-NNN	0.2-2.2	2.2-6.5	6.5-10.0
2-NNN	0.1-0.5	0.5-3.1	3.1-10.6
3-NNN	0.0-0.2	0.2-1.8	1.8-5.8
4-NNN	0.0-0.0	0.0-1.4	1.4-1.6

Additional catalyst needs for RFG 2 may involve catalyst variations that minimize the amount of gasoline heavy ends (T90 reduction). This 350-430°F fraction is highly aromatic and contains the majority of the S and N containing high molecular weight molecules. We may anticipate that it may be blended with LCO and the mixture hydrotreated. A portion will be recycled to the FCCU and the remainder used for low S content diesel fuel. The actual benefit of this requirement is questionable since the sulfur and aromatics concentration limitations address the negative aspects of heavy cat naphtha and it does help reduce the RVP of the gasoline. In any case, the catalyst consequences of T90 reduction appear to lie within the industry's current technology sphere.

Light olefin reduction in gasoline for RFG 2 may have catalytic impact since amylene is the principal target. Removal by distillation, however, and conversion into TAME or TAEE appears to be the most cost effective option when catalytic optimization in the gasoline stream using a 10-ring shape selective molecular sieve is attained. Amylene alkylation of all the C_5 olefins or of any of the excess olefins after oxygenate production would also be widely used (34).

Catalyst matrix changes (which may impact RFG 1 and 2 catalysts) relate mainly to pore structure accessibility, to acidity considerations and to how these variables affect the pre-cracking of molecules whose cracked fragments will react with the primary and secondary promoters present in the RFG catalyst. Matrices will be discussed in more detail along with resid catalysts.

FCC process considerations related to RFG 1 and 2 include Deep Catalytic Cracking (DCC), which enhances C_3 and iso C_4 and iso C_5 olefin yields, and high temperature (1050°F+) short contact time riser cracking (<2 sec) which enhances light olefin yields, including iso C_4 and iso C_5 olefins (35).

Future Resid Catalyst and Process Changes

The conversion of less than premium gas oils to transportation fuels in the FCCU has been a matter of fact for many years and the economic forces which drive this practice are not likely to subside in the future (36).

Common characteristics of resid feedstocks are high aromaticity and metals content and diminishing overall crackability which is directly related to diminishing hydrogen/carbon ratio. An outer FCC limit for today's technology in both catalyst and process is a feedstock with 9.0-10.0 Conradson Carbon Residue and 30-70 ppm total metals (Ni + V) (37, 38).

As illustrated in Figure 2, catalysts used in processing resids consist of a relatively high SiO_2/Al_2O_3 ratio type Y, or HSAY as a primary promoter together with an active matrix which may, in the future, be a shape selective large pore molecular sieve or a PILC (39, 40). Characteristics of the primary promoter are shown in Table 1 and 2. Given a stable high crystallinity, primary promoter, an effective resid catalyst must also contain an appropriately formulated matrix - pore structure and correct cracking activity. Correct pore structure and cracking activity refer to the matrix's ability to feed the primary promoter pre-cracked fragments (which it can effectively crack) with a selectivity that minimizes light gas and coke formation. Resid catalyst pore architecture has been recently described as a shrinking core of non-deactivated accessible sites surrounded by a progressing shell of large hydrocarbon molecules and metal contaminants (41). The matrix can thus take on a protective role in addition to its molecular supply function. In this instance, a matrix design which incorporates metals passivators (or sinks) primarily for nickel and vanadium is generally used (42). Nickel may be passivated using oil soluble antimony and bismuth compounds as additives and by matrix pore size distributions which encourage nickel agglomeration into catalytically inactive large nickel crystals (43). Vanadium is deactivated by reaction with catalyst components which interact with it forming stable vanadium (V) compounds. In this case, however, the catalyst component which is specific for vanadium is also reactive with SO_x, and in the presence of a high sulfur feedstock a competition is set up between the two functions usually resulting in vanadium trap deactivation. Group II titanates and oxides and minerals such as sepiolite fit into this category (44). The presence of an effective SO_x catalyst has been suggested as a method to minimize vanadium trap deactivation in high sulfur feedstock applications (45). The entire area of metals passivation will no doubt continue to be a critical area of R&D as long as resid feed processing is practiced.

The problems associated with cracking resid feeds to high value products involve catalyst and process innovation as well as environmental concerns related to the metals, sulfur and nitrogen present in the feed and the disposal of spent metals rich catalyst.

Illustrative data showing the effects of adding vacuum tower bottoms (VTB) to an average gas oil FCC feedstock are given in Table 4 (46).

The effects on product yields caused by adding 10 and 20% VTB to a gas oil feed are typical of virtually all operations in which a gas oil feed is mixed with a more refractory feed. That is:

- Conversion drops
- The C_3 through gasoline yields decrease
- LCO and HCO yields increase primarily due to the lower conversion levels
- Coke increases substantially
- Octanes increase: RON more than MON primarily from olefin increases
- LCO and HCO properties improve due to the lower conversion of these fractions into lighter products.

However, with the appropriate financial incentives (i.e., the delta cost between a resid feed and gas oil feed), resid processing makes sense and a large portion of the FCC field revolves around it.

TABLE 3. RATIO OF CRACKING TO HYDROGEN TRANSFER AS A FUNCTION OF UNIT CELL SIZE

U.C.S.	Total Al Sites	Ratio C/H-t $\left(\dfrac{\# NNN}{\# NNN}\right)$		
		0+1+2	0+1	0
		3+4	2+3+4	1+2+3+4
24.25	1.1	∞	10.0	2.7
24.30	6.4	31	8.1	1.2
24.40	17.1	4.3	1.7	0.3
24.55	33.2	3.5	0.85	0.2

TABLE 4.   EFFECT OF RESIDUE CONTENT ON FCC YIELDS

		VTB	
FEED	GAS OIL	+10%	+20%
C/O	BASE	-1.0	-1.0
CONV: VOL%	80.2	77.3	75.8
C_2-: WT%	2.6	3.0	3.2
TOT. C_3'S: VOL%	11.6	10.5	10.0
TOT. C_4'S: VOL%	18.3	16.1	15.2
GASO: VOL%	62.3	60.4	58.6
LCO: VOL%	14.8	16.1	16.2
HCO: VOL%	5.0	6.6	8.0
COKE: WT%	3.7	4.7	6.1
GASO. PROPS:			
RON	90.8	91.4	91.6
MON	78.3	78.3	78.8
OLEFINS:	33.0	35.0	42.9
VOL%			
LCO PROPS:			
GRAVITY °API	18.9	19.3	22.7
CETANE NO.	<20	20	26
HCO PROPS:			
GRAVITY °API	1.2	1.7	6.8

The effect of increasing feed H/C ratio is given in Table 5 where a comparison of selectivity changes caused by mildly and severely hydrotreating a West Coast feedstock is shown (47).

Feed hydrotreating is one of the most straightforward ways to approach the ideal product distribution described before: light olefins, isomeric gasoline, low S high octane number LCO, and controlled coke yield. Unfortunately, this mode of operation requires high capital and operating cost if sufficient hydrogen is unavailable at the refinery.

Table 5 is reasonably typical of the yield changes seen when the feedstock properties are altered (this time in the opposite direction shown in Table 4 when VTB caused a depression in H/C ratio). Here an increase in H/C ratio caused by mild and severe hydrotreating is shown to yield:

- A feedstock similar to naphthenic feeds in API gravity and aromaticity but with low nitrogen and S contents similar to high quality paraffinic feeds.
- Substantial removal of polynuclear aromatics (PNA) which would convert preferentially to coke.
- Increased conversion with significantly higher C_3, C_4 and gasoline yields.
- Lower LCO but more importantly, significantly lower HCO due to the higher conversion possible at constant coke.
- More aromatic gasoline and heavier LCO and HCO caused by the higher conversion.

An important benefit other then the increased yields of C_3 through gasoline at constant coke, is the low sulfur and nitrogen content of the hydrotreated feed. This benefit would carry through to lower sulfur cat gasoline and LCO and low SO_x emissions from the regenerator stack - both environmental pluses.

Process/hardware controls and equipment are also more straightforward since sulfur control is easier and the need for flue gas scrubbing or for an SO_x removal catalyst can usually be eliminated. In addition, the unit can be operated in partial or complete CO combustion since excess oxygen is not required for an SO_x transfer catalyst. A catalyst cooler can be eliminated in single stage regenerators if the Conradson Carbon is no higher than 2 (48).

When residual feeds are hydrotreated, the hydrogen uptake is usually low (300-600 scf/B) due to the high cost of hydrogen and the difficult processing conditions required (low LHSV and moderate reactor temperatures). In Table 6, we show a projected product distribution that would be possible if control of the feed H/C ratio was unlimited.

A Feedstock to give this product distribution would contain about 14.75 wt% Hydrogen.

TABLE 5.   EFFECT OF HYDROTREATING (HT) ON FCC YIELDS AT CONSTANT COKE

	AS RECEIVED	700 PSI HT	1800 PSI HT
WEST COAST			
FEEDSTOCK PROPERTIES			
GRAVITY: °API	23.0	25.9	27.7
ANILINE PT: °F	154	164	173
N_2: PPM/WT	2500	1560	334
S: WT%	1.3	0.13	0.06
ST'D PNA SATURATION: WT%	-	35	66
FCC PILOT PLANT YIELDS			
CONV: VOL%	51.0	65.0	77.0
TOT. C_3'S: VOL%	6.0	8.0	10.1
TOT. C_4'S: VOL%	8.4	10.0	11.9
GASO: VOL%	36.0	52.5	63.5
LCO: VOL%	29.0	22.0	16.5
HCO: VOL%	20.0	13.0	6.5
PRODUCT QUALITIES			
GASOLINE - RON	90.8	91.3	91.2
GASOLINE - MON	76.8	78.4	78.4
GRAVITY: °API	49.6	50.4	51.4
ANILINE PT: °F	92	84	73
BROMINE NO.	85	42	28
LCO			
GRAVITY: °API	27.0	23.0	20.5
ANILINE PT: °F	120	95	78
HCO			
GRAVITY: °API	19.5	14.0	9.0
ANILINE PT: °F	170	145	132

Catalyst and process control can only go so far in establishing selectivity patterns from various feedstocks but all else being equal, the high yields of products desired for RFG 1 and 2, LCO and petrochemicals will be ultimately dependent on the feedstocks hydrogen content and catalyst and process selectivities. Economics will dictate how far a refiner goes but more hydrogen in the feed allows more of the ideal products to be made.

Resid Catalysts

Given that catalysts for resid operations must be metals tolerant as described earlier and hydrothermally stable to withstand generally higher temperature regenerations than gas oil catalysts, the chief remaining resid catalyst issues relate to catalyst architecture. Since a variety of large molecules will be present in the resid feed, most of which cannot be accommodated directly in the primary zeolite Y promoter, it follows that ideal selectivity and maximum activity can only be achieved by a catalytically "proper" setting of active pores which selectively crack large molecules to feed the primary Y promoter's active sites. Many workers have described this "proper" setting (41, 49). On average, a matrix structure of about 100-120 m^2/g total surface area with a mixture of meso and macro pores of average pore diameter <60Å and 200-400Å, respectively, seems adequate (50). However, this pore structure, which is derived usually from a gamma type alumina, should deactivate at a rate more or less parallel to the deactivation of all other active components - the primary promoter and the secondary promoter (e.g., ZSM-5) if present (51). If this is not the case, a changing array of products results. It is known that the zeolite/matrix surface area ratio (Z/M) is important in controlling product distributions (52). At Z/M ranging from 1 to 4, it has been shown that bottoms yields increase, C$_3$/C$_4$ yields decrease, and LCO decreases while gasoline yields increase, and coke and dry gas decrease. Optimum Z/M ratios are dependent upon feed characteristics, both molecular size profiles and H/C ratios, the process configuration itself (how much coke can be burned) and the product slate desired.

In our opinion, there remains substantial opportunity in the resid catalyst area particularly in controls of matrix activity by additional pore size tailoring with added acid site strength control, both Bronsted and Lewis.

Resid Processes

Resid processes are currently offered by UOP, Stone and Webster/IFP, M.W. Kellogg and Exxon. Shell Oil has developed their own process but it is no longer available for third party licensing (53). UOP and Ashland Oil collaborated on the RCC$^\bullet$ process, Total Petroleum worked with both Stone and Webster and IFP while Mobil Oil is a partner to M.W. Kellogg though Phillips Petroleum was the developer of the original HOC process.

The chief design features of these units are summarized in Table 7.

TABLE 6. IDEALIZED PRODUCT DISTRIBUTION FROM
AN FCC UNIT FOR RFG OPERATIONS

PRODUCT	WT% OF FEED	WT% HYDROGEN
C_3^-	10.0	14.3
C_4^-	10.0	14.2
C_5^+ Gasoline*	50.0	≈ 15.7
n-Cetane	25.0	15.0
Coke	5.0	6.0

* C_5^+ Gasoline assumed to be all isomers (no aromatics)

TABLE 7. RESID FCCU DESIGN FEATURES

Feature	UOP	Stone & Webster/IFP	Kellogg	Exxon
Feed Injection	Premix	High Energy	Atomax	Venturi Design
Riser Terminator	Vented Riser or Ballistic Separator	Ramshorn-Axial Cyclone	Closed Coupled Cyclones	Closed Coupled Cyclones
Regeneration Scheme	2 Stage Coupled	2 Stage Independent	Single Stage	Single Stage

Feed injection technology is the most critical feature since poor contacting at the base of the riser leads to higher delta cokes and coking in the reactor and vapor transfer line. Higher delta cokes require more heat removal and therefore higher weight percent coke yields to achieve the same conversions. Fine atomization of the feed and good coverage of the riser cross-section are both necessary in any good feed injection system.

Recent studies have shown a significant amount of post riser cracking can occur in the dilute phase of an FCCU. Residence times as long as 40 seconds have been measured in the dilute phase of some units. Much of this uncontrolled cracking is thermal in nature, resulting in the loss of gasoline and the creation of extra unwanted dry gas and coke.

Two types of catalyst-vapor separators have been developed. Inertial separators replaced simple tee's at the end of the riser. These use the momentum difference of the rapidly moving vapor and catalyst to cause the separation. Putting a cyclone on the end of the riser (rough cut cyclones) was the other logical approach. This was followed by connecting the rough cut cyclone to the secondary cyclone. While this reduces the residence time, operational upsets can cause large amounts of catalyst to be carried over to the main fractionator.

Newer systems such as axial cyclones provide faster separation of the vapor and catalyst than conventional cyclones and are designed to minimize the operational problems associated with direct or close coupled systems. The amount of gas entrained in the diplegs of the new systems is less than 3% of the total vapors compared to as much as 30% in rough cut cyclones and 6-15% for regular cyclones (54). This entrained gas spends long residence times in the reactor/stripper and severely overcracks.

Regeneration is carried out in either one or two stages. The single regenerators use a catalyst cooler when Conradson Carbon exceeds 2 wt% of the fresh feed in order to protect the catalyst from excessive deactivation. Staged regeneration allows the generation of CO rather than CO_2 and acts as a catalyst cooler. This is far more effective when the regenerators are independent of one another since more CO can be generated. A separate catalyst cooler is not required in this design until Conradson Carbon exceeds 6 wt%. Since most of the moisture produced from the burning of hydrogen on the coke is removed in the first stage, a second stage regenerator can operate at much higher temperatures without deactivating the catalyst (55).

In the future, we may expect added improvements for resid as well as gas oil processing in feed distribution through advanced dispersion nozzle design, riser designs that include not only short (<2 second) contact times but rapid oil/catalyst disengaging to further minimize non-selective cracking and multi-stage strippers to recover more hydrocarbons and reduce the hydrocarbon load to the regenerator. New 2-stage regenerator and catalyst cooler designs will appear for resid units along with fast fluidized bed regenerators which will reduce unit inventories (56). Refinery operation centered around the FCC unit may appear to approach a specialty chemical operation coupling with petrochemicals to produce light olefins including perhaps ethylene from the FCCU (See Figure 2) and moving aromatics from FCC gasoline to the petrochemical plant (57). Increases in ether production will probably balance the gasoline losses from both reducing the T90 of the gasoline and removing all or part of the aromatics in accordance with the EPA Complex Model that will be implemented in March 1997.

Equilibrium catalyst disposal from FCC units, particularly those operating on resid feeds may require different procedures in the near future. Toxic Characteristic Leaching Procedure (TCLP) tests may show future catalysts as unacceptable for land fill. In this instance, chemical treatments such as DEMET or magnetic separation procedures (MagnaCat⁻) may need to supplement cement applications for dealing with high metals content FCC catalysts (58,59,60). The use of spent catalyst in asphalt or brick manufacturing has also been successfully practiced (61). Since high metals content hydroprocessing catalysts have been regenerated for years and methods have been developed for metals recovery from these materials, a number of environmentally friendly opportunities should exist for handling spent FCC (62).

SUMMARY

Catalysts

The FCC area has been full of surprises over the last 50 years but carbenium ion cracking catalyzed by acid sites generated from coordination of A1 (IV) and Si (IV) will likely persist into the foreseeable future. However, within the SiO_2/Al_2O_3 world there are many areas which warrant additional study: molecular sieves with both larger pores than 12-rings and smaller pores than 10-rings, PILCs, controlled acidity meso and macro porous structures.

In addition, there is an ever growing need to selectively crack large molecules and many of these are extremely easy to crack. At the present time we tend to overcrack them. Acidity control needs to be exercised on both the molecular sieves and matrix and will be dependent on the reaction conditions and feedstocks processed.

Structural control of catalyst systems should be integrated with acidity control. Much of the data generated on primary and secondary promoters show the benefits of structural control. Extension of structural control and acidity control to selectively crack large molecules would, no doubt, be beneficial.

Process

Can we ideally make only light olefins, isomeric gasoline, n-cetane and coke? Without high H/C ratios in the feed, it is virtually impossible. However, the die is cast for fossil fuels in the transportation sector and only molecules of the above types will have little or no environmental problems. Feedstock hydrogen contents of 14.74 wt% are necessary to approach product distributions that are environmentally friendly but excess hydrogen to accomplish this is generally unavailable.

We are presently in a situation where reformer hydrogen yields are fixed or may be dropping while demands for feed or product hydrotreating are increasing (57). More imaginative ways of increasing hydrogen availability are needed and a number of these; PSA, cryogenics, membrane technologies for FCC dry gas, hydrocracker and hydrotreater off-gas, have been implemented in refineries.

Separate risers are usually avoided to minimize unit cost and complexity. Feeds are segregated depending on their coking tendencies and their crackabilities and placed at different locations in the same riser. The use of secondary, smaller pore zeolites are used to selectively crack linear molecules without another riser. A second reactor only will be employed when a portion of the feed needs very different reaction conditions (i.e., pressure, temperature or catalyst/oil).

Feed atomization is extremely important since poor atomization can result in as much as an 8 volume percent loss in gasoline with gas oils and may make operations on residual feeds impossible (63). As contact times shorten, feed atomization will be an even more critical control point since it will affect average reaction temperature and contact time.

Environmental

Environmental concerns are an umbrella over the entire area of FCC catalysis. Attention to environmental issues is important from basic catalytic process science through refinery operation. A formal climate of communication between EPA and industry has been established (Reg Neg) where mutual problems can be posed and joint solutions can be discovered (64). It is important to the consumer and the refining industry that arbitrary decisions not be taken since they are unlikely to represent the best technical or economic solutions to environmental needs.

REFERENCES:

1. Oil & Gas Journal Special Report, December 20, 1993, 37.
2. Avidan, A. A.; "Origin, Development and Scope of FCC Catalysis", in FCC: Science & Technology, J. S. Magee & M. M. Mitchell Jr., Eds., Elsevier, Amsterdam, 1993.
3. Schwatz, A. B.; U.S. Patents 4,072, 600 and 4,093,536 (1978).
4. "U.S. Ethylene Growth to 2002 Will Rely on Heavier Feedstocks", Oil & Gas Journal, December 27, 1993, 99.
5. Oil & Gas Journal, October 8, 1984, October 14, 1985.
6. Scherzer, J.; "Correlation Between Catalyst Formulation and Catalytic Properties in FCC: Science & Technology, J. S. Magee & M. M. Mitchell, Jr., Eds., Elsevier, Amsterdam, 1993.

7. Pine, L. A.; Maher, P. J.; Wachter, W. A.; J. Catal. 1984, 85, 486.
8. Wachter, W. A.; "The Role of Next Nearest Neighbors in Zeolite Acidity and Activity", in Theoretical Aspects of Heterogeneous Catalysis, J. Moffat Ed. Van Nostrand, Rheinhold, New York, 1990, 110.
9. de Jong, J. I.; Ketjen Catal. Sym. 1986, Paper F-2.
10. Otterstedt, J. E.; Zhu, Yan-Ming; Sterte, J.; Appl. Catal., 1988, 38, 143.
11. Dwyer, F. G.; Degnan, T. F.; "Shape Selectivity in Catalytic Cracking" in FCC Science & Tech., J. S. Magee & M. M. Mitchell, Jr., Eds., Elsevier, Amsterdam, 1993.
12. Vaughan, D. E. W.; "Complexity in Zeolite Catalysts: Aspects of the Manipulation, Characterization and Evaluation of Zeolite Promoters For FCC", FCC: Science & Tech., J. S. Magee & M. M. Mitchell, Jr., Eds., Elsevier Amsterdam, 1993, 85-88.
13. Ibid., 83.
14. Letzsch, W. S.; Ashton, A. J.; "The Effects of Feedstock on Yields and Product Quality", in FCC: Science & Tech., J. S. Magee & M. M. Mitchell, Jr., Eds., Elsevier Amsterdam, 1993, 455-64.
15. "The Clean Air Act and the Refining Industry", UOP Brochure, September 1, 1991, 3.
16. Dale, C.; Hackworth, J. H.; Shore, J. M.; Ostrich, J.; Oil & Gas Journal, October 25, 1993, 66-69.
17. Ibid., 69-75.
18. "New High Conversion Etherification Process Tested and Patented", Oil & Gas, December 27, 1993, 98.
19. Skeel, G. W.; Breck, D. W.; Proc. 6th Zeol. Conf., Reno, Nevada, 1983.
20. Keyworth, D. A.; Reid, T. A.; Kreider, K. R.; Yatsu, C. A.; Zoller, J. R.; "Controlling Benzene Yield", AM93-49, NPRA, March, 1993.
21. Reference 15, 16-17.
22. Magee, J. S.; Moore, J. W.; "Mechanisms of Product Yield and Selectivity Control with Octane Catalysts" in Fluid Catalytic Cracking, ACS Symposium Series 375, M. L. Occelli, Ed., 1988, 87-100.
23. Ref. 6, 178.
24. Ref. 15, 19.
25. Young, G. W.; Suarez, W.; Roberie, T. G.; Cheng, W. C.; AM 91-38, NPRA Annual Meeting, March 1991.
26. Biswas, J.; Maxwell, I. E.; p. 63, 197, Appl. Catal. (1990).
27. Evans, R. E.; Quinn, G. P.; "Environmental Considerations Affecting FCC", FCC: Science & Technology, J. S. Magee and M. M. Mitchell, Jr., Eds., Elsevier, Amsterdam 1993, 582.
28. O'Connor, P.; Gevers, A. W.; Humphries, A. P.; Gerritsen, L. A.; Desai, P. H.; in ACS Symposium Series 452, M. L. Occelli, Ed., 1990, 318.
29. Hettinger, W. P.; in ACS Symposium Series 375, M. L. Occelli, Ed., 1987, 308.
30. Wilson, C. P. Jr.; Carr, B.; Ciapetla, F. G.; U.S. Patent 3,395,103, 1968.

31. Magee, J. S.; Blazek, J. J.; "Preparation and Performance of Zeolite Cracking Catalysts", Zeolite Chemistry and Catalysis, J. A. Rabo, Ed., ACS Monograph 171, 1976, 655-656.

32. Ref. 6, 156.

33. Ref. 12, 93-96.

34. Ref. 15, 24.

35. Li, Z. T.; Shi, W. Y.; Pan, R. N.; Jiang, F. K.; Sym. on Adv. in FCC, ACS Petrol. Div. Preprints, 1993, 38, 581-3.

36. Mitchell, M. M. Jr.; Hoffman, J. F.; Moore, H. F.; "Residual Feed Cracking Catalysts", in FCC: Science & Technology, J. S. Magee, M. M. Mitchell, Jr., Eds., Elsevier, Amsterdam, 1993, 293-294.

37. Ref. 2, 37.

38. Ref. 36, 295.

39. Davis, M. E.; Saldarriaga, C.; Montes, C.; Garces, J.; Crowder, C.; Nature, 1988, 698, 331.

40. Figueras, F.; Catal. Rev. Sci. Eng., 1988, 457, 30.

41. O'Connor, P.; Humphries, A. P.; in Sym. on Adv. in FCC, ACS Petrol. Div. Preprints, 1993, 601-602.

42. Ref. 6, 149.

43. Nielsen, R. H.; Doolin, P. K.; "Metals Passivation", in FCC: Science & Technology, J. S. Magee, M. M. Mitchell, Jr., Eds., Elsevier, Amsterdam, 1993.

44. a.   Occelli, M. L.; in "Fluid Catalytic Cracking: Role in Modern Refining", (M. L. Occelli, ed.), ACS Sym. Sers., Vol. 375, A.C.S. Washington, D.C., 1988, 1.

    b.   Occelli, M. L.; Stencel, J. M.; in "Zeolites as Catalysts, Sorbents, and Detergent Builders", (Y. E. Karge and J. Weitkamp, Eds.), Elsevier, Amsterdam, 1989, 127.

    c.   Occelli, M. L.; Stencel, J. M.; in "Fluid Catalytic Cracking II Concepts in Catalyst Design", (M. L. Occelli, Ed.), A.C.S Sum. Sers. Vol. 452, A.C.S. Washington, D.C., 1991, 252.

    d.   Ng, H. N.; Calvo, C.; Can J. Chem, 1972, 50, 3619.

45. Ibid., p. 375-78. Ref. 43, 375-78

46. Ritter, R. E.; Young, G. W.; Davison Catalagram No. 68, 1984, 1.

47. Ref. 31, 658-659.

48. Valeri, F.; "New Methods For Evaluating Your FCC", Katalistiks 8th Annual FCC Symposium, Budapest, June 1987.

49. Dai, P-S. E.; Neff, L. D.; "Effect of Secondary Porosity on Gas Oil Cracking", Sym. on Adv. in FCC, ACS Petrol. Div. Preprints, 1993, 38, 594-76.

50. Ref. 36, 307.

51. Palmer, J. L.; Cornelius, E. B.; Applied Catal., 1987, 35, 217-235.

52. Wear, C. C.; Mott, R. W.; Oil & Gas Journal, July 25, 1988, 71.

53. Ref. 36, 294-295.

54. Silverman, M. A.; 4th International Conf. on Circulating Fluid Beds, Sommerset, PA, August 1-5, 1993.

55. Oil & Gas Journal, October 4, 1982 and October 11, 1982, Dean, Mauleon and Letzsch.
56. Upson, L. L.; Henler, C. L.; Lomas, D. A.; in "FCC: Science and Tech.", J. S. Magee & M. M. Mitchell, Jr., Eds., Elsevier, Amsterdam, 1993, 436.
57. Ref. 2, 38-39.
58. Ref. 27, 578, 582, 584.
59. Elvin, F. J.; AIChE Annual Mtg., November 1993.
60. Goolsby, T. L.; Moore, H. F.; Mitchell, M. M. Jr.; Kowalczyk, D.; Letzsch, W. S.; Campagna, R. J.; AIChE, Spring Meeting, 1993, 64.
61. Oil & Gas Journal, November 18, 1991, Rodolphe Schmitt.
62. NPRA Q&A, 1984 Meeting, 164-165.
63. Ref. 56, 431.
64. Ref. 15, 6.
65. Sohn, J. R.; DeCanio, S. J.; Lunsford, J. H.; O'Donnell, D. J.; Zeolites, 1986, 6, 225
66. Pine, L. A.; Maher, P. J.; Wachter, W. A.; J. Catal. 1984, 85, 466.

RECEIVED July 13, 1994

INDEXES

Author Index

Affiliation Index

Subject Index

A

Production: Meg Marshall
Indexing: Deborah H. Steiner
Acquisition: Anne Wilson
Cover design: Bob Sargent

Printed and bound by Maple Press, York, PA

Highlights from ACS Books

Good Laboratory Practice Standards: Applications for Field and Laboratory Studies
Edited by Willa Y. Garner, Maureen S. Barge, and James P. Ussary
ACS Professional Reference Book; 572 pp; clothbound ISBN 0–8412–2192–8

Silent Spring Revisited
Edited by Gino J. Marco, Robert M. Hollingworth, and William Durham
214 pp; clothbound ISBN 0–8412–0980–4; paperback ISBN 0–8412–0981–2

The Microkinetics of Heterogeneous Catalysis
By James A. Dumesic, Dale F. Rudd, Luis M. Aparicio, James E. Rekoske,
and Andrés A. Treviño
ACS Professional Reference Book; 316 pp; clothbound ISBN 0–8412–2214–2

Helping Your Child Learn Science
By Nancy Paulu with Margery Martin; Illustrated by Margaret Scott
58 pp; paperback ISBN 0–8412–2626–1

Handbook of Chemical Property Estimation Methods
By Warren J. Lyman, William F. Reehl, and David H. Rosenblatt
960 pp; clothbound ISBN 0–8412–1761–0

Understanding Chemical Patents: A Guide for the Inventor
By John T. Maynard and Howard M. Peters
184 pp; clothbound ISBN 0–8412–1997–4; paperback ISBN 0–8412–1998–2

Spectroscopy of Polymers
By Jack L. Koenig
ACS Professional Reference Book; 328 pp;
clothbound ISBN 0–8412–1904–4; paperback ISBN 0–8412–1924–9

Harnessing Biotechnology for the 21st Century
Edited by Michael R. Ladisch and Arindam Bose
Conference Proceedings Series; 612 pp;
clothbound ISBN 0–8412–2477–3

From Caveman to Chemist: Circumstances and Achievements
By Hugh W. Salzberg
300 pp; clothbound ISBN 0–8412–1786–6; paperback ISBN 0–8412–1787–4

The Green Flame: Surviving Government Secrecy
By Andrew Dequasie
300 pp; clothbound ISBN 0–8412–1857–9

For further information and a free catalog of ACS books, contact:
American Chemical Society
Distribution Office, Department 225
1155 16th Street, NW, Washington, DC 20036
Telephone 800–227–5558

Bestsellers from ACS Books

The ACS Style Guide: A Manual for Authors and Editors
Edited by Janet S. Dodd
264 pp; clothbound ISBN 0–8412–0917–0; paperback ISBN 0–8412–0943–X

The Basics of Technical Communicating
By B. Edward Cain
ACS Professional Reference Book; 198 pp;
clothbound ISBN 0–8412–1451–4; paperback ISBN 0–8412–1452–2

Chemical Activities (student and teacher editions)
By Christie L. Borgford and Lee R. Summerlin
330 pp; spiralbound ISBN 0–8412–1417–4; teacher ed. ISBN 0–8412–1416–6

Chemical Demonstrations: A Sourcebook for Teachers,
Volumes 1 and 2, Second Edition
Volume 1 by Lee R. Summerlin and James L. Ealy, Jr.;
Vol. 1, 198 pp; spiralbound ISBN 0–8412–1481–6;
Volume 2 by Lee R. Summerlin, Christie L. Borgford, and Julie B. Ealy
Vol. 2, 234 pp; spiralbound ISBN 0–8412–1535–9

Chemistry and Crime: From Sherlock Holmes to Today's Courtroom
Edited by Samuel M. Gerber
135 pp; clothbound ISBN 0–8412–0784–4; paperback ISBN 0–8412–0785–2

Writing the Laboratory Notebook
By Howard M. Kanare
145 pp; clothbound ISBN 0–8412–0906–5; paperback ISBN 0–8412–0933–2

Developing a Chemical Hygiene Plan
By Jay A. Young, Warren K. Kingsley, and George H. Wahl, Jr.
paperback ISBN 0–8412–1876–5

Introduction to Microwave Sample Preparation: Theory and Practice
Edited by H. M. Kingston and Lois B. Jassie
263 pp; clothbound ISBN 0–8412–1450–6

Principles of Environmental Sampling
Edited by Lawrence H. Keith
ACS Professional Reference Book; 458 pp;
clothbound ISBN 0–8412–1173–6; paperback ISBN 0–8412–1437–9

Biotechnology and Materials Science: Chemistry for the Future
Edited by Mary L. Good (Jacqueline K. Barton, Associate Editor)
135 pp; clothbound ISBN 0–8412–1472–7; paperback ISBN 0–8412–1473–5

For further information and a free catalog of ACS books, contact:
American Chemical Society
Distribution Office, Department 225
1155 16th Street, NW, Washington, DC 20036
Telephone 800–227–5558